Pseudo-Aristoteles (Pseudo-Alexander), *Supplementa Problematorum*

Peripatoi

Philologisch-historische Studien zum Aristotelismus

Herausgegeben von

Wolfgang Kullmann,
Robert W. Sharples, Jürgen Wiesner

Band 20

Walter de Gruyter · Berlin · New York

Pseudo-Aristoteles (Pseudo-Alexander), *Supplementa Problematorum*

A new edition of the Greek text
with introduction and annotated translation

Edited by
Sophia Kapetanaki and Robert W. Sharples

Walter de Gruyter · Berlin · New York

♾ Printed on acid-free paper which falls within the guidelines
of the ANSI to ensure permanence and durability.

ISSN 1862-1465

ISBN-13: 978-3-11-019140-0
ISBN-10: 3-11-019140-7

Bibliographic information published by the Deutsche Nationalbibliothek

The Deutsche Nationalbibliothek lists this publication in the Deutsche Nationalbibliografie; detailed bibliographic data are available in the Internet at http://dnb.d-nb.de.

Printed in Germany

Printing and binding: Hubert & Co., Göttingen
Cover design: Christopher Schneider, Berlin

Contents

Introduction

1. The texts and their attribution

1.1. The evidence of content

In 1841 Ideler published two books of *Medical Puzzles and Physical Problems* (Ἰατρικὰ ἀπορήματα καὶ φυσικὰ προβλήματα) attributed to Alexander of Aphrodisias, the commentator on Aristotle who flourished around 200 A.D.[1] In 1859 Usener published two more books which he argued were books 3 and 4 of the same collection.[2] But in 1857 Bussemaker had appended to his edition of the pseudo-Aristotle *Problems* as contained in Bekker's edition three extra books of *Problemata inedita* – i.e. *hitherto* "unpublished" – the first two of which are the same as the two edited by Usener.[3]

The extent to which any of these problems is "medical" needs to be heavily qualified. The topics selected are indeed predominantly concerned with the functioning of animal bodies; but we are definitely not dealing with any sort of therapeutic manual, and the driving force is intellectual curiosity.

---

1 A new edition of these is currently being prepared by Carl-Gustaf Lindqvist of Gothenburg, to whom we are grateful for advice and the opportunity to exchange information and ideas.

2 Usener was working on transcripts of MSS entrusted to him for editing by Brandis (Usener 1859, iii).

3 We use Bussemaker's numbering for the books, in accordance with the majority of scholars – cf. Flashar 1962b, 362 n.1 – and in order to recognise the "Aristotelian" character of some of the texts. However, we use Usener's numbers for the individual problems within the books; Bussemaker did not give individual numbers to problems that are preserved in only one of the MSS he himself used, presumably regarding them as late additions in this MS, but they in fact appear in other MSS too, as Usener discovered, and in any case a single continuous sequence of numbers is desirable. Ideler's books 1 and 2 are throughout distinguished from Bussemaker's by using i. and ii. for the former, 1 and 2 for the latter. Since it is inappropriate to continue to call problems first edited in 1857 *Problemata inedita*, we have used the title *Suppl(ementa) Probl(ematorum), Supplementary Problems*.

For example, the two problems concerned with nausea are (2.105) why some people feel nausea and hunger at the same time, and (2.106) why the vomit of those who are sea-sick on ships is bilious while other vomit is phlegmatic. We have not been able to find any parallels for discussion of these issues in the medical literature, which discusses nausea and vomiting rather in the practical contexts of disease and of antidotes to poisoning. Moreover, most of the problems that are "medical" are about why certain symptoms occur; there is some discussion of why certain diet is suitable for certain conditions, but relatively little about practical treatment.

Although Bussemaker included his three books in an edition of Aristotle, he none the less distinguished in his introduction between those items which he regarded as "Aristotelian" and those which he regarded as "Alexandrian". It has indeed been generally recognised that Bussemaker's three books contain material of two different types.[4] A new preface separates the major part of book 2, introduced as a discussion of "common symptoms", from the first 38 problems. The problems in the major part of book 2 and the first 45 problems of book 3 – the latter indeed being preserved in only one MS, separate from books i and ii Ideler and 1 and 2 Bussemaker, and with no explicit attribution of their own[5] – bear a marked similarity in manner to the pseudo-Aristotle *Problems* included in Bekker's edition.[6] They are divided into sequences on specific topics, some of which include a high proportion of problems also found in the Bekker collection while others do not.[7]

---

4 Usener 1859, xii notes that Theodore Gaza, who in 1453 translated extracts from books i-ii, 1 and 2.1-38 (below, §2.1.17), already rejected the attribution to Alexander himself. – A cautionary note is appropriate: material of this sort is by its nature liable to attract additional items, and inferences from the character of some or even most of the items in a particular collection cannot be used to draw conclusions about individual items within it with any degree of certainty.

5 3.46-49, together with two further problems added in an appendix by Usener, are a collection of odds and ends that occur in the non-standard sequence only (see below), four of the six being extracts from or paraphrase of Olympiodorus' commentary on the *Meteorologica*. For Bussemaker's 3.50-52 see below, at n.46.

6 In this book '[Ar.] *Probl.*' will always indicate the collection included in Bekker's edition, as opposed to our *Supplementary Problems*.

7 Although there are in books i-ii sequences on related topics – thus the first ten problems are all concerned either with hair, or with eunuchs, or with both, and the next eleven are concerned with physical reactions to emotion, or in one case to age – these two books are in general much less systematically arranged than the latter and major part of book 2. The lack of system is noted by Flashar 1962b, 366. Book 1 is so short that it is difficult to conclude very much about arrangement within it.

In the majority of the MSS book 1 begins with a preface which Dietz edited from MS V, where it occurs rather as a preface to Hippocrates' *Aphorisms*. The view that this was in fact its original context has however been challenged by Flashar.[8] The book itself contains 22 miscellaneous problems, some medical and some concerned with food, wine and the like. Book 2 Usener starts with 38 miscellaneous physiological problems. Both in book 1 and in the first 38 problems of book 2 there are relatively few parallels either with the [Aristotle] *Problems* in Bekker's edition or with the [Alexander] *Problems* in books i-ii Ideler.

After 2.38 there is a new preface introducing a discussion of "common symptoms", defined as those that can occur at any age.[9] The list corresponds, though not exactly, to the topics of 2.39-126. The remaining problems of book 2 (127-192) are concerned with four-footed creatures. 2.127-137 are on four-footed animals in general, 2.138-161 on pigs, and 2.162-192 on horses, asses and mules. In MS K, 2.127-192 are labelled as a separate book ("book 5") with the title "by the same author (i.e., by implication, Alexander; but this may be artificial because book 3 and the start of book 4 are missing in this MS): different questions and solutions on quadrupeds"; Valla, who used K, followed it in making 2.127ff. a separate book with a separate title.[10] As the table shows, there are some, but not many, points of contact with Aristotle's zoological works; to those listed there one may add two where a point made by Aristotle is used to answer a problem not itself stated by him.[11] There are also strikingly *few* points of contact with the secondary zoological traditions collected in Constantine Porphyrogenitus' summary of Aristophanes of Byzantium and other sources.[12]

---

8 Dietz 1834, 244.16-245.29; Flashar 1962a, especially 412. See however below, n. 149. MS V is dated to the first half of the 16th century by Hunger 1969, 101; it was owned by Johannes Sambucus.

9 Flashar 1962b, 361 argues that this preface is "obviously" evidence for a compiler's arrangement of the material, rather than, as Bussemaker supposed (1857, xix), of a change to a new source; but it may reflect the change to a new source even if added by the compiler rather than already present in the source-material.

10 Below, §2.1.16. In PT Book 2 has the subscription "medical *ἀπορίαι* concerning living creatures and quadrupeds": Usener 1859, x.

11 2.132 appeals to the claim, made at Ar. *GA* 5.6 786a35ff., that the colour of animals depends on their diet; 2.151 to the claim that females are less courageous than males except in the case of bears and leopards (Ar. *HA* 9.1 608a34).

12 2.145 is parallel to Ar. Byz. 2.570. 2.170 takes up Ar., *HA* 6.18 572a13-17, cited by Aelian *NA* 4.6 and thence by Ar. Byz. 2.613, but adds an explanation not found in the other texts.

Parallels with other texts.
(Figures with no brackets indicate similarity in both question and answer, those in square brackets the same question but a different answer.)

		parallels to [Ar.], *Probl.* Bekker	parallels to [Al.], *Probl.* Ideler	other Peripatetic parallels
1.1-22	medical, foodstuffs		1 (5%)[13]	
2.1-38	physiology	3 (8%); [1] (3%)[14]	3 (8%); [2] (5%)[15]	Aristotle, *GA* 1.19, *IA* 4: [2] (5%).[16]
2.39-53	dizziness, etc.	10 (67%)[17] (8 from book 33)	1 (7%); [1] (7%)[18]	Thphr., *Vert.*: 3 (20%)[19]
2.54-2.82	hair and nails	1 (3%)[20]	3 (10%); [4] (14%)[21]	Aristotle, *HA*, *GA* 5.3: 3 (10%); [1] (3%)[22]
2.83-97	voice and hearing	12 (80%)[23]	1 (7%)[24]	Aristotle, *GA* 5.2: 1 (7%)[25]
2.98-104	smell	6 (86%)[26]		
2.105-26	nausea, excretion and purging	1 (5%); [1] (5%)[27]	1 (5%)[28]	Thphr., *Vert.*: 1 (5%)[29]
2.127-192	four-footed animals	2 (3%); [1] (2%)[30]	2 (3%);[31] [2] (3%)[32]	Aristotle, *HA*, *PA*, *GA*: 7 (11%)[33]; [4] (6%)[34]
3.1-8 Buss.	plants and bread	3 (38%)[35]		Theophrastus, *HP*, *CP*: [3] (38%)[36]
3.9-29	wine and olive-oil	2 (10%)[37]	[1] (5%)[38]	Theophrastus, *CP*: 1 (5%)[39]
3.30-41	various foods			Aristotle, *GA*: 1 (8%)[40]
3.42-45	miscellaneous	2 (50%)[41]		

13 1.4 ~ the second part of ii.18 Ideler.

14 A similar solution to that in 2.17 is differently expressed in [Ar.], *Probl.* 8.8 and 8.33. The problem in 2.27 is discussed much more fully in [Ar.], *Probl.* 8.9. 2.34 is a much simplified version of [Ar.] 23.21 and 23.37. 2.32 has the same question as [Ar.], 13.7, but a different answer.

15 2.2 ~ i.31 (different answer), 2.5 ~ i.8 (more detailed here); 2.16 ~ i.33 (shorter here); 2.22 ~ i.55 (different answer); 2.34 ~ i.55.

16 2.28 ~ Ar., *GA* 1.19 726b7-11. 2.38 ~ *IA* 4 705b30.

17 2.40 = [Ar.] 33.15; 2.41 =33.8 (+ 31.1); 2.45 = 33.1; 2.46 = 7.6; 2.47 = 33.4 (different words); 2.49 = 6.4; 2.50 = 33.9; 2.51 = 10.18/33.10; 2.52 = 33.12; 2.53 = 33.13. 2.49 corresponds both to [Ar.] 6.4 and to Theophrastus *De vertigine* 2+13, but more closely to the former. *De Vertigine* is also the source for Cassius 27 and [Ar.] 3.9, 3.20.

18 i.34 is a shorter version of 2.46; 2.51 has the same question as i.144 but a different answer. 2.40, 2.41 = Cassius 44, 45.

19 2.39, 2.42, 2.43 (2.39 a fuller version of Theophrastus).

20 2.77 ~ [Ar.] *Probl.* 10.22.

21 2.77, 2.79, 2.81 have the same answers as [Al.] i.1, ii.19, i.10 (but 2.81 much fuller). 2.54, 2.58, 2.75, 2.80 have the same questions as [Al.] i.145, i.146, i.147, i.2/i.13 but different answers.

22 2.66, cf. 2.73, is a summary of passage in Ar. *GA* 5.3; 2.78 is an expanded version of a passage there; 2.80 has a similar argument to *GA* 5.3 but in a different context. 2.82 seems to be an explanation of why lice occur most on the heads of children, stated by Ar. *HA* 5.31 557a7, though the form of the problem seems to have been garbled.

23 2.83, 84, 85, 87, 89, 90, 91, 93, 94, 95, 96, 97 = [Ar.] *Probl.* 11.18; 11.19; 11.56; 11.57; 11.32; 11.45; 11.46; 11.20; 11.51/11.23; 11.61; 11.3; 11.41/48.

24 2.87~ [Al.] i.148.

25 2.97.

26 2.98, 99, 100, 101, 103, 104 = [Ar.] *Probl.* 13.9; 13.8; 13.2; 12.6 (shortened version here); 12.9 + 12.4; 12.5.

27 2.122 = [Ar.] *Probl.* 5.28. 2.126 has a similar question to [Ar.] 1.41, but a different answer. The theme of 2.107 is discussed in [Ar.] 3.9, 3.20, but Theophrastus is clearly the source.

28 2.107 ~ [Al.] ii.33; the latter is fuller.

29 2.107 = Theophrastus, *Vert.* 12.

30 2.172, 189 = [Ar.] *Probl.* 10.29, 10.27. 2.167 has the same question as [Ar.] 10.9 but a different answer.

31 2.150/156 abbreviated in second part of [Al.] i.141; 2.190 ~ [Al.] i.150, similar answer in different words.

32 2.153 ~ [Al.] i.140: 2.172/189 ~ i.151.

33 2.130 = Ar. *HA* 8.28 606b24ff.; Richter 1885, 30. 2.136 ~ Ar. *PA* 4.10. 2.139 draws on *HA* 8.21; 2.145 on *HA* 6.28 (also in Ar. Byz.); 2.158 on *PA* 3.1 (but also contradicts *HA* 2.5); 2.173 is an expanded version of *GA* 2.8 748a28ff.: Richter 1885, 31. 2.192 ~ *HA* 6.23, *GA* 2.8.

34 2.142 ~ *HA* 2.1, *GA* 5.8; 2.170 ~ *HA* 6.18 (also in Ar. Byz.); 2.181 ~ *HA* 6.23. 2.151 is imperfectly preserved in the MSS, but may take its problem from *HA* 6.28.

35 3.6, 3.7, 3.8 = [Ar.] *Probl.* 20.34, 21.5, 21.6 (all shortened).

36 3.3 ~ *CP* 2.15.1; 3.4 ~ *HP* 2.2.11, *CP* 2.14.2; 3.5 ~ *CP* 5.5.1.

37 3.11 is similar to [Ar.], *Probl.* 25.8, but not on quite the same issue, and actually cites Aristotle. On [Ar.] *Probl.* 27.4 and *Suppl. Probl.* 3.14 see below, n.54.

38 3.22 is on the same topic as [Al.] ii.70, but partly in disagreement.

39 3.12: general topic discussed by Theophrstus in *CP* 6 and elsewhere.

40 3.38 draws on Ar. *GA* 3.2, but adds the (striking) remark that a chick *has been observed* to change ends within the egg.

41 3.42 = [Ar.] *Probl.* 24.8; 3.45 = [Ar.] 23.40.

As for Bussemaker's third book, problems 1-45, which form the major part of this book, are found *only* in one MS, A, where they are separated from i-ii Ideler and 1-2 Bussemaker (all there attributed to Alexander) and are not themselves attributed to any author. None of them appear in the non-standard tradition. There is at least a question, therefore, whether these problems should be edited as part of the same collection at all.[42] We have nevertheless included them, being influenced by the similarity in manner between 3.1-45 and 2.39-192 Usener,[43] and by the presence in 3.9-29 of a possible Theophrastean connection.[44]

The remaining problems in Bussemaker's third book are a collection of odds and ends that occur only in MSS with a non-standard sequence of problems, attributed to Alexander. 3.46-47 Bussemaker occur in six MSS, BDHJLW.[45] 3.48-49 are excerpts from Olympiodorus' commentary on the *Meteorologica*; 3.48 occurs in four MSS, DGYo; 3.49 in DG only, some distance after 3.48. Usener added as an appendix two texts which in fact belong with 3.48 and 49 Bussemaker, all four being on meteorology; the only reason these two texts were not included by Bussemaker is that they occur in MSS Go, which he did not use, but not in D, which he did. The first of Usener's appendix texts is parallel to, though not verbatim quotation of, Olympiodorus' commentary; the second is again an extract from it.

3.50-52 Bussemaker are, as was pointed out by Westerink,[46] Michael Psellus, *De omnifaria doctrina* §§118, 119 and 109 respectively. The first two are concerned with hunger and thirst, the third with the evil eye. They occur in three MSS, EXc, at the end of what in these MSS is called book 3, which begins with some items from book 1 Bussemaker, but then consists mainly, as also do "books 4 and 5" in these three MSS, of material from the pseudo-[Galen] *Definitiones medicae.* 3.50-52 Bussemaker are thus much less securely linked with other pseudo-Alexander material than is the case with most of our texts; it looks rather as if Bussemaker included them

---

[42] In another MS, Florence, Laurentianus gr. 75.13, book i Ideler and the first part of book ii Ideler are followed, without distinction, by three problems of Cassius and then by problems 10 to 121 of the "Hippocratic problems", known from eight other MSS and of which the incipits are listed by Jouanna 1997; cf. Guardasole 2003, and Garzya and Masullo 2004, 19.

[43] 3.11 Bussemaker, which is similar to [Ar.], *Probl.* 25.8, but not on quite the same issue, includes the comment "thus Aristotle", which is at first sight, as Bussemaker 1857, xix pointed out, difficult to reconcile with the view that 3.1-45 were originally "Aristotelian"; but the comment could be a gloss.

[44] See Sharples 2004, and below at n.54.

[45] Rochefort 1950, 6-7 suggested that 3.46-7 are not by Alexander.

[46] Westerink 1948, 6.

simply because he did not recognise their origin and they seemed to him to be possible 'Alexander' problems (unlike the medical definitions preceding and following them, presumably). We have not therefore included them in this edition; the proper place for recording the readings of EXc in these problems is in an edition of Psellus.[47] The question arises whether we should have included the material from Olympiodorus as well; after some hesitation we decided to do so, partly because it is integrated into our other material in the non-standard tradition in a way that the material from Psellus is not, and partly because Usener's first appendix item is not actually verbatim quotation from Olympiodorus. 3.53-54 Bussemaker, taken by him from Valla's translation and printed in Latin, presumably on the basis that he supposed that the Greek did not survive,[48] are in fact simply the second and third parts of [Al.], *Probl.* i.88, which Valla had divided into three sections.[49]

What is striking, as the table above shows, and as was already noted by Bussemaker (1857, xviii) is the extent to which *some* sections in 2.39-126, unlike 1 and 2.1-38, are parallel to the [Aristotle] *Problems* included in Bekker's edition.[50] And, conversely, on certain topics the proportion of those *Problems* also to be found in our book 2 is not entirely negligible.[51] It seems clear that there is a real connection between these *Problems* and our

---

[47] Westerink used E, known to him from Bussemaker: he does not seem to have been aware of X or c. There are indeed parallels between these three texts and other collections; Linke notes (1880, 50 n.6) that 3.50 is parallel to Macrobius, *Sat.* 7.13.1-5 and Plutarch, *Quaest. conv.* 6.1 686e-687b, and (1880, 51 and n.4) that 3.51 is parallel to Macrobius, *Sat.* 7.12.18-19 and to the second explanation in Plutarch, *Quaest. conv.* 6.3 689a-690b. It would clearly be a mistake to assume that the texts collected by Psellus in *De omnifaria doctrina* should be regarded just for that reason as sharing a common history. Nevertheless, in editing the texts we must be guided in part by the contexts in which they have reached us.

[48] At least initially; he realised the mistake when his edition was already in press (Bussemaker 1857, xviii n.1).

[49] Bussemaker 1857, x regarded 3.46-48, 50-52 as "Alexandrian", 3.53-54 as "Aristotelian", and 3.49 as later in date, which is ironical given that 3.53-54 actually come from the [Alexander] problems and that 3.48 and 49 both come from Olympiodorus.

[50] Flashar 1962b, 364 notes that 31 (i.e. 16%) of the 198 problems in [Al.], *Probl.* i-ii also occur in the [Aristotle], *Problems*; that is a considerably lower proportion than that which applies in *some* of the sections of our book 2.

[51] Of 18 problems in book 33 of the [Aristotle] *Problems*, 8 appear in 2.39-53 here; of 62 problems in book 11, twelve appear in 2.83-97; of 25 problems in books 12 and 13, 6 appear here.

collection in these areas at least, and that the latter is to some extent excerpted from the former. This raises the question whether those sections of 2.39-126 which do *not* have numerous parallels in the Bekker *Problems* may have been related in a similar way to some other body of material now lost, and specifically whether this may have been some of the lost works of Theophrastus. But that is a topic that has been discussed elsewhere.[52] Similar questions also arise about the material on animals in 2.127-192,[53] and about 3.9-29.[54] 2.104 is a single statement of an explanation which, as Marenghi 1991, 134 notes, appears *twice* in the [Aristotle] *Problems*, at 12.5 in connection with good odours and at 13.12 in connection with bad ones. It seems more likely that it has been duplicated there as part of an editorial process than that our text has combined two different ones.

Flashar has argued that the *Supplementary Problems* differ from the [Aristotle] *Problems* in that, even where there is a sequence on a specific topic, the answers are specialised and no general *Weltbild* becomes apparent.[55] The extent to which there is a real difference is in our view not

---

52 In Sharples 2004. It must be admitted, indeed, that such a reconstruction is not comparing like with like, for it supposes that the [Aristotle] *Problems* drew on Theophrastus, among other sources, and that the *Supplementary Problems* drew sometimes on the [Aristotle] *Problems* and sometimes on Theophrastus directly. One way of avoiding this asymmetry would be to suppose that both sets of *Problems* are themselves extracts from a single larger collection, as was supposed by Richter 1885, 44-47. See however below, at nn.65-69. – Another possibility, that *Supplementary Problems* 2.39-192 and 3.1-45 Bussemaker are simply extra books of the [Aristotle] *Problems* that have become detached, seems to be ruled out by the extent of overlap between the former and the latter; there are indeed cases of repetition of the same problems in books of the [Aristotle] *Problems*, but not on a comparable scale.

53 Flashar 1962b, 361 and n.2.

54 See Sharples 2004. [Ar.], *Probl.* 27.4 explains the alleged fact that brave men like wine more by the argument that wine, being hot itself, quenches thirst better than water does, and states that the reason for *this* being so has been given elsewhere. As Bussemaker 1857, xix and Flashar 1962b, 694 note, *Suppl. Probl.* 3.14 has precisely this as its question; why does wine, which is hot, quench thirst more than water, which is cold? (The answer is unfortunately corrupt in the text.) Flashar consequently suggests that 3.14 was originally part of [Ar.], *Probl.* as well.

55 Flashar 1962b, 362. Bussemaker 1857, xviii on the other hand had argued that both in subject-matter and in style 2.39-192 were “very like” the [Aristotle] *Problems*, 1 and 2.1-38 very like [Al.], *Probl.* i and ii Ideler. Blair 1999, 177 comments that i-ii Ideler are concerned with “dispensation of knowledge rather than active manipulation of principles” which characterises the Bekker

easy to judge; certainly in both collections there are appeals to principles of Aristotelian physics and biology in the solution of specific problems, and the extent to which a *systematic* physical theory becomes apparent may be more a matter of editorial arrangement than anything else. Of the three texts in the group 2.39-2.53 that have no parallel in the [Aristotle] *Problems* or in Theophrastus, 2.48 appeals to the orthodox Aristotelian doctrine that the purpose of respiration is to cool internal heat; 2.44 explains the pleasure of yawning and stretching in terms of *pneuma*, and there is certainly nothing alien to the early Peripatos in such a type of explanation. 2.85 and 2.95 contradict each other on whether voices are higher or deeper in winter, but so too, as Prantl 1851, 367 noted, do [Aristotle], *Problems* 11.56, 11.61 and 11.17, to which these are parallel. 2.92 on why the pan-pipes and shrill voices give the impression of an empty place draws on the connection of shrillness with distance present also in 2.93 = [Aristotle], *Probl.* 11.20. Striking is the remark, bracketed by Usener as a marginal comment, that the explanation in 2.78 of why greyness and baldness do not begin on the same parts of the head is "more to do with nature than with the discussion of problems".

If we consider parallels both to the [Aristotle] *Problems* and to other Peripatetic texts, there does therefore seem to be a difference in kind between 1 and 2.1-38 on the one hand, and the rest of 2 and 3.1-45 on the other. The latter do seem "Peripatetic", at least in terms of incorporating material from a tradition. Most of the problems in 2.1-38 are much shorter than those in the latter part of the book. 1 and 2.1-38 share the feature that the answer to the problem begins simply "Because" (*ὅτι*) whereas in 2.39-192 it begins with "Is it because" (*ἢ ὅτι*), as in the [Aristotle] *Problems*. And parallels with ancient medical literature seem different in character in the two groups of texts. In the discussion of purging and excretion in 2.105-126 there are parallels in the Hippocratic writings to a number of the alleged facts that provide the problems, though not to the solutions.[56] This section regularly uses for "excessively purged" the term *ὑπέρινος*, which was used by Hippocrates, Aristotle and Theophrastus,[57] but is not part of standard later vocabulary; it is one of the Hippocratic terms that Galen explains,[58] and it also occurs in the lexicographical tradition. In

---

Problems. On the latter see now Jacob 2004, 40-48.

56 Cf. 2.115, 116, 119, 124.

57 Ar., *GA* 3.1 750a29; Theophr., *HP* 9.14.2.

58 Galen, *Ling. Hippocr. explic.*, 19 148.15-149.1 Kühn. 2.6.2 appears to use the present tense of *κτᾶσθαι* in a passive sense, which seems to be a late usage. On the other hand, if *ποικιλότης* of all the MSS and Psellus is the original form in 2.184.2, rather than Usener's emendation *ποικιλία*, that is an argument for a late date at

the case of 1 and 2.1-38, on the other hand, while there are a few parallels to earlier medical texts,[59] the majority are to the writings of Galen[60] or even later.[61] 1.1 is concerned with the digestibility of the testicles of cockerels fed on milk, and 1.6 with the digestibility of fish that live among rocks; Galen refers to both together,[62] but the occurrence of the two topics both in his discussion and in our collection may just be coincidence.

Bussemaker regarded 1 and 2.1-38 as "Alexandrian", 2.39-192 and 3.1-45 as "Aristotelian", and Rose 1863, 216-219, argued that this was indeed the original attribution. That 2.39-192 are more "Aristotelian" in character than 1 and 2.1-38 was indeed recognised by Usener too.[63] But he contended that the "Aristotelian" material in 2.39-192 was used by "Alexander" – i.e. the compiler of the collection consisting of books i, ii, 1 and 2 – and that the order of the "Aristotelian" material was changed to conceal the "fraud". Usener's argument (1859, xii) for the "Alexandrian" origin of books 1 and 2 was that 1.17 referred back to ii.70 Ideler – which in any case on Bussemaker's and Rose's view says nothing about the original attribution of 2.39-192. Usener indeed asked (1859, xi) why, if Bussemaker was right, MS M (and PRT, for that matter) attributed 1 and 2.1-38 to Aristotle, and not just 2.39-192. But that argument cuts both ways, since Usener needed an account of how 1 as well as 2 came to be attributed to Aristotle in some MSS.[64]

---

least for this problem from the *second* part of book 2 (the form ποικιλότης appears to occur first in Origen).

59 2.20 ~ Plato, *Timaeus* (question and solution); 2.14 (question ~ Hippocrates, answer ~ Galen). See the notes ad loc.

60 1.1, 1.4, 1.6, 1.8, 1.18, 2.14, 2.19 (question and solution); 1.12 (question; solution ~ Stephanus, 7th cent. AD). See the notes ad loc. Cf. Filius 1997, 82-83 for the influence of Galenic medical ideas both on the [Alexander] *Problems* (i-ii) and on the Arabic version of the [Aristotle] *Problems*, which he suggests derives from a *Greek* rewriting of the original in late antiquity.

61 1.9 has an etymology for "spleen" from "attract" which occurs in the lexicographer Orion (5th cent. AD) and the medical writers Meletius (7th cent. AD or later) and Leo (9th cent.). See below, n.179 ad loc.

62 Galen, *Meth. med.* 7.6 (10 481.15-17 Kühn), 8.5 (10 575.17 Kühn), *Comp. med. secundum locos* 2.1 (12 518.1-2 Kühn).

63 Usener 1859, ix-xi; indeed he accepts Bussemaker's argument (1857, xviii; Prantl 1858, 205) that in citing "Aristotle" [Al.], *Probl.* i.141 is actually citing *Suppl. Probl.* 2.150 and 156.

64 The fact that Gaza's translation of problems attributed to Alexander includes only material from i, ii, 1 and 2.1-38 (the last three all being treated by Gaza as part of his "book 2") is not evidence that he found only 1 and 2.1-38, but not 2.39-192, attributed to Alexander. For the MS that Gaza used was linked with

Richter 1885, 40[65] explained the differences in character between 2.1-38 and 2.39ff. as reflecting the different approaches to a single body of Peripatetic material of two, or perhaps three,[66] excerptors. (He strangely disregards book 1 Bussemaker completely: 33.) One of the difficulties with Richter's analysis is his tendency to explain the present state of our material in terms of the result of a process of excerption modifying what was originally a single homogenous set of discussions, and – influenced by the relation of some of the Bekker *Problems* to the surviving works of Theophrastus – to regard problem literature in general as itself resulting from the extraction of material from more continuous works and its recasting in problem form; while the latter process certainly took place in the third century B.C. and continued to take place later,[67] it seems that allowance should also be made for the composition of new problems originating in school-discussion rather than in earlier literary sources,[68] and for the development of collections by a process of accretion rather than selection.[69]

---

the NOSbi group (below, §2.1.17). N in its present state ends with 2.22, jumping to the middle of Cassius 1, but this is the result of mechanical loss, and the relation of N to Oi suggests that it originally contained the whole of book 2 (below, §2.1.12-13). NOi treat the whole of book 2 as by Alexander, at least by implication; Oi, like Gaza, do not treat book 1 as a separate book from [Al.], *Probl.* ii, and none of NOi regard 2.1 as the start of a new book. No extant Greek MS, as far as we are aware, in its numbering of the problems treats 2.39 as the start of a new book.

65 Cf. Susemihl 1885, 1482.

66 The third excerptor being responsible for 2.127ff.

67 Cf. below, n.90.

68 One may compare the *Quaestiones* attributed to Alexander of Aphrodisias; some of these may be derived from passages in his lost commentaries, but others definitely are not. Cf. Sharples 1994, 111 n.23, and references there.

69 Richter allows that this process played a part: 1885, 40. But he refers it to the 11th and 12th centuries A.D.; the process of accretion may at an earlier date too have played a greater part than his general analysis recognises. Sandbach 1982, 224 considers the question whether differences between the [Aristotle] *Problems* and Plutarch's citations of Aristotle are to be explained by Plutarch's having used an original version from which the extant collection of [Aristotle] *Problems* derives, or conversely an expanded version deriving from these.

1.2. The secondary tradition

That parts at least of our material were attributed to Aristotle in antiquity is suggested by a number of references and parallels. As Roselli 1979 noted, Oxyrhynchus papyrus 2744, from the 2nd century A.D. quotes a fuller version of problem 2.156, on why pigs like things with a bad smell, as "Aristotle in the *Aporemata*". Luppe 1970 argued that the text in the papyrus which contains this quotation is a commentary written by Didymus Chalcenterus in the first century B.C. Even if this is not so, at any rate we have evidence that this problem not only existed but also belonged to some collection attributed to Aristotle already in antiquity. However, Flashar suggests (1962b, 367) that Plutarch, *De fac. in orb. lun.* 19 932bc (= Aristotle fr.210 Rose^3) refers to ii.46 Ideler; and if this is correct we have an example of an "Alexander" problem which was earlier attributed to Aristotle. This suggests that 2.156 too *could* have been part of a collection which was attributed to Alexander at a later stage in its history.[70]

That oil causes water to slip over it is asserted in 3.47, in the context of oil calming the waves, and also in Bekker *Problem* 32.11, but in the specific context of what happens inside the ears of divers. Flashar 1962b, 360 n.1 notes that Plutarch, *Quaest. Nat.* 12 915a, where Aristotle is cited for this point, shares the reference to divers, and argues that it probably refers to the Bekker *Problem* rather than to our 3.47. However, the question in Plutarch, and the point for which Aristotle is cited, is the general one raised in 3.47 rather than the specific point about divers in the Bekker *Problem*; and in fact Plutarch goes on to attack the claim of those (plural) who say that the same explanation applies to divers. So it looks as if the reference may be to 3.47 after all. Moreover Plutarch, *De prim. frig.* 13 950b, where Aristotle is cited as saying that the oil makes the *wind* slip over the waves, seems to be a reference to *Suppl. Probl.* 3.29.[71] And the citation of Aristotle at Athenaeus 14.72 656ab (cf. Rose 1863, 219) is, as Bussemaker (1857, xix) and Flashar (loc. cit.) note, closer to 3.43 than to Aristotle, *Meteor.* 4.3 380b21; it shares with the former but not with the latter the point that roasting after boiling does not improve food.[72] Flashar rightly observes that such citations prove

---

[70] True, Plutarch only gives a brief (and corrupt) reference to ii.46, while the papyrus quotes the actual text of 2.156. See also below, n.457 on 2.145.

[71] Cf. Flashar 1962b, loc. cit.

[72] However, as Sandbach 1985, 8-9 notes, Athenaeus says that what is roast is dryer than what is boiled, while the *Meteorologica* and our text both say the opposite, the *Meteorologica* explaining that the greater heat of the surrounding fluid draws out the moisture in the thing being boiled.

nothing about the antiquity of the entire books in which they are now found; that is indeed true, but they do suggest that the problems referred to formed part of some collection attributed to Aristotle (since they can hardly have circulated in isolation), even if we cannot rule it out that material from this was subsequently incorporated into an "Alexander" collection.[73] A *terminus ante quem* for the existence of some of the texts is also provided by parallels with book 7 of Macrobius' *Saturnalia*; Linke 1880, 52-54 noted 24 such parallels in [Alexander], *Probl.* i and ii, arguing that only four of these are close enough to suppose that Macrobius might have been dependent on "Alexander", and that we should therefore rather think of a common source. There are also parallels between Macrobius and *Suppl. Probl.* 2 and 3, but these are fewer in number.[74]

Evidence that the first part of *Suppl. Probl.* 2 was at one time attributed to Aristotle is provided by the medieval Latin collection which Rose edited as the *Problemata Bambergensia,* attributed to Aristotle in the oldest, late eighth-century MS.[75] 23 of *Suppl. Probl.* 2.1-38 have close, though often abbreviated, parallels in this collection. Of the remaining 20 *Problemata Bambergensia*, nine are parallel to problems of Cassius the Iatrosophist and one to [Al.], *Probl.* i.99.[76] Lou Filius, who is preparing an edition of Arabic versions which include material from the first part of book 2, in the standard order, interspersed with material found in the *Problemata Bambergensia,*

---

[73] Other parallels are less close. 3.17, asking why bronze corrupts wine but not water, mention that wine *and oil* stick to the bronze; Rose points out (1863, 219) that Plutarch, *Pyth. Orac.* 3 395F cites Aristotle as explaining the fact that oil creates rust (patina) on bronze by arguing that other liquids disperse the rust but the oil because of its density does not do so. The doctrine is similar, but the point at issue different. See also below, n.528 on 3.9.

[74] Linke 1880, 54 n.2. 2.4, on eunuchs, is parallel to Macrobius 7.7.8, and Macrobius 7.12.9-10, on wine and honey, is closer to *Suppl. Probl.* 3.22 than to [Al.], *Probl.* ii.70. However Macrobius 7.10.13, also on eunuchs, is closer to i.8 and i.97 than to 2.5; Macrobius 7.10.5, on baldness at the front of the head, is closer to i.2 than to 2.80; and Macrobius 7.6.9 contradicts 2.12, the former arguing that wine causes trembling because it is cold, the latter that it does so because it is hot.

[75] Bamberg, Staatsbibl. cod. med. 1; ed. Rose 1863, 666-676, and Stoll 1992, 68-75. For the dating of the MS cf. Stoll 1992, 13-14; the collection, also known as the *Problemata vetustissima*, additionally appears, without the attribution to Aristotle, in two 12th-century MSS, Bruxellensis 2419-2431 and Berolensis lat. qu. 198 p.14 (Rose 1863, 606; Stoll 1992, 31). Cf. Rose 1863, 219; Flashar 1962b, 359 n.1.

[76] Rose 1863, 666-676; Stoll 1992, 31.

Cassius[77] and the "Bekker Problems", has kindly informed us that in all the Arabic MSS that give an attribution these texts are attributed to Aristotle rather than to Alexander.[78] Since it is unlikely that either the Arabic or the Latin is directly dependent on the other, this suggests that there existed a collection in Greek which was the common source of the Arabic and the Latin versions. It is impossible to be certain exactly what it contained, because the presence of a problem in one collection but not in the other could be due either to addition in the former or to omission in the latter; but the similarity in content between the Arabic and the Latin collections does suggest that at some stage in late antiquity 2.1-38 existed as all or part of a separate collection which was distinct both from 1 and from 2.39-192 Usener, and which was attributed to Aristotle. Flashar 1962b, 366 has argued that the prologues of [Alexander], *Problems* ii and of *Supplementary Problems* 1[79] are so similar in theme that it is difficult to suppose that the two books originally formed parts of a single collection.

There are also parallels between texts from book 2 and the medieval Latin *Salernitan Questions*,[80] all but one coming from 2.1-38 and involving texts also included in the *Problemata Bambergensia.* Although the problems correspond, the answers in the *Salernitan Questions* are different to a greater or lesser extent.[81] In at least one case our Greek text is compressed to a point where it is only comprehensible with the aid of the Latin.[82]

---

77 There is little correspondence, as to which Cassius problems they include, between the Arabic version on the one hand, and the "non-standard" sequence of problems attributed to [Alexander] in the Greek on the other.

78 Cf. Filius 1999, xiii, xv-xvi: also Meredith-Owens 1955-56, Daiber 1970, 545 and Ullmann 1970, 94-95. [Alex.], *Probl.* i and ii, on the other hand, are attributed to Alexander in a Syriac version (Daiber 1970, 545-6; Filius 1999, xvi.).

79 This indeed depends on the view that the preface to book 1 does indeed belong to it; see nn.8, 149. The preface is not actually included in Bussemaker's edition, even though it is present in all the MSS which contain the whole of book 1; its absence in A is probably only due to damage (below, §2.1.2). It is included by Usener. Bussemaker was no doubt influenced by Dietz' view that the prologue was not originally written for this book.

80 Noted by Lawn 1979; we are also grateful to Charles Burnett for drawing our attention to specific parallels here.

81 2.2 ~ *Quaest. Salern.* Ba63; 2.4 ~ P43; 2.5 ~ P40; 2.8 ~ W3 (the answer may in fact be the same here, but 2.8 is too compressed for it to be clear); 2.12 ~ P21; 2.13 ~ B228; 2.19 ~ B294-5; 2.39 ~ Ba39.

82 See below, n.239 to 2.19.

1.3. The evidence of the MSS

1.3.i. MSS with the standard sequence

The Greek MSS of the *Supplementary Problems* fall into two groups.[83] Those in the first group, the "standard sequence", contain the texts, or excerpts therefrom, in the order in which they appear in the modern editions, allowing for minor variations. Of this group five MSS (MPRTm), of which the earliest (P) is thirteenth century, attribute books i and ii to Alexander and books 1 and 2 to Aristotle.[84] In M, but not PRT, books 1-2 *precede* books i-ii, though book 1 is none the less described as "Selections from Aristotelian medical and physical problems, *book 3*". Two further MSS (k, 14th-15th century, and w, 16th century) have only part of book 2, and book 1, respectively, in both cases attributed to Aristotle. Two MSS from the 14th and 15th centuries (NS) attribute book 1 to Alexander and treat books 1 and 2 as a single book;[85] seven more implicitly attribute all four books to Alexander, either by not distinguishing the books (COgi)[86] or as a result of omissions (AFK).[87] One MS (e, 16th century) apparently attributes book 1 to *both* authors, and book 2 to Aristotle (see further below); one dated to 1325-1334 (b)[88] attributes book 1 to Alexander and leaves book 2 unattributed. In MS V from the 16th century the prologue to book 1 appears without attribution to any author. And finally, three MSS

---

83 For details of the MSS, and for our conjectural stemma, see below, 82-84.

84 See below, nn.149 and 209.

85 In N book 1 is explicitly attributed to Alexander, and book 2 follows on from book 1 with no break in the numbering. In S book 1 is attributed to "the same", i.e. Alexander, and books 1 and 2 are not distinguished. See Garzya 1998, 86.

86 C contains only excerpts, from all of books i, ii, 1 and 2, in a single sequence with sequential numbering and no recognition of the book divisions. MS 1 originally formed part of N; see below, §2.1.11. O and i run ii, 1 and 2 together, as does g (the translation by Gaza, above n.35. On O and i see also below, §2.1.12). On the letter from Trebizond which Cranz attributes to Gaza and regards as evidence for the latter's use of a now lost MS see below, §2.1.17.

87 A has the beginnings of both book 1 and book 2 missing as a result of damage (below, §2.1.2). F jumps – with no break in numbering – from near the end of book ii to the middle of book 2, and thus is missing the beginning of book 2; of book 1 it has only the preface, *following* book 2, and attributed to *Alexander*. K is missing the whole of book 1 and the first part of book 2. The translation by Giorgio Valla, based in part on K, attributes material from books i, ii and 2 to Alexander.

88 So dated by Garzya 1998, 86.

(EXc), of which the earliest (X) is again 13th century, and which we will label "collection A", attribute to Alexander five books, of which books 1 and 2 are i and ii from the standard sequence, with omissions; book 3 is a selection from the first part of book 1 in the standard sequence followed by a selection from [Galen], *Def. med.* in question-and-answer form and then by three problems from Psellus (above, §1.1); books 4 and 5 in these MSS are material from *earlier* in [Galen] *Def. med.* with a selection from Palladius' *Synopsis de Febribus*[89] intervening in book 4.[90]

The nature and complexity of the tradition does not make drawing up a conjectural stemma easy, but if we hypothetically rule out contamination between MSS except in cases where there is a specific reason to suppose it, there are good reasons – in the sense of omissions of phrases or of entire problems – to suggest that PRTw at least are not dependent on any other extant MSS, and that therefore it cannot at any rate be *demonstrated* that the attribution of books 1 and 2 to Aristotle is secondary and that to Alexander primary.[91]

At the start of book 1 MS e has "Section from the second book of Alexander's *Problems*: Aristotle on the peculiar characteristic (sic) of animals", with a marginal note (no doubt resulting from comparison with S: below, §2.1.15) identifying this rather as the *third* book of Alexander's *Problems*. In both e and m book 2 is attributed to Aristotle but said to be book 3 of the medical problems, book 4 of the physical. Book 1 also has the title "Aristotle on the peculiar characteristic of animals" in MPRTmw.[92]

---

89 Identified by Wiesner in L; see below.

90 Usener 1859 ix, not recognising the material as a version of pseudo-Galen, remarks that Alexander's name was so distinguished that it came to be applied to "such pitiful little questions". With the conversion of the *Definitiones medicae* to question-and-answer form one may contrast the presentation of i.85-99 Ideler in the form of statements rather than problems in Oxford, Barocc. 131 404v-405, noted by Pontikos 1992, 2-6 (cf. Sharples 1998a, 393), and the conversion of problems to statements in Psellus' summary (see below). See also below, n.98, on BL.

91 See below, §2. Usener 1859, xi indeed argues that "the good MSS" attribute all four books to Alexander. – Whereas ekw are the only MSS with the standard sequence that contain books 1 and 2, or parts thereof, without books i and ii, there are 7 more MSS known to us that contain the whole of books i and ii, and another 16 with excerpts therefrom, without books 1 and 2. Books i and ii themselves are always attributed to Alexander in the Greek MSS tradition, never to Aristotle.

92 With the book number 3 associated with this title in PRTmw, but not in M. See below, n.149.

But "peculiar characteristic of animals" seems to fit the contents of the latter part of book 2 (127ff.)[93] much better than book 1. This could be explained if the latter part of book 2 was originally so described, and this title was then transferred to the start of book 1 so that a description of all the contents of book 1 and 2 should appear at their start.[94] For one possible explanation of the subsequent variations is then that some copyists will have taken the dual title as a sign that *all* the material from this point on was Aristotle collected by Alexander, and chosen to attribute both 1 and what follows, including book 2 when they came to it, to Aristotle (ζη); others will have treated the reference to Aristotle as incidental and attributed book 1, and following it book 2, to Alexander, explicitly or implicitly (εκ). However, it may well seem implausible that the late MS e should have preserved the dual form of the title where no other extant MS has done so; the dual title could be the result of contamination. And, while the title "the peculiar characteristics of animals" fits the latter part of book 2 – at least from 2.127 onwards – better than it does book 1 as a whole, it is not, read in a medical sense as pertaining to diet, inappropriate to the *first* problem in book 1, which is "Why are the testicles of cockerels fed on milk large and easy to digest?" (cf. also 1.6).[95]

That the attribution to Alexander, as well as that to Aristotle, is early as far as the Greek tradition is concerned is suggested by the fact that MS o, which contains material from the "non-standard tradition" (see below) at 62v-64v, and i.85-99 Ideler in the form of statements rather than problems at 404v-405 (above, n.90), also contains, at 441v-446, a paraphrase of much of books i-ii and 1-2 (and the *Problems* of Cassius) attributed to Psellus.[96] In

---

93 'On the peculiar characteristic of animals' is also the Greek title of the work by Aelian commonly referred to as *On the Nature of Animals*. On the relation between our text and Constantine Porphyrogenitus' collection see above, 3 and n.12.

94 This would admittedly require also that *two* originally separate titles – one indicating the start of "Aristotelian" material at 2.39, the other the start of the discussion of animals at 2.127 – had been collapsed into one.

95 Flashar 1962a, 409 n.2 suggests that the title indicated book 1 and the first part of 2 by contrast with 2.39ff. on *common* symptoms. But 1 and the first part of 2 are not obviously as a whole about particular animals.

96 Edited, from this MS, by Duffy 1992, who recognises that [Alexander] is the source: cf. his preface, xxxvii-xxxviii. According to Wilson 1978, 171 the whole of 397v-446v in o are material assembled by Psellus as source material for his *De omnifaria doctrina*. o is dated to 1250-1275 by Bodleian Library 1966, 22. See also Garzya and Masullo 2004, 16.

this paraphrase the texts are considerably abbreviated, and their expression as problems removed by omitting the initial "Why?" and subsequent "Is it that?" so that "Why is such-and-such the case? Is it because ...?" becomes just "Such-and-such is the case because ...". The sequence in this MS is based on the standard ordering, but dislocated in a way that seems to bear no close relation to that in any other known MS.[97] Nevertheless, the fact that Psellus (who lived from 1018 to 1097) drew on all four books i, ii, 1 and 2 suggests that they had already been united into a single corpus by his time. He gives no indication that he thinks he is drawing on material from more than one author; this is an argument, though admittedly an argument from silence, for his having found the material in such a form that he could regard it as all in some sense from a single author, and, given that books i and ii are always attributed to Alexander, that can only have been Alexander. If there was no attribution of books 1 and 2 to Alexander in αβ we would have to suppose that Psellus and ε both failed to notice that they were dealing with a collection two books of which were attributed to Alexander and two to Aristotle.

1.3.ii. MSS with non-standard sequences

Further evidence for the attribution to Alexander being early is provided by a further group of MSS which contain material from some or all of the four books i, ii, 1 and 2 in non-standard sequences. As will be argued later in this Introduction (below, §2.2) these can be traced back to two MSS, one extant (J), the other (γ) hypothetically reconstructed. The first group ("book 2" in BL, and DHJW) is not really relevant to the present discussion, for it includes nothing from books 1 and 2, and only two problems which were included by Bussemaker in his third book, as 3.46 and 3.47, but occur *only* in this group of MSS.[98] The second group (DGYjo)[99] however includes

---

97 1.1-22, 2.1-160 (2.45 and 2.46 follow 2.73; 2.47 and 2.156 are omitted), i.101-152, 2.162-3, ii.1-76, 2.161, 1 praef., 2.164-192, Cassius 1-18.

98 In BL what is presented as "book 2" is a version of part of the non-standard sequence, while what is presented as "book 1" is excerpts from books i, ii, 1 and 2 in the standard order, all attributed to Alexander. These two MSS, which we will label "Collection B", have the same "books 4 and 5" as do EXc, and no "book 3" at all, but instead Avitzianus' treatise *De urinis* (= Ideler 1841-1842, 2 286ff.) See Moraux et al. 1976, 433-434.

99 Together with two MSS, Vienna, med. gr. 44 and Venice, App. gr. IV.16, which include only items from [Al.], *Probl.* i. The relevant pages in o are 62v-64v: see

material from 1 and 2, as well as i and ii; and the important point for the present argument is that in most of these "non-standard" MSS *all* the material is attributed to Alexander of Aphrodisias.[100] The "non-standard" MSS include some which are earlier than all or most of the "standard" MSS; J is dated to 1075-1085 by Rochefort,[101] G is 12th century and Xo are 13th. The non-standard sequence includes some of the problems of Cassius, which frequently accompany the [Alexander]/[Aristotle] *Problems* in MSS with the standard sequence as well.[102]

There are some places in the non-standard sequence where it seems as if there has, by comparison with the standard sequence, been some attempt at grouping texts by theme, even if in a very localised way.[103] But there are

---

above, at n.96. D, which is relatively late, includes sequences from both groups; it can be demonstrated (see below, §2.2) that this is because it draws successively on different exemplars.

100 The exceptions are W (14th century), which has "some wise person" in the title, and what appears to be "'Ἀλεξάνδρου Βασιτίου" in the margin, and j (15th century), in which part of the non-standard sequence is attributed to "Alexander of Alexandria". On j cf. Ruelle 1892, who argues on the basis of abbreviations in the MS that the "fraud" – one might rather say, the confusion? — goes back to a MS of the 13th or 14th century.

101 Rochefort 1950, 4. Omont 1886-98, pt.3 301, had dated this MS to the 12th century.

102 Of 22 MSS known to us, directly or from Garzya 1994 and Garzya and Masullo 2004, containing all or some of the *Problems* of Cassius, 13 (ACKMNPRSTbm*p*w) contain all or part of the *Supplementary Problems* in the standard sequence; two more, Florence, Laurentianus gr. 75.13 (above, n.42) and Paris, BN, ancien gr. 2230, contain some problems of Cassius along with part of [Al.], *Probl.* i-ii; and DGYo contain Cassius as part of the non-standard sequence. (The three MSS that contain the *Problems* of Cassius but not the *Supplementary Problems* are Milan, Ambr. gr. 23 (A 92 sup.), Paris, anc. gr. 1943, and Vatican, gr. 1614; also Vat. gr. 1398, but that is a copy of the printed edition of Cassius. Garzya and Masullo 2004, 15-20.) Conversely, of the MSS of our standard tradition containing books 1 and 2 (EXc, with only a few problems from book 1, are hardly relevant here) only four (Oeik) do not contain material from Cassius. This suggests, what is in any case clear from the presence of Cassius material in the non-standard tradition, that the Cassius problems and ours were transmitted together from an early date. The association in the *Problemata Bambergensia* (above) of problems from Cassius with material from 2.1-38 specifically may require separate explanation.

103 In one early MS (o) i.1, on hair, follows 2.56 on hair and 2.59 on nails. In two MSS (GY) i.50, on cold, is followed by 2.24 and 2.27, on the same theme.

also places where a thematic sequence seems to have been broken up;[104] and in general the "non-standard" sequence does not show the degree of systematic organisation we find at least in book 2 in the standard tradition. Typically, where there is a group of related problems in the "non-standard" sequence, it consists of extracts from a *longer* group of problems on the same topic in the "standard" sequence.

The possibility that the non-standard sequence is primary, the standard sequence having developed from it by rearrangement and accretion, can be ruled out on the grounds that such a rearrangement would have had to create *both* the groupings by approach and type of discussion *and* the groupings by theme which we find in the standard tradition; that one or the other might have been introduced may be possible, but that both should be seems unlikely. The additions that would have been required would also have been considerable; the "non-standard" MSS with the greatest number of problems are DGYo, with respectively 146, 113, 157, and 70 problems (we omit those from Cassius), whereas the four books of the standard sequence total 442 problems in all. And the presence in the non-standard sequence of some problems from Cassius is a further argument for its secondary nature; if the standard sequence was created from the non-standard ones the Cassius problems would have to have been carefully removed. If on the other hand, as seems likely, the non-standard sequences are the result of selection from and rearrangement of the standard one, it seems reasonable to suppose that the compiler of the second sequence found all his material already attributed to Alexander, rather than that he attributed to Alexander material some of which he found attributed to Alexander and some to Aristotle.

At this point one might wonder whether it is the attribution to Aristotle, rather than that to Alexander, that is itself a late and erroneous addition to the tradition. There are however two arguments against this. First, with the stemma as we have reconstructed it (see below, §2.1.1), the addition would have to have been made independently in ζ and in η. And, secondly, there is, as we have seen, evidence that some of our material was attributed to Aristotle at a much earlier date; so that if it did appear only at a late stage in the transmission, it was in fact a revival of an earlier attribution.

---

[104] Thus in the late MS D i.26-8 Ideler, which are concerned with wisdom, follow i.11-i.21 Ideler, on physiological reactions to emotion, but are separated from them by 2.59, on fingernails. In another group of MSS (GYo) i.28, on wisdom, comes between i.9 and i.5, both on eunuchs. In BDJLW 3.46, on eagle's feathers, comes between ii.68, on reproduction, and ii.25-26, on the effects of cold; and the latter are followed by 3.47, on the effect of oil on water, which is followed in turn in D by i.24-5, on itching. The early MS o places Cassius 53, on fever, between Cassius 49 and i.13 Ideler, both on physiological reactions to emotion.

While 2.39 ff. are more Aristotelian in character than 1 and 2.1-38, 2.1-38 were attributed to Aristotle at some point before the eighth century. Perhaps 2.1-38 and 2.39ff. were then joined together, and attributed as a whole to Aristotle; subsequently, perhaps, they were united with book 1, already attributed to Alexander. The attribution to Alexander might then explain how these books came even later[105] to be connected with [Al.], *Probl.* i-ii, both in the Greek MSS transmission at the stage where it becomes directly accessible to us, and also in the excerpts made by Psellus and by the compiler of the non-standard sequence which includes material from *Suppl. Probl.* 1 and 2.

1.4. The title

The title appears as *Natural Puzzles and Medical Problems* (φυσικὰ ἀπορήματα καὶ ἰατρικὰ προβλήματα) in MSS EX (books ii and 1, the start of i being lost), N (book 1), c (books i and 1) and m (book 1); this is usually in the genitive depending on *Selections* (ἐκλογαί), the exceptions being N(1) and c(i); in c(1) ἐκλογαί is added in the margin. The reverse order *Medical Problems and Natural Puzzles* (ἰατρικὰ προβλήματα καὶ φυσικὰ ἀπορήματα) appears, with ἐκλογαί, in M(1), em(2) and Tmw (subscription to 1). F(i) has just *Natural Puzzles* (ἀπορήματα φυσικά). The pairing of the terms is exchanged, giving *Medical Puzzles and Natural Problems* (ἰατρικὰ ἀπορήματα καὶ φυσικὰ προβλήματα), in the heading to book ii in KNOSbi (KOi have ἐκλογαί, Nb do not); this exchanged pairing is unique to this group of MSS. With some hesitation we have adopted *Natural Puzzles and Medical Problems: Selections* as the most widely attested title.

A separate question concerns the title *Solutions to certain Puzzles* (ἐπί τισιν ἀπορίαις λύσεις). This is the title for book i in NOSbi, for "book 1" (in the standard sequence) in BL (above, n.98), and in the non-standard MSS GY. *Puzzles and Solutions* (ἀπορίαι και λύσεις) is the title for book i in K, and in the non-standard MSS H, o, and Vienna med. gr. 44.[106] It would appear that these titles are secondary and have arisen by confusion with the different collection of *School-puzzles and Solutions on Nature* (φυσικαὶ

[105] Even later, because of the similarity of the prefaces to ii Ideler and 1 Bussemaker – if indeed the latter was already attached to book 1. See above, at n.79.

[106] Of the other non-standard MSS in the BHJLW group, J and BL "book 2" have simply *On Nature* (φυσικά), W (like C) has *Enquiries* (Ζητήματα). j has *On Natures* (περὶ φύσεων), D *Natural Puzzles* (φυσικαὶ ἀπορίαι).

σχολικαὶ ἀπορίαι καὶ λύσεις) more reliably attributed to Alexander or at least to his school.[107]

1.5. The connection with Alexander and the date of the collections

There is no reason to suppose that the "Aristotelian" problems in 2.39ff. have any more connection with Aristotle himself than do the better-known problems edited by Bekker; some or all of them may well derive from the early Peripatetic school, and Aristotelian authorship of individual items cannot perhaps be excluded, but we have no way of telling.[108] What however of the possible connection of any of this material (including books i and ii) with Alexander of Aphrodisias himself? We do find orthodox, or reasonably orthodox, Aristotelianism not only in 2.39-192 but also in some parts of 1 and 2.1-38 as well. 1.3 argues that milk is concocted blood, as at Aristotle, *GA* 4.8 777a7ff. The allegory in 1.17 explains Pan's love of Echo by saying that the heavenly sphere desires harmonious movement; the term used for desire, *ephesis*, is that used by the genuine Alexander for the desire of the heavens for the Unmoved Mover,[109] and the idea that the motion of the heaven is *itself* the object of the desire of the heaven has recently been advanced as an interpretation of Aristotle by Broadie 1993.[110]

However, there are a number of considerations against Alexander's authorship of i-ii, 1, or 2.1-38.

(1) Flashar 1962b, 363 argues that the prologue to book 1, which says that Hippocrates was a gift of divine providence to human beings, cannot be by Alexander because it follows a Stoic doctrine of providence which Alexander himself rejected.[111] Flashar argues for similar Stoic influence also in the reference in 1.16 to the influence of the stars on human beings at their birth and in bringing about plague.[112] The idea that the heavens have an influence on the sublunary is not in fact incompatible with Alexander's views, provided it is not understood as determining outcomes completely or

---

[107] Edited by Bruns 1892.

[108] Louis 1991, 149-155 argues that book 10 of the [Aristotle] *Problems* is by Aristotle himself; Flashar 1962b, 503 that it is closer to him than the rest. (We owe these references to John Dudley).

[109] Alexander, *Quaestio* 1.1 3.16-4.2, 1.25 40.11-26.

[110] See Sharples 2002, 5-7.

[111] The fact that there is a marginal note in MS M saying "this is like Aristotle" is hardly a counter-argument. Cf. Flashar 1962b, 362-363.

[112] Already cited as un-Aristotelian by Prantl 1851, 367 and n.72.

as involving providential concern for individuals;[113] Alexander's attitude to astrology in the sense of the influence of natal horoscopes is a topic of debate.[114] However, the view of providence in the prologue at least is difficult to reconcile with Alexander's. Prantl (1851, 376), followed by Flashar (1962b, 366) argue that the argument in i.26 that small people are wiser because their souls are concentrated into a smaller space suggests a conception of the soul as a material substance in its own right, which does seem incompatible with Alexander's hylomorphic views.[115] 2.8 explains temporary loss of vision when moving from an illuminated place to a dark one by saying that the darkness conceals "this substance (*ousia*) which enables vision". It is unclear from this text alone whether what is envisaged is the Platonic or Stoic theory of a visual ray emerging from the eye, or the theory of Aristotle's *De Anima* that the transparent is matter for light. However, Filius 1997 has shown that both the Arabic version of the [Aristotle] *Problems,* and the [Alexander] *Problems,* have a distinctive theory of visual *pneuma* deriving from psychic *pneuma*, the latter being contrasted with vital *pneuma* as in Galen; and *Probl. Bamberg.* 10, which is parallel to 2.8, refers to visual *pneuma* explicitly. It seems that 2.8 is a compressed and hence unclear version of the original explanation, and that that did indeed involve something issuing from the eye.[116] Aristotle himself sometimes speaks in terms of something issuing from the eye, rather than of the *De Anima* theory;[117] however, Alexander himself loyally maintains the *De Anima* theory even when (as in the *Meteorology*) Aristotle's text seems to imply something different.[118]

---

[113] Cf. *Quaest.* 2.3, with Genequand 1984, 114 and 2001, 18; Sharples 1987, 1188; Fazzo and Zonta (1998) 41, 63-68. Whether Alexander would accept that the influence of the heavens is *solely* responsible for the occurrence of plague – if the *Problem* indeed implies that – may be more questionable.

[114] See Fazzo 1988, and Sharples 2001, 526-9.

[115] Even if Alexander's view of soul were regarded as epiphenomenalist, the language of the *Problem* would not seem appropriate. On Alexander's conception of soul see Caston 1997, 347-54. We may note that [Ar.], *Problem* 30.3 argues that small heads go with greater practical wisdom.

[116] [Alex.], *Probl.* ii.37 solves the same problem by appealing to the darkness making the visual πνεῦμα dense and turbid, or else to the suddenness of the change. Prantl 1851, 361 n.54 (cf. 376) saw this as an indication of an un-Aristotelian materialism; he also notes the references to our natural heat "fleeing" and the like at [Alex.] *Probl.* i.12, i.130 and ii.22, comparing [Ar.], *Problem* 27.10.

[117] Cf. Sharples 2003, 217-220 and references there.

[118] Ibid. 218 n.73.

(2) Conversely, Rose argued (1863, 219) that books i and ii show Neoplatonist influence. Rose cites a number of specific passages but does not discuss them in detail; not all seem equally telling, but there are enough both to rule out Alexander's authorship of the whole of these books and to show, when combined with point (1), that the collection draws on more than one source. i.87 refers to "the divine Plato" – as does the prologue to *Suppl. Probl.* 1; but so too does Alexander himself in the genuine part of his commentary on the *Metaphysics*, at 18.3. However, i.89 advocates asceticism, i.121 appeals to Plato's theory of recollection, and the prologue to book ii asserts the impassivity of soul. ii.46 mentions – without endorsement, but also without any distancing of the author from the ideas expressed – how people clash metal objects in order to drive away δαίμονες during eclipses because the light of the heavenly bodies is no longer available to do so. ii.47 refers to seven and ten as perfect numbers; but David Sedley has recently emphasised that the idea of significant numbers is found in the Aristotelian corpus too.[119] ii.53 discusses the evil eye, and ii.61 refers to air as like a cord or rope between earth and heaven. ii.67, which begins as a discussion of why oil and water do not mix, lists the types of mixture which do not apply to the relation between soul and body – which is not dissimilar to what Alexander himself does at *DA* 15.5ff. and *Mantissa* 115.33ff., 116.5ff. – but then goes on to describe πνεῦμα as a mixture of corporeal and incorporeal, which recalls *Timaeus* 35a, to describe its functioning in the body in a similar way to Galen, and to describe the spherical shape of the head in terms reminiscent of *Timaeus* 44d.

(3) The allegorical explanation of myth that we find in 1.17 (in fact combining allegorical and etymological interpretation), and also, as Usener 1859, xi notes, in i.87, is not characteristic of the genuine Alexander; neither is the statement at the end of 1.12 that an egg resembles the universe because it is spherical, made of the four elements, and shares in the life-giving faculty. The attribution of life-force to the world as a whole, and indeed the reference to the universe as composed of only *four* elements, are Stoic or Platonic rather than Aristotelian. The argument in 1.19 that the sphere is the most perfect shape is not in itself un-Aristotelian (though I am not aware that Aristotle applies it, as here, to the shape of plant seeds), but it does not seem in the manner of either Aristotle or Alexander to express it by saying the sphere is the most perfect shape *in god's eyes*.

---

[119] Sedley 2000, 331 n.7. The statement in ii.47 that "according to Aristotle and the philosophers" the number ten encompasses things subject to coming to be and passing away may cause some surprise (a reference to the categories?).

(4) There are also divergences from Aristotelian doctrine on specific points, which are strikingly in contrast with Alexander's own tendency to defend Aristotelian orthodoxy even when subsequent discoveries, e.g. in anatomy, had made it hard to do so.[120] (Significantly, these divergences occur in book 1 and the first part of book 2, rather than in the second, "Aristotelian" part of book 2.) Some of the divergences themselves show the influence of later medical thought. 1.1 recognises the existence of "female testes", i.e. ovaries, known to Hellenistic and later medicine but not to Aristotle.[121] 1.18, distinguishing between parts of the body formed from seed and those formed from blood, the latter being able to regenerate but the former not, contradicts Aristotle, for whom *all* parts including sinews are formed from blood (*GA* 2.6 743a18); a distinction like that in our text is however found in Galen.[122] Other divergences from Aristotle include 2.5, which explains the higher voices of eunuchs quite differently from the way in which Aristotle does himself,[123] and 2.10, which attributes the menstrual blood of women to their less active life, whereas Aristotle attributes it to their natural coldness and weakness (*GA* 1.19 726b31). 2.24 says that snow burns because it is compressed and extremes (here, of hot and of cold) are alike; Aristotle, *Meteor.* 4.4 382b8, followed by Alexander in his commentary (*In Meteor.* 202.28ff.), says rather that cold is *said* to burn because it concentrates *heat*. It rather looks as if Aristotle's explanation may have been misunderstood.[124]

(5) The general intellectual level of i-ii, 1 and 2.1-38 does not seem characteristic of Alexander[125] – though Bruns could only explain some sections of the *Quaestiones*, which do seem more securely connected with

---

120 Notably in his argument at *DA* 96.28-9 that the eyes, ears and nose are connected through the head to the heart as the common sense-organ.

121 Below, n.162 on 1.1. The term our text uses is not Herophilus' δίδυμοι, but ὄρχεις, used for ovaries e.g. by Galen, *De venarum arteriarumque dissectione*, 2 810.2ff. Kühn.

122 See below, n.204 on 1.18.

123 Below, n.216 on 2.5.

124 For disagreement with the Bekker *Problems* see also below, n.458 on 2.146.

125 Flashar argues (1962b, 366) that in books i-ii the impression is of *ex cathedra* statements cast in problem-and-answer form, rather than of the results of genuine school discussion and debate, and questions whether such a collection can genuinely derive from Alexander. This is reflected in the use, which Flashar notes, of "I say that" in place of the characteristic "Or rather" (which still seems as an appropriate translation of ἢ, in spite of the comments of R.B. Todd at *BMCR* 1995, 05.05) of the [Aristotle] *Problems*. See also above, 9.

Alexander, by regarding them as notes dictated for teaching purposes,[126] and Richard Sorabji has suggested that some of the same collection read like student essays.[127]

Leaving aside details of the argument of individual problems, a wider question is raised by the attribution to Alexander of Aphrodisias of a collection which includes a reference to medicine in its title, even though the content is "medical" only in a relatively broad sense. The fact that Alexander is included in one list compiled in late antiquity of those "distinguished" in medicine[128] may not mean all that much. The fact that he drew upon Aristotle's physiological and zoological works in his own work *On the Soul*, as Accattino and Donini have emphasised,[129] does show some interest in biological matters.[130]

There is another medical work attributed to Alexander, the treatise *On Fevers*. Usually his authorship of this has been denied, on the grounds that he wrote no medical works apart from the spurious *Medical Problems*; and his authorship of the latter has been denied on the grounds that he wrote no medical works apart from the spurious *On Fevers*. This is plainly a case where the argument becomes weaker the more works it is applied to. The most recent editor of *On Fevers*, Tassinari, came to the conclusion, following a suggestion of Verbeke, that this treatise was probably written by someone lived at approximately the same time as Alexander of Aphrodisias, who had a particular interest in philosophical analysis, rather than in medical treatment, and who held distinctively Peripatetic views concerning natural philosophy and the status of forms.[131] Tassinari did not himself go as

---

126 Bruns 1892, ix-x.

127 Personal communication: cf. Sharples 1990, 4 and n.16.

128 Cyril, *Lexicon* (5th cent. A.D.), in Cramer, *Anecdota Graeca* 4 196 = Diocles, fr.12 van der Eijk; see van der Eijk's comments in his 2000-2001, vol. 2, 16.

129 Accattino and Donini 1996, xvi f.

130 Thillet 1984, xlvii-xlviii, while rejecting the attribution of books i-ii to Alexander, notes, in the context of the possible relation between Alexander and Galen, that Alexander at *In An. pr.* 39.31ff. uses an example from medicine not present in Aristotle's original text.

131 Tassinari 1994, vi. Cf. Verbeke 1945, 202-203, arguing against Wellmann that the author was not simply a member of the Pneumatic school of medicine, but one who showed considerable Peripatetic influences. Verbeke suggests that the author was, not merely another Alexander (so Tassinari) but another Alexander *of Aphrodisias*. And we do now know that Alexander the Commentator's father was also named Alexander and was described by his son as a philosopher: see Chaniotis 2004.

far as accepting the conclusion to which these facts might seem to point, namely that the author *was* Alexander. Nevertheless, if the treatise *On Fevers* is in fact by an unknown namesake of Alexander of Aphrodisias, roughly contemporary with him and also influenced by Peripatetic doctrine, we may have a candidate for the authorship, or at least the compilation, of parts of our material as well. The question will be whether the distinctively non-Peripatetic features in *some* of the texts are sufficient to exclude that hypothesis. It may however be that we are simply dealing with just one more example of the tendency for texts to be ascribed to famous individuals.[132]

As we have seen, the existence of individual items from our collections is attested in the first and second centuries AD. The collection from which 2.1-38 derive existed early enough for a selection from it to be translated into Latin by the eighth century AD, and also into Arabic; this collection also included material from the problems of Cassius, whose date is uncertain.[133] It seems likely that books i, ii, 1 and 2 had been gathered into a single collection by the time of Psellus in the eleventh century AD. Rose argued (1863, 216-219) that an author "later than Theophilus the commentator on Hippocrates" (i.e. Theophilus Protospatharius, 9th/10th cent. AD) compiled a collection of five books (indicated, he argues, by subscriptions in cod. Paris. anc. gr. 1943 of Cassius and in our P and R) containing three books of problems attributed to Alexander (= i Ideler, ii Ideler, 1.1-2.38 Bussemaker) and one attributed to Aristotle (2.39-192 Usener) followed by a fifth containing the problems of Cassius.[134] If the early attribution of 2.1-38 to Aristotle was not simply an error, it seems, as argued above, more likely that such a collection would have anticipated the book divisions we now find, 2.1-38 having been joined to 2.39-192 *before* book 2 as a whole was linked to book 1.

Richter argued (1885, 47) that the *Supplementary Problems* were collected "in the Alexandrian [i.e. Hellenistic] period"[135] and are *earlier* than

---

[132] For another probable example cf. the glossary edited in Kapetanaki and Sharples, 2000.

[133] On the date of the collection attributed to Cassius and whether it is to be attributed to Cassius the African (5th century AD) see Flashar 1962b, 367-368 and n.2; Garzya and Masullo 2004, 13. We are grateful to Wolfgang Kullmann for drawing our attention to the importance of this issue. See also above, n.102.

[134] Is the fact that both "Collection A" and "Collection B" attribute five books to Alexander (even though in Collection B book 3 is missing, and book 3 in Collection A consists mainly, and books 4 and 5 in both Collection A and B entirely, of pseudo-Galen) a pure coincidence? Or do the five books somehow represent the substitution of different material for the original five?

[135] This is Richter's own formulation. Flashar (1962b, 364 n.1), citing Susemihl's

the *Problems* included in Bekker's edition, which he regarded as later in date because of the disorganised nature of the collection (Richter 1885, 41-44). The parallels with later medical theory (above, at nn.60-62) may cast doubt on a Hellenistic date for 1 and 2.1-38 at least. Flashar dates the *Supplementary Problems* to "not long after Alexander" (*sc.* of Aphrodisias: Flashar 1962b, 364, cf. Stoll 1992, 30 and Filius 1997, 83 n.17) and connects them with the medical school of Alexandria (Flashar 1962a, especially 413), seeing Christian influence in the reference to divine providence as making Hippocrates "nature incarnate" (*τὴν φύσιν σαρκώσας*) in the prologue to book 1 (line 7).[136] Usener (1859, xii; followed by Linke 1880, 54 n.2) however regarded the compiler of books 1 and 2 as Byzantine in date.

The fact that our material seems at some stage to have existed as at least three different collections – book 1, book 2.1-38 and book 2.39-192 – means that it becomes less significant whether it was from an early stage transmitted together with the Bekker *Problems*, unless we suppose that 2.39-192 might have been transmitted with these, the other sections being added subsequently. It seems more likely that where our texts and the Bekker *Problems* occur together in the MSS, this reflects the choices of copyists rather than a shared transmission from antiquity. In fact the number of cases where the *Supplementary Problems* and the Bekker *Problems* are found together is very limited. The 33 MSS of the Bekker *Problems* mentioned by Ruelle 1922, Marenghi 1961 and Marenghi 1991, 69-71 and 62 n.1, include only four (NOSb) of the 21 MSS which include all or part of books 1-2 in the standard sequence, only six of the further 23 MSS which include only books i-ii or parts thereof in the standard sequence, and only two (BL) of the MSS with non-standard sequences.

report of Richter, says "soon after the Alexandrian period"; Susemihl himself (1885, 1483) reports Richter as saying "in the late Alexandrian period".

136 Sluiter 1994, 272 suggests the term may not be exclusively Christian. She considers, but rejects, the possibility of interpreting *σαρκώσας* rather as "having strengthened".

2. Relations between the MSS

2.1. MSS with the standard sequence

(i.e. ACEFKMNOPRSTVXbceiklmw, and BL in their book 1.)[137]

2.1.1. MPRT

M does not derive from any of the following MS with the standard sequence: **A:** 2.68.6, 2.88.5-7, 2.139.11-12, 2.147.3-4, 2.153.3-4; **C:** content, 1.9.5-6, 1.22.11-12, 2.13.4; **EXc:** content; **F:** content, 2.106.4, 2.112.5-6, 2.113.5-6, 2.136.11, 2.137.4-5, 2.139.4,18-19, 2.140.3, 2.145.2, 2.173.3-4, 2.175.4-6; **K:** content, 2.78.14, 2.139.5, 2.150.2, 2.154, 2.156, 2.173.3-4; **NOil:** 1.3.10, 1.6.3, 1.17.33, 1.18.5, 1.20.6, 2.8.2-3, 2.78.14, 2.139.5, 2.150.2, 2.173.3-4; **P:** 1.10.3, 1.13.5, 1.14.1, 1.15.9, 1.18.4, 2.45.2, 2.51.1, 2.52.2, 2.55.7, 2.81.2-4, 2.112.5-6; **S:** 2.8.2-3, 2.78.14, 2.130.1 (unless a correction), 2.150.2, 2.173.3-4; **T:** date, 1.17.30,40, 2.5.1, 2.17.1, 2.24.1, 2.42.8, 2.45.2, 2.57.5-6, 2.59.1, 2.76.2-3, 2.77.7, 2.82.2, 2.147.2; **V:** date, content; **b:** 2.8.2-3, 2.78.14, 2.173.3-4; **e:** date, 2.51.1; **k:** content, 2.5.2-3, 2.15.1, 2.17.1-3; **m:** 2.45.1, 2.50.10, 2.51.2,4, 2.57.5, 2.58.2,4, 2.68.3-6; **w:** date, content.

M does not derive from any MS with a non-standard sequence: contents, but also: **B:** date, 2.39ff.pref.4-5, 2.39.1-3,4, 2.40.8-9; **D:** date, 2.27.1, 2.56.8, 2.72.1,2, 2.73.5-6; **G:** 1.17.1-44, 2.27.1, 2.73.5-6, 2.74.5,6-9, 2.90.5-13, 2.98.2; **L:** 2.39.1-3,4, 2.40.8-9; **Y:** 2.27.1, 2.73.5-6, 2.89.4-5, 2.90.5-13, 2.98.2; **o:** 1.17.1-44, 2.56.4-9, 2.74.5,6-9.

P does not derive from any of the following MSS with the standard sequence: **A:** date, 2.68.6, 2.81.15, 2.88.5-7, 2.139.11-12, 2.147.3-4, 2.153.3-4; **C:** date, content, 1.9.5-6, 1.22.11-12, 2.13.4; **E:** date, content; **F:** date, content, 2.106.4, 2.113.5-6, 2.136.11, 2.137.4-5, 2.139.4,18-19, 2.140.3, 2.145.2, 2.158.11-12,14-17, 2.173.3-4, 2.175.4-6; **K:** date, contents, 2.78.14, 2.81.4-33, 2.90.11-12, 2.107, 2.139.5, 2.150.2, 2.154,

[137] In what follows it should be recognised that arguments for relations between MSS based on their dates are only as certain as those datings themselves. Where passages relating to a group of MSS are discussed together (e.g. NOil) it should be understood that each reference relates only to the MSS extant in each case. When MSS are described as agreeing against others, the readings are more similar in the MSS said to agree, but not always identical.

2.156, 2.158.11-12, 14-17, 2.173.3-4; **M:** date, 1.2.17, 2.8.3, 2.9.2, 2.17.2, 2.19.2, 2.39.1, 2.70, 2.81.4-33, 2.90.11-12, 2.107, 2.108.8, 2.125.11, 2.128.1,3, 2.131, 2.158.11-12,14-17, 2.188.1-3; **NOil:** date, 1.6.3, 1.17.33, 1.18.5, 1.20.6, 2.8.2-3, 2.39.8, 2.78.14, 2.81.4-33, 2.90.11-12, 2.119.7-8, 2.131, 2.139.5, 2.150.2, 2.158.11-12,14-17, 2.173.3-4; **R:** date, 2.70, 2.81.4-33, 2.90.11-12, 2.107, 2.128.3, 2.131, 2.158.11-12,14-17; **S:** date, 2.8.2-3, 2.39.8, 2.70, 2.78.14, 2.81.4-33, 2.90.11-12, 2.107, 2.130.1 (unless correction), 2.150.2, 2.158.11-12,14-17, 2.173.3-4; **T:** date, 1.17.30,40, 2.5.1, 2.17.1, 2.24.1, 2.42.8, 2.45.2, 2.57.5-6, 2.59.1, 2.76.2-3, 2.77.7, 2.82.2, 2.147.2; **V:** date, content; **X:** content; **b:** date, 2.8.2-3, 2.19.2, 2.39.8, 2.70, 2.78.14, 2.81.4-33, 2.90.11-12, 2.107, 2.119.7-8, 2.131, 2.158.11-12,14-17, 2.173.3-4; **c:** date, content; **e:** date, 1.2.17, 2.19.2, 2.81.4-33, 2.90.11-12, 2.107, 2.119.7-8, 2.131, 2.158.11-12,14-17; **k:** date, content, 2.5.2-3, 2.15.1, 2.17.1-3; **m:** date, 1.2.17, 2.45.1, 2.50.10, 2.51.2,4, 2.57.5, 2.58.2,4, 2.68.3-6; **w:** date, content.

P does not derive from any MS with a non-standard sequence: content, but also: **B:** date, 2.39ff.pref.4-5, 2.39.1-3,4, 2.40.8-9, 2.81.4-33; **D:** date, 2.27.1, 2.56.8, 2.72.1,2, 2.73.5-6; **G:** 1.17.1-44, 2.27.1, 2.73.5-6, 2.74.5,6-9, 2.90.5-13; **L:** date, 2.39.1-3,4, 2.40.8-9, 2.81.4-33; **Y:** date, 2.27.1, 2.73.5-6, 2.89.4-5, 2.90.5-13; **o:** 1.17.1-44, 2.56.4-9, 2.74.5,6-9.

R does not derive from any of the following MSS with the standard sequence: **A:** 2.68.6, 2.88.5-7, 2.139.11-12, 2.147.3-4, 2.153.3-4; **C:** date, content, 1.9.5-6, 1.22.11-12, 2.13.4; **E:** content; **F:** content, 2.106.4, 2.112.5-6, 2.113.5-6, 2.136.11, 2.137.4-5, 2.139.4,18-19, 2.140.3, 2.145.2, 2.173.3-4, 2.175.4-6; **K:** date, content, 2.78.14, 2.139.5, 2.150.2, 2.154, 2.156, 2.173.3-4; **M:** date, 1.2.17, 2.8.3, 2.9.2, 2.17.2, 2.39.1, 2.108.8, 2.125.11, 2.188.1-3; **NOil:** date, 1.3.10, 1.6.3, 1.17.33, 1.18.5, 1.20.6, 2.8.2-3, 2.78.14, 2.139.5, 2.150.2, 2.173.3-4; **P:** date, 1.10.3, 1.13.5, 1.14.1, 1.15.9, 1.18.4, 2.45.2, 2.51.1, 2.52.2, 2.55.7, 2.81.2-4, 2.112.5-6; **S:** 2.8.2-3, 2.78.14, 2.130.1 (unless correction), 2.150.2, 2.173.3-4; **T:** date, 1.17.30,40, 2.5.1, 2.17.1, 2.24.1, 2.42.8, 2.45.2, 2.57.5-6, 2.59.1, 2.76.2-3, 2.77.7, 2.82.2, 2.147.2; **V:** date, content; **X:** content; **b:** 2.8.2-3, 2.78.14, 2.173.3-4; **c:** date, content; **e:** date, 1.2.17, 2.51.1; **k:** content, 2.5.2-3, 2.15.1, 2.17.1-3; **m:** 1.2.17, 2.45.1, 2.50.10, 2.51.2,4, 2.57.5, 2.58.2,4, 2.68.3-6; **w:** date, content.

R does not derive from any MS with a non-standard sequence: content, but also: **B:** date, 2.39ff.pref.4-5, 2.39.1-3,4, 2.40.8-9; **D:** date, 2.27.1, 2.56.8, 2.72.1,2, 2.73.5-6; **G:** 1.17.1-44, 2.27.1, 2.73.5-6, 2.74.5,6-9, 2.90.5-13, 2.98.2; **L:** 2.39.1-3,4, 2.40.8-9; **Y:** 2.27.1, 2.73.5-6, 2.89.4-5, 2.90.5-13, 2.98.2; **o:** 1.17.1-44, 2.56.4-9, 2.74.5,6-9.

T does not derive from any of the following MSS with the standard sequence: **A:** 2.68.6, 2.81.15, 2.88.5-7, 2.139.11-12, 2.147.3-4, 2.153.3-4;

C: content, 1.9.5-6, 1.22.11-12, 2.13.4; **EXc:** content; **F:** content, 2.106.4, 2.112.5-6, 2.113.5-6, 2.136.11, 2.137.4-5, 2.139.4,18-19, 2.140.3, 2.145.2, 2.158.11-12,14-17, 2.173.3-4, 2.175.4-6; **K:** content, 2.78.14, 2.81.4-33, 2.90.11-12, 2.107, 2.119.7-8, 2.139.5, 2.150.2, 2.154, 2.156, 2.158.11-12,14-17, 2.173.3-4; **M:** 1.2.17, 2.8.3, 2.9.2, 2.17.2, 2.19.2, 2.39.1, 2.70, 2.81.4-33, 2.90.11-12, 2.107, 2.108.8, 2.119.7-8, 2.125.11, 2.128.1,3, 2.131, 2.158.11-12, 14-17, 2.188.1-3; **NOil:** 1.3.10, 1.6.3, 1.17.33, 1.18.5, 1.20.6, 2.8.2-3, 2.39.8, 2.78.14, 2.81.4-33, 2.90.11-12, 2.119.7-8, 2.131, 2.139.5, 2.150.2, 2.158.11-12,14-17, 2.173.3-4; **P:** 1.10.3, 1.13.5, 1.14.1, 1.15.9, 1.18.4, 2.51.1, 2.52.2, 2.55.7, 2.81.2-4, 2.112.5-6, 2.119.7-8; **R:** 2.70, 2.81.4-33, 2.90.11-12, 2.107, 2.119.7-8, 2.128.3, 2.131, 2.158.11-12,14-17; **S:** 2.8.2-3, 2.39.8, 2.70, 2.78.14, 2.81.4-33, 2.90.11-12, 2.107, 2.119.7-8, 2.130.1 (unless correction), 2.150.2, 2.158.11-12,14-17, 2.173.3-4; **V:** content; **b:** 2.19.2, 2.39.8, 2.70, 2.81.4-33, 2.90.11-12, 2.107, 2.119.7-8, 2.131, 2.158.11-12,14-17, 2.173.3-4; **e:** 2.8.2-3, 2.19.2, 2.51.1, 2.81.4-33, 2.90.11-12, 2.107, 2.119.7-8, 2.131, 2.158.11-12,14-17; **k:** content, 2.15.1, 2.17.1-3; **m:** 1.2.17, 2.45.1, 2.50.10, 2.51.2,4, 2.57.5, 2.58.2,4, 2.68.3-6; **w:** content.

T does not derive from any MS with a non-standard sequence: content, but also: **B:** 2.39ff.pref.4-5, 2.39.1-3,4, 2.40.8-9, 2.81.4-33; **D:** 2.27.1, 2.56.8, 2.72.1,2, 2.73.5-6; **G:** 1.17.1-44, 2.27.1, 2.73.5-6, 2.74.5,6-9, 2.90.5-13, 2.98.2; **L:** 2.39.1-3,4, 2.40.8-9, 2.81.4-33; **Y:** 2.27.1, 2.73.5-6, 2.89.4-5, 2.90.5-13, 2.98.2; **o:** 1.17.1-44, 2.56.4-9, 2.74.5, 6-9.

MR and PT each form a pair closer to each other than MR is to PT. MR is closer to the KNOSBi group than PT are. MR agree with PT against KNOSbil at 1.2.21 (R^{pc}), 1.17.12 (correction in N?),33, 2.8.2-3, 2.22.3, 2.28.2 (not P), 2.41.2,5, 2.42.6 (*δὴ δεῖ*), 2.49.1, 2.50.13, 2.51.7, 2.65.7, 2.75.2, 2.78.14, 2.83.2, 2.88.3, 2.89.5, 2.90.5, 2.104.2, 2.110.8 (*οὐ*), 2.111.1, 2.136.6 ($P^{ac}T^{ac}$). Both at 1.16.4 and at 2.53.1 MPRTem have *ὑπὸ* and BLNOSbil *ἀπὸ*; this suggests that an exemplar (ε, in our reconstruction) may have been such as to be misread in a consistent way. MR agree with KNOSbi against PT 1.2.8,11, 1.3.4 (but N^{pc}Sb agree with APT: as the result of a correction?), 1.14.5, 1.17.26, 2.42.6 (*πάσχοντα*), 2.45.2, 2.50.2, 2.58.2, 2.81.4-33, 2.87.4, 2.96.2, 2.111.2, 2.112.2, 2.113.2, 2.114.9, 2.136.3, 2.178.2, 2.190.2.

2.107 and 2.131 are misplaced in Oi and omitted in MRSb; Usener 1859, v, argued that they were in the margin in the archetype of MROS. This could suggest that MR belong to the NOSbi group. But there is an alternative explanation for the presence of 2.107 and 2.131 in Oi; see below, §2.1.13. PT agree with SbN^{pc} against MROeim 1.3.4; P agrees with OSbi against MRTem, 2.28.2, and with KMORSbei against AFT, 2.147.1. MT,

NOP and RS diverge at 1.17.42, and MRTSb and NOP at 1.18.6; but this proves little. MP agree against all other MSS, 2.156.3. MPRT share the description of book 2 as *Φυσικὰ προβλήματα* (*φυσικὰ* om. P); in A and F the title is in a missing part.

That MPRT form a single group would be suggested if they agreed against *both AF and* the NOSbi group. AMPRTe^{ac}m agree against F and KOSbi at 2.183.4, but this could be independent correction in F and KOSbi. MPRT agree against NOi and BL, 1.13.2, 1.16.4, 1.17.33, but that may just show BLNOi are linked. At 2.143.2 it appears that *ἔχει* was missing from the ancestor of the AFPT group and was restored by P alone of this group; that MRe have the word in a different place from KOSbi*p* may show either that it was omitted in a common ancestor of AFPT MRe and restored in different places in P and in MRe, or that it was present in a common ancestor of KOSbi*p* MRE but was transposed in the latter.

However, MR share the omission of 2.81.4-33 with all but APTl^2. This in itself might not show MR are not linked with PT; independent omission by MR and others (in this over-long problem) is possible. But only APTl^2 include 2.70 (the sequences containing 2.70 and 2.81 are missing from F); only AFPT do not have omissions at 2.90.11-12 and 2.158.11-12,14-17, and only AFOPTi include 2.107 and 2.131. (On the inclusion of this material in Oil see further below, §2.1.13) AFPT agree against other MSS at 2.113.2, 2.114.9, 2.140.3 (by including *ὗς*), 2.149.9, 2.152.2, 2.189.8; APT at 1.2.11, 2.79.2, 2.138.1-2 (and cf. F), 2.143.1; AFT (and *p*) at 2.119.7-8 (where it is perhaps more likely that we are dealing with omission due to homoioteleuton in the other MSS than that a clause has been added by AFT; F's *οὐ* in 7 must be an error arising either from dittography of *οὖσα* or from the scribe's eye jumping to line 8, but M's *οὐ may* be a remnant of the omitted clause). There may seem too much coincidence here if PT are not linked more closely with AF than with MR. In the *Problems* of Cassius Garzya and Masullo 2004, 21-22, place our APTm*p*w in one family, KNRSb in another (cf. Garzya 1994, 186-187).

The question then arises whether AFPT, MR and KNOSBi should be regarded as three co-ordinate branches of the tradition or whether a common ancestor should be postulated for MR KNOSbi.

There are two possible explanations for the presence of material in AFPT which is absent both from MR and from KSb. Either (i) it was present originally and was omitted from the latter two groups – which is easier to explain if they share a common ancestor – or (ii) it was an addition to the ancestor of AFPT (β), as Bussemaker's numbering suggests he thought, and not part of the original tradition at all. Psellus' paraphrase *p* omits 2.70, 2.81.4-33, 2.107 and 2.131; but its relation to the other MSS is unclear (below, 2.1.21), and so it is not clear that this is independent evidence that this material was missing from the archetype. In the case of 2.81 the material absent from most MSS seems to be an integral part of the problem, more likely to have been omitted because of its length than to have been added in the medieval period to a shorter original text, and the omission in *p* does not begin in quite the same place as in the other MSS, though this may simply reflect *p*'s being a summary. The length of 2.131 might make it too a likely candidate for omission. 2.107 is paraphrase of Theophrastus, and could have been added from Theophrastus' text, but it is not alone among our *Problems* in deriving from Theophrastus.

Moreover, in two places – 2.8.3 and 2.19.2 – phrases found only in PT (and m – below, §2.1.7; AF do not include these problems) appear to be reflected in the Latin of the *Problemata Bambergensia.* In 2.19.2 it is conceivable that this is coincidence, additions having been made to clarify the argument, but in 2.8.3 it seems that the source of the Latin had something after ἐπιμείνασα which *aperiens porus* is an attempt to translate, and ἐπιμείνασα in Re may also be a remnant of the clause present in its entirety only in PTm.

At 2.103.3 AP (and Y) agree with [Ar.] against all our other MSS in reading καταφέρεται rather than καταφαίνεται. It is possible indeed that AP have been corrected from Aristotle and that the archetype of *our* Problems had καταφαίνεται; but πόρρω φερόμενον in Psellus' paraphrase may suggest that he too found καταφέρεται here. *If* the archetype read καταφέρεται, the number of places where corruption need have occurred is lessened if MR and KNOSbi share a common ancestor.

The question whether the material found in AFPT is part of the original tradition, and the question, which in part depends on it, whether MR and KNOSbi share a common ancestor not shared by AFPT, are distinct from the question whether PT should be grouped with MR or with AF. The answer to the latter question, at least, is clear; it is that PT belong with AF. For the question of the origin of the material shared by this group of MSS has no bearing on its relevance to the question of the affinities between them.. And the linking of PT with AF rather than MR makes it more difficult to argue that the attribution of books 1 and 2 to Aristotle was an error that occurred just once in a common ancestor of MPRT. For F attributes the preface to book 1 to Alexander. (To be sure, F has the preface in isolation from the rest of book 1, which may make the attribution to Alexander less significant than it would otherwise be.) As for the questions whether the material only in AFPT (and Oil^2) is part of the original tradition, and whether MR and KNOSbi share a common ancestor not shared by AFPT, we are inclined to answer these in the affirmative, but without sufficient certainty to incorporate the latter point into our stemma.

Pe may be corrected from [Aristotle] at 2.51.1 (*ἄλλων*). See §2.1.7 below. M is dated to the second half of the 15th century by de Andres 1987, 126; it was copied by Constantine Lascaris in Messina (ibid. 127). P and R were both owned by Bessarion; P is dated to the middle of the 13th century, and R to the beginning of the 14th, by Mioni 1985, 390 and id. 1981, 371 respectively. T contains a significant number of errors and omissions.

2.1.2. A

A does not derive from the following MSS with the standard sequence: **C:** date, content, 1.9.5-6; **E:** content; **F:** content, 2.106.4, 2.112.5-6, 2.113.5-6, 2.136.11, 2.137.4-5, 2.139.4,18-19, 2.140.3, 2.145.2, 2.158.11-12,14-17, 2.173.3-4, 2.175.4-6; **K:** date, content, 2.78.14, 2.81.4-33, 2.90.11-12, 2.107, 2.119.7-8, 2.139.5, 2.150.2, 2.154, 2.156, 2.158.11-12,14-17, 2.173.3-4; **M:** date, 2.70, 2.81.4-33, 2.90.11-12, 2.107, 2.108.8, 2.119.7-8, 2.125.11, 2.128.1,3, 2.158.11-12,14-17, 2.188.1-3; **NOil:** date, 1.3.10, 1.6.3, 2.78.14, 2.81.4-33, 2.90.11-12, 2.119.7-8, 2.139.5, 2.150.2, 2.158.11-12,14-17, 2.173.3-4; **P:** 2.81.2-4, 2.90.12, 2.112.5-6, 2.119.7-8; **R:** 2.70, 2.81.4-33, 2.90.11-12, 2.107, 2.119.7-8, 2.128.3, 2.158.11-12,14-17; **S:** 2.78.14, 2.81.4-33, 2.90.11-12, 2.107, 2.119.7-8, 2.150.2, 2.158.11-12,14-17, 2.173.3-4; **T:** date, 2.76.2-3, 2.82.2, 2.90.12, 2.147.2; **V:** date, content; **Xc:** date, content: **b:** 2.78.14, 2.81.4-33, 2.90.11-12, 2.107, 2.119.7-8, 2.158.11-12,14-17, 2.173.3-4; **e:** date, 2.81.4-33, 2.90.11-12, 2.107, 2.119.7-8, 2.158.11-12,14-17; **m:** 2.68.3-6; **w:** date, content.

A does not derive from any MS with a non-standard sequence: content, but also: **BL:** 2.81.4-33, and date (B); **D:** date, 2.73.5-6; **G:** 2.73.5-6, 2.74.5, 6-9, 2.90.5-13, 2.98.2; **Y:** 2.73.5-6, 2.89.4-5, 2.90.5-13, 2.98.2; **o:** 2.74.5, 6-9.

A very often errs in breathings and spelling, far more often than we have been able to indicate without overloading our apparatus; itacism and other phonetic spellings (e.g. *αι* for *ε* and vice versa) are frequent. This is however a different issue from the general quality of its text; it sometimes preserves good readings. We have cited A as agreeing with other MSS even where it differs from them in spelling and accentuation, provided that comparison with its errors of this sort elsewhere makes clear which variant it intended.

Bussemaker 1857, viii argued that A originally had books 1-2 in full and that pages have been lost (cf. Usener 1859, v); it now has only 1.1-12, with the preface and first part of 1.1 and the end of 1.12 missing, and 2.66-end, again with the beginning of 2.66 missing. On the history and condition of the MS see Garzya and Masullo 2004, 17-18.

2.1.3. F

F does not derive from the following MSS with the standard sequence: **A:** 2.88.5-7, 2.139.11-12, 2.147.3-4, 2.153.3-4; **C:** date, content; **EX:** content; **K:** date, 2.90.11-12, 2.107, 2.119.7-8, 2.139.5, 2.150.2, 2.154, 2.156; **M:** date, 2.90.11-12, 2.107, 2.108.8, 2.119.7-8, 2.125.11, 2.128.1,3, 2.188.1-3; **N:** date; **O:** date, 2.90.11-12, 2.107, 2.119.7-8, 2.139.5, 2.150.2; **P:** 2.119.7-8; **R:** 2.90.11-12, 2.107, 2.119.7-8, 2.128.3; **S:** 2.90.11-12, 2.107, 2.119.7-8, 2.130.1 (unless correction), 2.150.2; **T:** date, 2.90.12, 2.147.2; **V:** content; **b:** 2.90.11-12, 2.107, 2.119.7-8; **c:** date, content; **e:** date, 2.90.11-12, 2.107, 2.119.7-8; **i:** date, 2.90.11-12, 2.107, 2.119.7-8, 2.139.5, 2.150.2; **l:** date; **w:** date, content.

F does not derive from any MS with a non-standard sequence: content, but also: **G:** 2.90.5-13, 2.98.2; **Y:** 2.89.4-5, 2.90.5-13, 2.98.2.

F agrees with APT against all other MSS in not having the omission at 2.90.11-12; it agrees with AFOPTi alone in preserving 2.107 and 2.131, and it seems likely that these problems were marginal additions to an ancestor of Oi (below, §2.1.13). F also agrees with APT against other MSS at 2.140.3 (ὗς), 2.143.2(P^{ac}), 2.149.9 (F has a corrupt reading but shares word-order of APT), 2.152.2. Although it agrees with the majority against APT at 2.143.1, F is, it seems, to be linked with APT.

F agrees with PT against A, 2.93.5; with AT against P, 2.147.1; with P against AT, 2.112.5-6 (omission from homoioteleuton). At 2.90.12, first FPT agree against A, then FP agree against AT. There do not seem to be strong grounds for linking F either with A or with PT more closely than the with the other. See also below, §2.1.13.

2.1.4. V

Of our texts V only contains the prologue to book 1. It agrees with Tw at 1 prol.20 and with MP at 1 prol.24.

2.1.5. MR

On the relations between MR and other MSS see above, 2.1.1. The question arises whether M could derive from R. The title that appears at the start of book 1 in M is the same that appears at the end of the book in R, but the extra title that appears before problem 1.1 in M does not come from R. The readings of M at 1.17.42, 2.8.2-3, 2.45.10, 2.48.1, 2.77.1 (shared with K) and 2.166.9 are presumably corrections. Other agreements between M and the KOil group may be the result of coincidence, as at 2.75.4 and 2.147.3. At 2.156.3 MP agree against all other MSS. Harder to reconcile with dependence of M on R, however, is the fact that at 2.165.7 M and e^{ac} apparently both reflect a corrupt or damaged exmplar, whereas R has restored the correct text (below, §2.1.9). It is probably safer to regard M and R as siblings.

2.1.6. Tw

w is close to T (1.1.15, 1.10.5(T^{ac}), 1.3.8, 1.14.5, 1.15.9 (where $Tm^{pc}w$ agree with DGY only in what is probably a conjecture in μm^{pc}, given that m^{ac} has the vulgate reading and p a totally different one), 1.16.4, 1.17.26, 42, 1.20.3,5, 1.21.4), and Tmw have in common the title *Aristotle's medical problems and natural problems, selections, book 3* at the end of book 1, in addition to the title *On the peculiar characteristic of animals* at the beginning. But w does not depend on T: at 1.11.3 *πλήττει πάλιν σφοδρῶς τὸν ἐφκέφαλον* is present in Aw Buss. Us., missing in MNOPRSTbegi, and present as a marginal addition in m. At 1.17.30 *ἄκουε δὲ καὶ περὶ τοῦ πανὸς*, and at 1.17.40 *τὸ δὲ κάτω προβατοειδὲς* are present in all the MSS except T, including w. And T does not depend on w: since w contains only book 1, it seems unlikely that T is a descendant of it.[138] Garzya notes the closeness of T and w in Cassius (1994, 188; Garzya and Masullo 2004, 22). Capocci 1958 notes the closeness of w to m in Cassius and in the pseudo-Alexander/pseudo-Aristotle problems. w belonged to Carlo Strozzi (Capocci 1958, 255-256; Garzya and Masullo 2004, 18).

2.1.7. m

m does not derive from any of the following MSS with the standard sequence: **EXc:** content; **F:** 2.175.4-6; **K:** 2.65.7; **M:** 2.8.3, 2.9.2, 2.17.2, 2.19.2, 2.39.1, 2.125.11, 2.188.1-3; **NOi:** 1.3.10, 1.6.3, 1.17.33, 1.20.6, 2.8.2-3; **P:** 1.10.3, 1.13.5, 1.14.1, 1.15.9, 1.16.4, 1.18.4, 2.45.2, 2.51.1,

---

[138] Unless what we now have in w is only a fragment of what it originally contained from our texts, and there seems nothing to suggest this. Book 1 at 33-43 is preceded by Cassius the Iatrosophist at 1-32, and followed by works by Galen at 48ff. (Capocci 1958, 254); Capocci notes (255) that fols. 44-47 are blank, and argues that the MS as we have it is made up of several bound together, and that these blank pages are ones left at the end of what was originally an independent MS.

2.52.2, 2.55.7; **R:** 2.8.3; **T:** date, 1.16.4, 1.17.30,40, 2.5.1, 2.17.1, 2.24.1, 2.42.8, 2.45.2, 2.57.5-6, 2.58.6, 2.59.1; **V:** date, content; **b:** 2.19.2; **e:** date, 2.8.3, 2.19.2, 2.51.1; **k:** content, 2.5.2-3, 2.15.1; **w:** date, content.

m does not derive from any MS with a non-standard sequence: content, but also: **B:** date, 2.19.2, 2.39ff. pref.4-5, 2.39.1-3,4, 2.40.8-9; **D:** date, 2.27.1, 2.56.8; **G:** 1.17.1-44, 2.27.1; **L:** 2.19.2, 2.39.1-3,4, 2.40.8-9; **Y:** 2.27.1; **o:** 1.17.1-44, 2.56.4-9.

m agrees with MPRT rather than NOSbil: 1.2.21, 1.4.5, 1.16.4, 1.17.33, 1.18.5, 2.8.2-3, 2.28.2, 2.41.2,5, 2.42.6, 2.49.1, 2.51.7, 2.53.1, 2.65.7. m agrees with PT: 1.2.8, 1.14.5 (apart from the breathing: correct reading in PTmw, possibly due to emendation), 1.17.26, 2.8.3, 2.19.2, 2.42.6, 2.45.2, 2.50.2, 2.178.2, 2.189.8. At 1.10.5 m^{ac} agrees with PT^{pc} and NOSbi, and m^{pc} with T^{ac} and MRew; possibly ζ had both readings and we are dealing with the recording of recognised variants. At 2.42.8 T agrees with m^{pc}. Tmw have in common the title *Aristotle's medical problems and natural problems, selections, book 3* at the end of book 1, in addition to the title *On the peculiar characteristic of animals* (also in P amd MRe) at the beginning. m agrees with MR against PT at 1.2.11, 1.3.4, 1.16.4; but these latter agreements do not seem sufficient ground for detaching m from the PT branch.

At 1.17.42 m, which with MTw alone had the correct reading *φθορᾶς*, has been "corrected" to agree with *φοβερᾶς* in RSbe. Presumably the transmitted text was the impossible *φοβερᾶς* present in two separate branches of the tradition, Sb and Re, and *φορᾶς* in $MTm^{ac}w$ and *φθορᾶς* in BL NOi and P are emendations; m will then have changed the reading to agree with the transmitted text.

m apparently incorporates a reading from [Aristotle] at 2.51.2 *εὐχερεῖς*; Pe similarly at 2.51.1 *ἄλλων*. It is odd that Pe seem to have ignored [Aristotle] where m followed, and vice versa.

The arrangement of the material from our texts in m is strange. Problem 2.68 breaks off partway through on fol.31r, to be followed by the problems of Cassius from 21 (from its beginning, but wrongly numbered as 20) onwards, ending on 38v. But on fols. 78r11-78v10, in a different hand (Garzya 1994, 183; Garzya and Masullo 2004, 15-16) the first part of 2.68 is repeated verbatim, without attribution, followed without a break by the latter part of 2.174 and then by 2.175-192, followed in turn by Cassius 1-21 (Garzya loc. cit.). The fact that this latter unattributed section does not begin or end at a page break shows that the separation of this group of texts from the rest is not the result of mechanical dislocation in m itself; and the fact that both 2.68 and Cassius 1.21 are copied *twice* suggests that we are not dealing with the results of simple mechanical dislocation at an earlier stage in the tradition either. The way in which m has arranged the material could be explained on the following (highly conjectural) hypothesis; if m's exemplar had lost 2.68b-2.174a inclusive, the scribe of m might have come to the bottom of the page in the examplar where 2.68 broke off, realised that something was wrong, and stopped copying our text altogether, going on instead to Cassius as the next text in his exemplar concerned with human physiology (2.174-192 being on horses, and clearly out of place). But subsequently it was thought that the omitted material should not be discarded altogether, and 2.68a was copied again by a different hand, this time with 2.174b-192, the repetition of 2.68a presumably resulting from a desire not to start in the middle of a problem. (2.68 could after all in itself be supposed to relate to horses.) This admittedly does not explain why the earlier part of the MS jumps from 2.68a to Cassius 21 rather than to Cassius 1. But if the omission of 2.68b-174a is due to loss of material in the exemplar, then we must suppose either that ζ was mutilated after Pμ were copied from it, or that there was an intermediary between ζ and m. See further below, §2.1.16, on K.

m was owned by Giorgio Valla (Garzya 1994, 183; Garzya and Masullo 2004, 15). See further below, 2.1.16.

2.1.8. k

Louis 1986, 197-8, notes that k contains a collection of extracts from various sources but especially from book 2 Bussemaker (= 4 Usener). He also notes that on fol.166 there is a table of contents which suggests that what is in this MSS is only part of what was once a much larger collection. In fact only fols. 96r-98r contain problems from the first part of book 4 Usener; the remainder of 98r-166r is material from various medical texts, principally Alexander of Tralles' *Therapeutica*, with major dislocations apparently due to loose pages being rebound in the wrong order, and the table of contents is simply the chapter headings of Alexander of Tralles, *Therapeutica* 2 335.5-365.4 Putschmann. k is dated to the 14th-15th century by Omont 1886-1898, pt.2 235.

MS k is highly erratic with many unique errors. It shares the omission at 2.5.2-3 with T^{ac} and *g*, so it is not the source of any other MS; and that it is the source of T is excluded by content and by the omission at 2.151. K's affinity seems to be with the MPRTem group (2.28.2, 2.41.2), and within this group it is from Re that there is least reason to dissociate it. 2.17.1 and 2.24.1 suggest that k is not a descendant of T, which is in any case ruled out on chronological grounds. At 2.20.1 k agrees with M in *πυρετῶν* against all other MSS, but the change is a natural one in the context, and against any link with M cf. 2.20.2. At 2.24.1 k does not share the note in Re.

2.1.9. e^{ac}

c^{ac} does not derive from any of the following MSS with the standard sequence: **A:** 2.68.6, 2.88.5-7, 2.139.11-12, 2.147.3-4, 2.153.3-4; **C:** 1.9.5-6, 1.22.11-12, 2.13.4; **EXc:** content; **F:** 2.106.4, 2.112.5-6, 2.113.5-6, 2.136.11, 2.137.4-5, 2.139.4,18-19, 2.140.3, 2.145.2, 2.173.3-4, 2.175.4-6; **K:** 2.78.14, 2.139.5, 2.150.2, 2.154, 2.156, 2.173.3-4; **M:** 2.9.2, 2.17.2, 2.39.1, 2.108.8, 2.125.11, 2.128.1,3, 2.188.1-3; **NOil:** 1.3.10, 1.6.3, 1.17.33, 1.20.6, 2.8.2-3, 2.78.14, 2.139.5, 2.150.2, 2.173.3-4; **P:** 1.10.3, 1.13.5, 1.14.1, 1.15.9, 1.16.4, 1.18.4, 2.45.2, 2.51.1, 2.52.2, 2.55.7, 2.81.2-4, 2.112.5-6; **R:** 2.128.3; **S:** 2.78.14, 2.130.1 (unless correction), 2.150.2, 2.173.3-4; **T:** 1.16.4, 1.17.30,40, 2.5.1, 2.17.1, 2.24.1, 2.42.8, 2.45.2, 2.57.5-6, 2.58.6, 2.59.1, 2.76.2-3, 2.77.7, 2.82.2, 2.147.2; **V:** content; **b:** 2.78.14, 2.173.3-4; **k:** 2.5.2-3, 2.15.1, 2.17.1-3; **m:** 2.45.1,2, 2.50.10, 2.51.2,4, 2.57.5, 2.58.2,4, 2.68.3-6; **w:** content.

e^{ac} does not derive from any MS with a non-standard sequence: content, but also: **B:** 2.39ff.pref.4-5, 2.39.1-3,4, 2.40.8-9; **D:** 2.27.1, 2.56.8, 2.72.1, 2.73.5-6; **G:** 1.17.1-44, 2.27.1, 2.73.5-6, 2.74.5,6-9, 2.90.5-13, 2.98.2; **L:** 2.39.1-3,4, 2.40.8-9; **Y:** 2.27.1, 2.73.5-6, 2.89.4-5, 2.90.5-13, 2.98.2; **o:** 1.17.1-44, 2.56.4-9, 2.74.5,6-9.

e^{ac} agrees with MPRT rather than NOSbil: 1.2.21, 1.4.5, 1.16.4, 1.17.33, 1.18.5, 2.8.2-3, 2.27.3, 2.28.2, 2.41.2,5, 2.42.6, 2.49.1, 2.50.13, 2.51.7, 2.53.1, 2.65.7, 2.78.14, 2.88.3, 2.90.5, 2.132.3. At 2.82.5 e agrees with OSi in an easy mistake. At 2.158.4 e^{ac} agrees only with K, but the mistake is again easy. At 2.56.3 it is remarkable that DGYe have the correct reading *τόπου* against all other MSS; in the case of e this is presumably emendation.

e^{ac} probably derives from the common source of MR. e^{ac} agrees with MR only, 2.21.3 (R^{pc}), 2.30.1, 2.44.1, 2.97.3, 2.106.4, 2.108.2 (R^{pc}), 2.114.9, 2.116.2, 2.128.2 (R^{pc}), 2.133.4, 2.136.6, 2.137.3 (R^{pc}), 2.139.22, 2.140.2,3, 2.143.2, 2.144.1-2, 2.147.3, 2.158.3, 2.160.6, 2.170.7, 2.181.3,4, 2.183.2, 2.189.12, 2.190.7, 2.192.1. e agrees with MR and T^{ac}m^{pc}w against PT^{pc}m^{ac} and NOSbi, 1.10.5; with MR and NOSbi against PTmw, 1.14.5. e^{ac} agrees with M only: 2.21.3, 2.33.2, 2.40.1, 2.50.4, 2.79.1, 2.116.5, 2.117.6, 2.176.2 (an obvious correction, but one not made by other MSS); with AMRT, 2.112.12; with AFMRT, 2.112.11; with R only: 2.8.2-3, 2.24.1, 2.93.5, 2.108.8, 2.126.5, 2.163.2; with PT, 2.76.5. At 2.165.7 R and other MSS have the correct *χρησομένοις*, M has *χρησίμοις*, and e originally wrote *χρησόμοις*; this could be explained by a damaged or abbreviated text in the exemplar of M and e, correctly restored by R if M and R are indeed siblings. On the double attribution of book 1 in e see above, §1.3.i.

The question arises whether e and m may not be more closely related than our stemma suggests, given that these are the two MSS which have a double Aristotelian title with two different book numbers at the start of book 2. The answer appears to be no, as far as their readings are concerned. e is not the source of m (above, §2.1.7), and m is not the source of e (above, this section). At 1.2.17 Mem alone have the error ὄξος for ὄζον; but this hardly counts for much against the other evidence and may be coincidence, especially as M is not the source of m (above, §2.1.7) or vice versa (§2.1.1). The treatment of book 2 as the third book, only in these two MSS, is more natural in e where book 1 has been referred to as the second book than it is in m. But contamination of m from e is ruled out on grounds of date.

Pe^1 may incorporate a reading from [Aristotle] at 2.51.1 (above, §2.1.7). e certainly adds a marginal reading from [Ar.] at 2.45.5.

e came to the Escorial from the library of Hurtado di Mendoza, who was the ambassador of Charles V to Venice (Miller 1848; Graux 1880, 208-9; de Andres 1965, 89-92).

2.1.10. KNOSbil

This group of MSS are closely linked by shared readings: 1.1.10, 1.2.21, 1.12.5-6, 1.16.4, 1.17.12,33, 2.8.2-3, 2.22.3, 2.28.2, 2.39.6, 2.41.2,5, 2.42.6, 2.49.1, 2.51.7, 2.53.1, 2.65.7, 2.75.2, 2.78.14, 2.83.2, 2.88.3, 2.89.5, 2.90.5, 2.126.5, 2.147.2, 2.149.4, 2.158.3 (where K seems to have emended the corrupt text), 2.170.7, 2.173.3-4, 2.181.2, 2.185.4, 2.188.3. At 1 prol. 21, for πάθος of the other MSS, N^{ac}Sb have πάθοι and $N^{pc}$$Oe^{pc}$i have πάθοις; possibly the corrector of N intended rather πάθος, but if so Oi misinterpreted the correction. NOSbi divide i.2 Ideler alike (in the case of S apparently as a correction at the time the numbering was added; information from Carl-Gustaf Lindqvist. Kl do not include this part of book i). At 1.17.12 N's addition of the negative is presumably an emendation. KNOSbi are linked by having ἰατρικὰ coupled with ἀπορήματα and φυσικὰ with προβλήματα in the title to book ii Ideler (above, §1.3; not included in l, but see below §2.1.11). This exchanged pairing of the terms seems not to occur in any of the titles for any of the books in other MSS. NOSbi are linked not only by having the exchanged terms in the title for ii Ideler, but also by having the form ἐπί τισιν ἀπορίαις λύσεις in the title for i. (K however has ἀπορίαι καὶ λύσεις). ἐπί τισιν ἀπορίαις λύσεις is also the form of the title in book 1 of BL, and in the non-standard MSS GY (but o and Vienna med. gr. 44 have ἀπορίαι καὶ λύσεις).

That the original text in l (as opposed to the addition of 2.70 and the greater part of 2.81; see below, §2.1.13) belongs to the KNOSbi group is suggested by 2.27.3, 2.37.2, 2.39.6, 2.41.2, 2.52.2. Oi agree with l at 2.39 pref. 5-6, 13, 2.39.9, 2.47.2, 2.50.2,11, 2.52.2, 2.55.6,7, 2.56.2,9, 2.68.6, 2.74.4; KOil at 2.69.2, 2.73.6.

The similarity between N, S and b is noted at Moraux et al. 1976, 65 and by Garzya 1998, 86-9, who argues (89) that this group had an uncial archetype; in Cassius problem 74 NSb share a lacuna at a point where a corrupt text, reflecting misinterpretation of an uncial MS, has been interpreted in different ways by our Rm, AT, and w (Garzya 1994, 188; Garzya and Masullo 2004, 23). Marenghi 1991 discusses four of our MSS in his edition of [Ar.]: Oxford, Corpus Christi MS gr. 113 (our O, his E) which he places in his family α, and Oxford, New College MS gr. 233, Venice, Marc. gr. 259 (982) and Bologna, MS gr, XXI.3635 (15) (our NSb; his DxB) which he places in his family δ. At 13.2 907b28 (Bekker) Marenghi has our b and our O agreeing in error against our N and our S (σώματι for στόματι), but this may just be coincidence. Hubert-Drexler 1952, xiii-xiv and Cherniss 1976, 10, place S (in the case of Hubert-Drexler) and b (in both) close to l in the tradition of Plutarch, *Platonic Questions*; see below, n.139.

KNOSbil do not derive from the following MSS with the standard sequence: **A:** 2.68.6, 2.69.3, 2.74.3-4, 2.76.4, 2.88.5-7, 2.139.11-12, 2.147.3-4, 2.153.3-4; **C:** content, 1.9.5-6, 1.22.11-12, 2.13.4; **EXc:** content; **F:** 2.106.4, 2.112.5-6, 2.113.5-6, 2.136.11, 2.137.4-5, 2.139.4,18-19, 2.140.3, 2.145.2, 2.173.3-4, 2.175.4-6; **M:** 1.2.17, 2.9.2, 2.17.2, 2.39.1, 2.108.8, 2.125.11, 2.128.1,3, 2.188.1-3; **P**: 1.10.3, 1.13.5, 1.14.1, 1.15.9, 1.16.4, 1.18.4, 2.45.2, 2.51.1, 2.52.2, 2.55.7, 2.81.2-4, 2.112.5-6; **R:** 2.128.3; **T:** date, 1.16.4, 1.17.30,40, 2.5.1, 2.17.1, 2.24.1, 2.42.8, 2.45.2, 2.57.5-6, 2.58.6, 2.59.1, 2.76.2-3, 2.77.7, 2.82.2, 2.147.2; **V:** content; **e:** date, 1.2.17, 2.51.1; **k:** 2.5.2-3, 2.15.1, 2.17.1-3; **m:** 1.2.17, 2.45.1,2, 2.50.10, 2.51.2,4, 2.57.5, 2.58.2,4, 2.65.1,5, 2.68.3-6; **w**: date, content.

KNOSbil do not derive from any MS with a non-standard sequence: content, but also: **B:** 2.39 pref. 4-5, 2.39.1-3,4, 2.40.8-9; **D:** 2.27.1, 2.56.8, 2.72.2, 2.73.5-6; **G:** 1.17.1-44, 2.27.1, 2.73.5-6, 2.74.5,6-9, 2.90.5-13, 2.98.2; **L:** 2.39.1-3,4, 2.40.8-9; **Y:** 2.27.1, 2.73.5-6, 2.89.4-5, 2.90.5-13, 2.98.2; **o:** 1.17.1-44, 2.56.4-9, 2.74.5,6-9.

OSbi do not derive from **K:** 2.127.1, 2.154, 2.156. **l** cannot derive from K, 2.74.4. **S** is not the source of KOil: 2.130.1, 2.150.2. **Oi** are not the source of KSbl: 2.31.1, 2.82.1 (*ζῶσιν*). KOi agree against Sb at 2.139.5, 2.148.5, 2.150.2,3, 2.186.2.

2.1.11. N and l

The material from our *Problems* in MS l is on inserted leaves in a different hand, beginning with 2.23 at the top of fol.18r (Wegehaupt 1918, 470).[139] The material from our *Problems* in N ends with 2.22 completed at the bottom of fol. 64v, after which the rest of book 2 has been lost and the text jumps to the middle of Cassius *Problem* 1 on what is now 65r. In N 2.22 is numbered 43 (the numbering continues from book 1 to book 2 Bussemaker without restarting, and 1.2 and 1.3 are both numbered as 2); in l 2.23 is numbered 44. The layout of the material in the two MSS is identical; large gaps are left in the line between problems, but each problem does not necessarily start a new line; the first letter of each problem is rubricated, and placed in the margin in cases where a new problem does start a new line; the rubricated numbers for the problems are faint and placed well out into the margin. The two MSS show a similar alternation between runs of ornate and runs of simple rubricated initials, and a sequence of the latter extends from 2.16 in N to 2.38 in l. The handwriting of the main text in the two MSS is very similar if not identical. The pages of N are now 21.7 x 14.3 cm (Hutter 1997, 151); those of l are now 21.9 x 14.2 cm (Wegehaupt 1918, 469). Fol. 64 of the New College MS is the first leaf of an original gathering of eight leaves the rest of which is now missing (information from Jürgen Wiesner); the material from our problems in l occupies eight leaves (18-25 inclusive), which are made up of three folded leaves (18-23) and two single leaves (24 and 25).[140] This is compatible with fol.64 of N and fol.24 of l being the two halves of the outermost leaf of the original gathering of which l 18-23 are

---

[139] MS l consists of material from several different MSS, the section (1-28c) into which our material has been inserted having once belonged to Patricius Junius: de Meyier 1955, 254-5. There is a question whether the material into which our pages have been inserted – Plutarch, *Platonic Questions* at fols. 1-10 and 26-28 – *also* came originally from N; it has the same watermark (on fols. 2 + 9, 4 + 7, and 28 + 28a) and the writing is very similar. But there is nothing to suggest that Plutarch's *Platonic Questions* ever formed part of N. Sb do contain the *Platonic Questions*, but after the problems of [Aristotle] which there follow those of [Alexander] and Cassius, whereas in N the loss of pages is from the end of [Alexander] and the beginning of Cassius. In N too the [Aristotle] *Problems* follow Cassius, but are followed in turn by [Plutarch], *Placita* and Aelian *On Animals.* Oi do not contain Cassius (above, n.102); K includes various works by Plutarch, but not the *Platonic Questions.* Fols.11-17 in N, on the other hand, are quite clearly an intrusion from a different source.

[140] De Meyier 1955, 255.

the inner sheets, and fol.25 of l being the first leaf of the next gathering the rest of which is lost.[141] The paper of fol.18 recto of l is noticeably discoloured compared with its verso and the following leaves; fol.25 is faded (especially the red marginal numbers and initial letters), rather than discoloured, on *both* sides compared with those that precede. This seems consistent with 18 recto, alone of what survives, having been for a time exposed, as the first outer page of a larger detached portion. Leaves 56-63 of N have the watermark Briquet 439, while 64 has no watermark (information from Jürgen Wiesner); leaves 18 + 23 and 20 + 21 of l have the same watermark, while leaves 19 + 22 and 24 have none[142] – which is compatible with fol.64 of N being the other half of fol.24 of l. It would appear that N was damaged with some violence; for the possible historical context see §2.1.12 below.

The copy of Bussemaker's edition in the Warburg Institute Library, London, was once the property of Wegehaupt and has MS annotations, presumably his, to 2.23-81 Buss., i.e. the texts included in MS l; presumably they were made in preparing his articles on this. They appear to consist mainly if not entirely of notes of Usener's readings and of the MSS used by Usener which contain each of these problems. Both Wegehaupt 1915 and Wegehaupt 1918 were published posthumously, the first comparing the readings of MS l with Usener's edition (though with some reference to Bussemaker) and the second with Bussemaker's; posthumous publication may explain both the fact that the two papers cover the same material, though the second is fuller, and the reference to Alexander in one title, Aristotle in the other. In the apparatus we have simply cited as "Wegehaupt" material that appears in some cases in both papers, in others only in the later one.

2.1.12. NOil

NOi agree against other MSS at 1.19.7 (where the reading of b may suggest how the error arose) and 1.20.6. At 1.3.4, 1.17.12 and 2.19.2 Oi disregard corrections in N; at 1 prol. 21 they follow the correction in N and possibly misinterpret it (above, §2.1.10). NOi (and S) all fail to distinguish book 2 from book 1.

---

141 We are grateful to Dr A.T. Bouwman of the Universiteitsbibliotheek Leiden for this suggestion.

142 We are grateful to Marlein van Raalte and to Dr Bouwman for investigating this for us.

If l is part of the missing portion of N, and Oi derive from N before its mutilation, Oi must derive from l. At 2.31.1 Oi have incorporated only part of the marginal addition in l, and at the wrong place. At 2.42.10 Oi's reading is a further modification of that in l.

1.12.4 and 2.7.2 suggest that O and i are not independent copies of N, as does the fact that both fail to distinguish book 1 from book ii Ideler, whereas N does so. At 2.16.3 it appears that O followed the correction in i; at 1.9.5 and 1.17.49 (the same error in both cases) and at 2.106.2, either i was copied from O before the latter was corrected, or O copies the error in i and then makes an (easy) correction. At 2.78.1 O differs from all other MSS including i, which suggests i is not a copy of O.

Wiesner 1978, 484 identifies the marginalia in N as by Demetrius Chalkondyles (cf. Garzya 1998, 86); there are also marginalia by Bessarion (private communication by Wiesner). N is dated to the third quarter of the 15th century, and before Bessarion's death in 1472, by Wiesner, who suggests (private communication) that it was brought to Oxford by William Grocyn. Hutter 1997, 151-3 suggests that it was written in Bessarion's circle in Rome in 1450 or soon after, and that it formed part of the library of Cardinal Reginald Pole (ob. 1558), Archbishop of Canterbury under Mary Tudor. i is dated to the end of the 15th century by Devreese 1945, 316, who also notes that the copyist was Caesar Strategos; cf. Vogel and Gardthausen 1966, 224. Wiesner and Victor 1971-2, 64 date O to c.1490, copied in Florence by Petros Hypsilas and brought to Oxford by William Grocyn.

2.1.13. The additions in (N)Oil

l shares 2.70, as well as the latter part of 2.81 (both in the second hand in l) with APT only (F is not extant here). 2.70 has been added partly at the bottom of fol.23v directly after 2.69, and partly in the top margin of fol.24r. 2.81.4-33 have been added starting in the space left at the end of 2.81.1-4 (Nl regularly leave considerable space between problems; in this case the space happens to be at the end of a line) and then in the margins of fol.25v. The text clearly continued into the margin of the next folio which is now lost. Wegehaupt 1918, 470, notes that these additions may be by the same scribe as the original text, but are shown to be later by the colour of the ink.[143] ii.63 Ideler has been added to the margin of fols. 50v + 51r of N in a

[143] De Meyier 1955 notes that marginal additions on fols. 19-21 of l are by a younger hand and attributes them to Junius (above, n.139; drawn to our attention by Dr

similar way to that in which 2.81.4-33 has been added to l with a similar (not identical) reference mark. Significantly, in both cases this addition has been made by writing over the marginal numbers relating to the main text. At 2.19.2 the same hand, apparently, has added a clause in the margin of N which otherwise appears only in PT (AF are not extant here) and in the *Problemata Bambergensia* (above, §2.1.1).

In 2.70, and generally in 2.81, l^2 agrees with PT rather than A (cf. Wegehaupt 1915, 96). At 2.81.18-19 l^2 shares the expanded text with T only, which prompts the question whether l^2 took 2.70 and 2.81.4-33 from T (or ζ or μ), or from β via δ; this has a bearing, though not perhaps a decisive one, on the degree of authority we should recognise for the expanded text in 2.81.18-19. In favour of the view that l^2 derived its additional material from δ is the fact that Oi also contain material apparently derived from δ (below), and that it is more economical to suppose that this material was all copied together from δ into the margin of N+l, rather than that the contamination of the lost part of N happened first and that by l^2 from T or μ independently and much later. On the other hand, if the expanded text of 2.81.18-19 was already present in β, it is odd that A and P should both have omitted it, and if 2.70 and 2.81.4-33 were copied from δ into N+l at the same time as 2.107 and 2.131, it is odd that Oi omit them.

Wegehaupt suggested that the marginal correction by l^2 at 2.78.13 (*μύκητες*) derives from the same source as the addition of 2.70 and 2.81.4-33 and shows the use of some good MS as a source; on the contrary we suspect it is due to (a good) conjecture. (That the correction is by the second hand is suggested by its placing in the extreme left-hand margin *outside* the addition of 2.81; it is conceivable however that the correction was made first in the extreme margin and the addition written in such a way as to avoid it.)

Oi are the only MSS apart from AFPT to preserve 2.107 and 2.131, but have them misplaced, after 2.117 and after 2.133 respectively. Usener explained this by their having been in the margin of the exemplar (above, §2.1.1). It seems likely that they were copied into the margin of the lost part of N. Oi in these passages agree with F against PT, 2.107.3; with F against APT, 2.107.1, 2.131.8,12,15,19,22; with FP against AT, 2.107.3-4, 2.131.18; with AFT against P, 2.107.2, 2.131.4,6. FOi's *συναναστπάσῃ* at 2.131.8 does not seem to make good sense, and the fact that F and Oi share this reading suggests that the source from which 2.107 and 2.131 reached an ancestor of Oi was δ rather than β.

Bouwman). These however are references in Latin to parallel passages in the [Aristotle] *Problems*, quite distinct from the marginal additions to the Greek text.

2.156 is omitted in K; in Oi it is placed before 2.165. But K also omits 2.154, so its omission of 2.156 may not be linked with the misplacing in Oi; and 2.156, also omitted in *p*, is a partial duplicate of 2.150. K could derive from the lost part of N. But if 2.107 and 2.131 were in the margin of N and entered Oi from there, their omission in K is more easily explained if K derives from the ancestor of N, or from N before the marginal additions were made.

2.1.14. Sb

Sb agree with MPRT, 1.18.5, 1.18.6 (not P; correction in Sb?), 1.20.6, 2.39.9; but much more often with NOi, 1.16.4, 1.17.12,33, 2.8.2-3, 2.22.3, 2.27.3, 2.28.2, 2.39.6, 2.41.2, 2.42.6, 2.49.1, 2.51.7, 2.53.1, 2.65.7, 2.75.2, 2.78.14, 2.83.2, 2.88.3, 2.89.5, 2.90.5. Sb do not derive from KNOi: date, 1.3.10, 1.6.3, 1.17.39, 1.18.6, 1.20.6, 2.139.5, 2.148.5, 2.150.2,3.

S and b generally agree. Exceptions: 1.17.15,49; 2.53.3, b agrees with PRTe^{ac}m*p*, presumably as a result of deliberate but misguided emendation. S cannot be source of b, 2.130.1 (unless correction), 2.150.2, and absence in S of division between books 1 and 2. If b is the source of S, S would have to have read b's γε as an abbreviation for γέγονε at 2.45.2 and to have re-corrected at 2.53.3. At 2.64.1, where it seems likely that those MSS that have the article *οἱ* (KMObi) do so as the result of an addition, it would be odd for S to have dropped it again. Hubert and Drexler 1959, xiv place S closer than b to 1 in Plutarch, *Platonic Questions*. At 2.156.5 S has a reading otherwise found only as a correction in b, but the text is clearly corrupt and the emendation could have been made independently; S does not follow a required correction in b at 2.166.4. It seems safest to regard Sb as siblings. Marenghi 1991, 65 indeed regards our SN as "twins" of b in the Bekker *Problems*.

S was owned by Bessarion. It is dated to the mid-14th century by Mioni 1981, 374; to the second half of the 14th century at Moraux et al. 1976, 65, where it is noted that the contents of S and of b as a whole are closely parallel. b was owned successively by the Sultan Mustapha I and by Count Luigi Ferdinando Marsili; Moraux et al. 1976, 66.

2.1.15. e corrected from S

Of the extensive marginal corrections in e, those at 1.17.15,17, 2.60.2, 2.87.1-2, 2.112.2, 2.139.3,4, 2.152.3-4, 2.156.5 (*φρυγῶν*) and 2.190.2 correspond with S alone, those at 2.82.2, 2.111.1, 2.115.6, 2.156.5 (*μεμολυσμένων*) and 2.166.4 (b^{ac}) with Sb alone, 1.17.49 with So alone, and only the following are not found in S: 1 pref. 21 (RNbi), 1.8.4 (no other MS), 2.4.1 (CT), 2.45.5,13 ([Ar.]), 2.55.2 (M), 2.65.1 (K ^{pc}MR^{pc}), 2.96.4 (no other MS), 2.139.7 (copied from 2.139.3), 2.179.4 (*τις* a misreading of *τε*, possibly from S?). At 1 prol. 21 the corrector of e has changed the correct *πάθος* apparently to *πάθοις*, the apparent reading of N^{pc} adopted by Oi, but may have intended *πάθοι*, the reading of N^{ac}Sb. At 1.17.22 Re have the same correction, possibly independently.

2.1.16. Kv

K is described by Kalinka 1909, 166-180; see also Usener 1859, v-vii; Cranz 1960, 131; Garzya and Masullo 2004, 16. Kalinka notes that an inscription at the end of fol. 133v shows the MS belonged to Valla, who either had it copied or purchased it soon after it was copied. It is likely that K has emended or supplemented the transmitted text: e.g. 2.64.1-2, 2.68.6, 2.77.1, 2.85.2, 2.86.1, 2.115.2, 2.151.5, 2.158.3, 2.189.10. Valla corrects K at 2.147.3, 2.169.2, 2.183.1, 2.188.3.

The arrangement of material in Valla's translation reflects the fact that this incomplete MS was his source; his books I (i.119-152), II (ii prologue, ii.1, 2.60-126) and V (2.128-192) correspond to the contents of K. His translation includes no book III or IV (in the original edition; in the 1501 edition we have used (*Problemata Aristotelis cu(m) dpulici [sic] translatio(n)e, antiqua. vz. [et] nouas. Theodori gaze: cu(m) expo(sitio)ne Petri Aponi ... Problemata Alexa(n)dri aphrodisei. Problemata Plutarchi*, per B. Locatellum: Venice, 1501) books V-VII are renumbered as III-V, and the 24th problem of book "III" is mistakenly numbered as 22, all subsequent numbers in this book being two less than they should be). Usener 1859, vii argues that Valla intended to translate books 1 and 2.1-59 as his books III and IV. Valla's book VI contains i.1-118, and his book VII, ii.2-75; these are the parts of books i and ii. found in m, which Valla certainly owned (Garzya 1994, 183) and which Usener 1859, vii and n.2 argued Valla also used. Usener noted that in K Cassius 1-19 follow 2.60-192, while in m Cassius 20 (but actually 21: see above, §2.1.7) follows 2.68, and therefore argued that a single MS has been divided up. Usener's view is rejected by Kalinka 1909, 180, and is ruled out by the overlap between material in K and that in m. There is also no affinity in our texts between the readings of K and those of m.

We have not attempted a critical edition of Valla's version itself. Consequently some of the readings we have noted may be due to the edition we have used rather than to Valla himself.

2.1.17. *g* (Gaza)

g's affinities are with the NOSbi group. He disagrees with APTm, 1.2.8; MRPTe, 1.13.3; M, 1.18.2, 2.17.2. He shares error or omission with BLNOi, 2.10.2; OSbil, 2.33.1; BL MRe OSbi, 2.19.2. He disagrees with BLOSbi at 1.17.12, but this is a question of interpretation. He agrees with E by omitting ἐνυγρανθέντα at 1.5.5, but this is chance due to the method of translation; at 1.20.5 he agrees only with P, but this again is a matter of interpretation. Marenghi 1991, 65 states that our N (his D) was used by Gaza in his translation of the Bekker problems; but Gaza's translation of our text was made in 1453 (Cranz 1960, 127-128), which is early for him to be using N (dated 1450-1472 by Wiesner, above §2.1.12), though not in principle impossible. Cranz 1960, 128 notes that Gaza has roughly the same items as N, but this is the accidental result of the damage to N rather than anything more significant: Gaza's translation of book 2 breaks off at 2.38, whereas we now know that N+l does not. It seems improbable that the MS Gaza used derived from a time before 2.39-192 were added to the collection at all. Gaza treated all the material he took from books 2-4 as part of his "book 2".

Cranz suggests that the MS Gaza used is the one mentioned in a letter written to Nicholas V from Trebizond, quoted by Müntz and Fabre 1887, 114: "tertius (of 4 codices) Aristotelis continet Problemata plura quam hucusque viderim, item Problemata Alexandri Aphrodisei ex eadem ut aiunt schola". The letter is discussed by Mercati 1925, 40-43, who rejects its attribution to Niccolò Perotti and connects it with the MS used by Gaza, while not holding that the letter was itself written by Gaza. But John Monfasani notes (personal communications; he has also consulted Hans Ramminger) that the letter, which is in Perotti's hand but does not seem to be composed by him, is highly problematic as evidence: Nicholas V is not named in the letter itself, which has no date; and "in Trapezunta", which is in any case grammatically odd – one would expect "from" – is no longer visible in the MS if it ever was. Monfasani adds that Trebizond would be a strange place to search for MSS, and that "if the author was writing from Trebizond on the Black Sea, I do not understand why he would tell the pope that he was sending the four libri mentioned via Cardinal Bessarion, who in the pontificate of Nicholas V was papal legate in Bologna with no special courier or diplomatic connection with Asia Minor."

Monfasani has further emphasised to us that it would be unwise to rely on Gaza's translations as evidence for the text, since Gaza paraphrases freely. On Gaza's cavalier attitude to texts see also Kraye 2003, 331. We have not in any case attempted a critical edition of Gaza's version itself, which would be a further task of a similar magnitude to the editing of the present Greek text, but have used the text in the 1562 Juntine edition of the complete Aristotle + Averroes, facsimile reprint 1962, 7 169r-204r. Some of the readings we have noted may therefore be due to the particular edition we have used rather than to Gaza himself.

2.1.18. C

C contains only excerpts, as follows: 1.9, 1.22, 2.2, 2.4, 2.3, 2.13, 2.15, 2.20, 2.22, 2.27, 2.28, 2.32, 2.38, 2.40, 2.41, 2.51, 2.66.

C does not appear to derive from any of the following MSS with the standard sequence: **MR:** 2.28.2; **P:** 2.51.1,5,7; **T:** date, 2.28.2; **e:** date, 2.28.2, 2.51.1; **m:** 2.28.1,2; **w:** date. C overlaps with A in its present state only in 1.9 and in part of 2.66; at 1.9.5 C agrees with ABELw against other MSS in having *ἔχει* (*ἔχει* A) rather than *ἔχῃ*, but the very fact that the reading (a pronunciation error) is shared with BL shows that it would be rash to infer any affinity between C and A on this basis.

C does not derive from any of the MSS with the non-standard sequence: the majority of the problems it includes do not occur in any of the non-standard MSS (there is considerable though not complete overlap in content with the *first* book in BL, but that belongs to the standard sequence), and the ordering in C follows the standard sequence.

C is linked by readings with the NOSbil group. At 2.22.3 it agrees with NOSbe^{mg}i against MPRTem; at 2.27.3 with OSbil against GMPRTYem; at 2.41.2 and 5 with BL OSbil against MPRTem; at 2.51.7 with OSbil against MPRTem. The hypothesis seems reasonable that it derives from ε.

2.1.19 EXc

Of our material these MSS of "Collection A" include only 1.5, 1.6, 1.9, 1.10, 1.12, 1.21 (and 3.50-52, but see above at n.16). They are linked by content, and by readings, e.g. 1.5.4, 1.12.2-5, 1.21.3,4. Since X is 13th century and EXc form a clear group, derivation of these MSS from any extant MS other than GJPo is excluded on grounds of date, and from GJo by content. EXc^{ac} agree with AGMRT^{ac}Yem^{pc}w against BDLNOPST^{pc}be^{mg}im^{ac}, 1.10.5 (though c has the reading of the latter group in the margin), but EXc agree with NOSbe^{pc}im against AGMPRTow at 1.12.5-6. Beyond this any placing of EXc in a stemma must be decided on the basis of the larger amount of material they include from books i-ii Ideler, if indeed the transmission there followed a similar route.

Ec agree against X, and then Xc against E, at 1.12.4. In 3.50-52 Buss. (= Psellus) Xc agree against E in *διψῶσι* at 332.6 Buss., but Ec against X in *τόδ' ὑγρὸν* for *δίυγρον* at 332.12. Usener 1859 viii remarked that D and E are "to be avoided" (*a quibus erat abstinendum*).

X was previously owned by Ulrich Fugger (Lehmann 1960, 455). c belonged successively to Jacques Mentel and J.B. Hautin.[144]

2.1.20. BL

BL ("Collection B") in items from 1-2, all in their book 1, have no significant agreements with other MSS against NOi, so they can here be linked with the source of the NOi group. BL agree with DNOPST^{pc}b$c^{mg}e^{mg}$im^{ac} against AEGMRT^{ac}XYcem^{pc}w, 1.10.5; with MRe and NOSbi against PTMw, 1.14.5; with N^{ac}OSbi, 1.17.12. At 1.20.6 they are closer to NOi than to other MSS. That BL do not derive from Go is shown by content and by 1.17.1-44; that they do not derive from D by content and by 2.72.2.

L does not derive from B: B is missing 2.37 and 2.39ff. pref. 4-5. Moreover, in the [Galen] *Medical Definitions* which form part of "Collection A" and "Collection B", where in def. 317 Kühn has *ὀστοῦ διακοπή*, EL have *ὀστοῦν διακοπὲν* but B *ἤγουν διακοπέν*. In a sequence of unidentified items after def. 395, B omits the section *τί ἐστι ἕλκος*, while E and L have it. In def. 419 *παρὰ τὸ σχῆμα ... παρὰ τὸ μέγεθος* is omitted by B but present in L. However, BL are very closely linked both by readings and by content. B can only derive from L if B has modified the erroneous reading of L at 1.17.26. Moreover, there are some passages in the [Galen] *Medical Definitions* that tell against the derivation of B from L: 326 line 3, *λεύκωμα* BE, om. L; 367 *βλεφάρου ... ἐκτροπῆς* BE, om. L. In the *Definitions* generally BL are closer to each other than either is to E, and E is closer to the text printed by Kühn than BL are. (BL have omissions relative to E in *Definitions* 367, 414 and 426-30; in 419 E omits *τόπον*, while L includes it, and B – see above – omits even more.)

"Collection B" occupies fols. 1-28 of B. Wartelle 1963 reports B as containing "problems of Alexander and Aristotle" at fols. 28-76; Louis 1986 indicates that this material begins with books 1-4 and 8 of the Bekker *Problems*, plus extracts from 6 and 7, but that fols. 45-70 are "an anonymous treatise which has nothing in common with the *Problems* of Aristotle". Louis further suggests (203-4) that this MS forms no part of the

---

144 Information from Pat Easterling and Vivian Nutton.

textual tradition of the Bekker *Problems*, but derives from "a specialised medical collection assembled in the early 14th century". The other items mentioned by Louis in B also appear in L, the material corresponding in position to fols. 45-70 in B being identified in the catalogue of L as extracts from Aelian, *De natura animalium*. However, in the light of what has been observed above about the texts from [Galen] it is probably safer to regard L as a sibling (or an uncle?) of B rather than as its parent.[145]

2.1.21. *p* (Psellus)

Psellus' report (*p*) is valuable in that Psellus himself wrote before any of our extant MSS, with the possible exception of J. It is however frequently a paraphrase rather than a full copy of the text, and we have treated it as such, reporting its readings from the edition by Duffy 1992 where they throw light on the MSS tradition of the full text, but not attempting a complete report of its entire content. We have reported the readings of *p* given in Duffy's apparatus, rather than his actual text, in which he corrects *p* from Usener's edition.

The extent to which Psellus abbreviates and modifies the text makes it difficult to determine the place of his source in the MS tradition. However, it is clear that (as its following the standard sequence would suggest) it is closer to the MSS with the standard sequence than to DGY (1.8.4, 2.56.3). *p* also agrees with other MSS against variants in EXc (1.5.4).

p agrees with AFMPRTbe against KOi, 2.150.2; with AMPRTe^{ac} against KOSb^{2}e^{pc}i, 2.161.3; with AFPT against other MSS, 2.87.4 (but this could be independent emendation); with AFT, 2.119.7-8 (where *p* may preserve the clause lost in other MSS, but without its final *ἱκανῶς*, or else a similar supplement has been made independently in AFT and in *p*; see above, §2.1.1); with APT, 2.190.2; with AT, 2.188.1. *p* agrees with (A)FMRSTb, 2.88.1; with MRem against NOSbi, 1.16.4 (om. PTw); with MRe, 2.108.2; with KMORSbei, 2.90.11-12 (but the omission may be due to homoioteleuton in the MSS and abbreviation in *p*), 2.97.6, 2.155.1. At 2.145.2 *p* agrees with KOi in having *γοῦν*; but the fact that Sb have *οὖν*

---

[145] Usener 1859, viii suggested that B was a sibling of L. For the dating of L to the second half of the 15th century see Pattie and McKendrick 1999, 197-8; cf. Diels 1970, 9. The part of L that concerns us may have been copied by Georgios Gregoropoulos or by Ioannes Rhosos; Wiesner in Moraux et al. 1976, 437, noting also that L was previously in the library of the Jesuit college at Agen (on the Garonne in Southern France). B was copied in 1503 by Manuel Gregoropoulos, the son of Georgios (Canart 1963, 75); Omont 1886-98, pt.2 158.

suggests that the agreement of *p* with KOi is coincidence. *p* shares error with m at 2.65.1, but this is coincidence. It is not clear that *p* can be attached with confidence to any one branch of the standard tradition, though it is with MR that it disagrees least.

p's omission of 2.156 can be explained by its being a duplicate of 2.150. It is not clear why *p* omits 2.47 or 2.183; 2.181 might have been omitted as a continuation of 2.180, and 2.189 as a duplicate of 2.172. At 2.158.11-12, 14-17 *p* does not have the expanded text found only in APT, but this could simply be due to *p*'s paraphrase being much compressed.

2.2. MSS with non-standard sequences

In the following table we have listed the entire contents of the non-standard MSS, even though most are from Ideler's books i and ii. The non-standard sequence in BHJLW is much closer to the standard order than is that in GYjo and the related MSS.[146]

BHJLW form a clear group starting with the sequence i.26-i.27-i.28-i.29-i.2 [lines 21-25 of the table]. GYjo and the related MSS Vienna, med. gr. 44 and Venice, App. gr. IV.16 form a group starting with the sequence i.29-i.10-i.7-i.8 [24-29]. In i.29 this latter group, uniquely, shares the reading ἀλόγων for ἀγρίων. MS D starts with an independent sequence [1-20], and then follows BHJLW as far as 3.47 Bussemaker [71]; it does *not* have GYjo's ἀλόγων in i.29 [24].[147] It then includes at [72-76] five items, i.24-i.25-i.57-i.61-i.62, all of which appear in the GYjo group *before* i.118 but not in the BHJLW group, before following the GYjo group from i.118 [77]. It looks very much as if the compiler of D has at this point switched from following a MS related to the BHJLW group to following one related to GYjo, and that the first thing he did was to go through his new exemplar from the beginning and insert problems which he had not already copied from the exemplar from the BHJLW group. If this is correct, it is strong evidence that D is the result of a late conflation of the two groups, not a unique direct survival from an archetype combining the two.

---

[146] What is listed and discussed in this section from BL is their "book 2", which contains of our texts only 3.46-47; book 1 in BL (for which see above §2.1.20) is a selection from all four of books i and ii Ideler and 1 and 2 Bussemaker in (more or less) the standard sequence. See Moraux et al. 1976, 433-434. – We are grateful to Carl-Gustaf Lindqvist for some corrections to the original version of this table.

[147] Information from Carl-Gustaf Lindqvist.

	‘i.29’ group											
							‘i.26’ group					
	Barocc. 131 62v-64v 13th cent. o	Paris anc. gr. 1268 12th cent. G	Vatican Palat. gr. 237 14th cent. Y	Vienna med. gr. 44 15th cent.	Paris, anc. gr. 2505 15th cent. j	Venice App. gr.IV.16 16th cent.	Paris, anc. gr. 2381 15th/ 16th D	Paris, suppl.gr. 690 11th cent. J	Vienna med. gr. 37 14th cent W	Paris 1630 disturbed order 14th cent H	London Harleian 6295 “book 2” 15th cent L	Paris anc. gr. 1884 1503 B
1							i.1					
2							i.2					
3							i.3					
4							i.4					
5							i.5					
6							i.6					
7							i.7					
8							i.8					
9							i.9					
10							i.11					
11							i.13					
12							i.14					
13							i.15					
14							i.16					
15							i.17					
16							i.18					

	o	G	Y	Vienna 44	j	Venice IV.16	D	J	W	H	L	B
17							i.19					
18							i.20					
19							i.21					
20							2.59					
21							i.26	i.26	i.26	i.26	i.26	i.26
22							i.27	i.27	i.27	i.27	i.27	i.27
23							i.28	i.28	i.28	i.28	i.28	i.28
24	i.29	i.29	i.29	i.29	i.29	i.29	i.29	i.29	i.29	i.29	i.29	i.29
25							i.2	i.2	i.2	i.2	i.2	i.2
26	i.10	i.10	i.10	i.10	i.10		i.10	i.10		i.10	i.10	i.10
27							i.12	i.12		i.12	i.12	i.12
28	i.7	i.7	i.7	i.7	i.7	i.7						
29	i.8	i.8	i.8	i.8								
30		i.9	i.9									
31	i.28	i.28	i.28		i.28	i.28						
32	i.5	i.5	i.5		i.5	i.5						
33	i.3	i.3	i.3									
34	i.23	i.23	i.23			i.23	i.23	i.23		i.23	i.23	i.23
35		i.24	i.24		i.24	i.24						
36		i.25	i.25		i.25	i.25						
37		i.31	i.31		i.31		i.31	i.31		i.31	i.31	i.31
38							i.34	i.34		i.34	i.34	i.34
39		i.35	i.35			i.35	i.35	i.35		i.35	i.35	i.35

	o	G	Y	Vienna 44	j	Venice IV.16	D	J	W	H	L	B
40							i.38	i.38		i.38	i.38	i.38
41							i.39	i.39		i.39	i.39	i.39
42							i.41	i.41		i.41	i.41	i.41
43							i.42	i.42		i.42	i.42	i.42
44							i.51	i.51			i.51	i.51
45		i.54	i.54				i.54	i.54		i.54	i.54	i.54
46	i.56	i.56	i.56		i.56		i.56	i.56		i.56	i.56	i.56
47	i.57	i.57	i.57									
48								i.58			i.58	i.58
49	i.61	i.61	i.61		i.61	i.61						
50		i.62	i.62		i.62							
51	i.63	i.63	i.63		i.63		i.63	i.63		i.63	i.63	i.63
52							i.65	i.65		i.65	i.65	i.65
53							i.68	i.68		i.68	i.68	i.68
54							i.74	i.74	i.74	i.74	i.74	i.74
55							i.89	i.89	i.89	i.89	i.89	i.89
56							i.97	i.97	i.97	i.97	i.97	i.97
57							i.102	i.102	i.102	i.102	i.102	i.102
58							i.115	i.115	i.115	i.115	i.115	i.115
59							i.117	i.117	i.117	i.117	i.117	i.117
60							i.123	i.123	i.123	i.123	i.123	i.123
61							i.127	i.127	i.127	i.127	i.127	i.127
62							i.128	i.128	i.128	i.128	i.128	i.128

	o	G	Y	Vienna 44	j	Venice IV.16	D	J	W	H	L	B
63							i.138	i.138	i.138	i.138	i.138	i.138
64							ii.13	ii.13	ii.13	ii.13	ii.13	ii.13
65							ii.16	ii.16	ii.16		ii.16	ii.16
66							ii.22	ii.22	ii.22	ii.22	ii.22	ii.22
67							ii.68	ii.68	ii.68	ii.68	ii.68	ii.68
68							3.46	3.46	3.46	3.46	3.46	3.46
69							ii.25	ii.25	ii.25	ii.25	ii.25	ii.25
70							ii.26	ii.26	ii.26	ii.26	ii.26	ii.26
71							3.47	3.47	3.47	3.47	3.47	3.47
72							i.24					
73							i.25					
74							i.57					
75							i.61					
76							i.62					
77	i.118	i.118	i.118				i.118					
78	i.122	i.122	i.122				i.122					
79	i.126	i.126	i.126			i.126	i.126					
80		i.127	i.127			i.127						
81		i.128	i.128		i.128							
82	i.133	i.133	i.133			i.133	i.133					
83	i.132	i.132	i.132				i.132					
84	i.134	i.134	i.134				i.134					
85	i.144	i.144	i.144			i.144	i.144					

	o	G	Y	Vienna 44	j	Venice IV.16	D	J	W	H	L	B
86	i.145	i.145	i.145			i.145	i.145					
87	i.149	i.149	i.149			i.149	i.149					
88	2.96											
89	2.99											
90	3.48	see below 160-4										
91	ii.48											
92	i.117											
93	i.115											
94	i.36											
95	i.6											
96	ii.71	see below 223-34										
97	ii.76											
98	1.12											
99	1.17											
100	1.22											
101	2.2											
102	i.16											
103	2.13											
104	i.26											
105	i.27											
106	ii.75											
107	Cass 55											
108	Cass 66											

	o	G	Y	Vienna 44	j	Venice IV.16	D	J	W	H	L	B
109	Cass 67											
110	Cass 80											
111	app.45U											
112	app.46U	see										
113	i.139	below										
114	2.74	204-19										
115	2.79											
116	2.84											
117	Cass 50											
118	Cass 49											
119	Cass 53											
120	i.13											
121	2.32											
122	i.79											
123	2.23											
124	2.24											
125	2.36											
126	2.37	see below										
127	2.38	178-87										
128	2.56											
129	2.59											
130	i.1	see below										

	o	G	Y	Vienna 44	j	Venice IV.16	D	J	W	H	L	B
131	i.21	197-8										
132	ii.26											
133	ii.22											
134	ii.7											
135	i.127											
136	i.128											
137	i.31											
138	i.2	i.2	i.2									
139	i.18	i.18	i.18									
140	i.38	i.38	i.38									
141		i.41	i.41									
142	i.42	i.42	i.42									
143		ii.7	ii.7				ii.7					
144	ii.14	ii.14	ii.14				ii.14					
145		ii.17	ii.17				ii.17					
146		ii.19	ii.19				ii.19					
147	ii.21	ii.21	ii.21				ii.21					
148		ii.22	ii.22									
149		i.102	i.102				i.102					
150		2.85	2.85									
151		2.89	2.89		2.89							
152		2.90	2.90									
153		2.95	2.95				2.95					

	o	G	Y	Vienna 44	j	Venice IV.16	D	J	W	H	L	B
154		2.96	2.96				2.96					
155		2.98	2.98									
156		2.99	2.99				2.99					
157		2.103	2.103				2.103					
158		Cass 47	Cass 47									
159		Cass 74	Cass 74									
160	see above 90-93	3.48	3.48				3.48					
161		ii.48	ii.48				ii.48					
162							Cass. 47					
164		i.117	i.117									
165		i.115	i.115									
166		ii.5	ii.5				ii.5					
167		ii.13	ii.13				ii.13					
168		ii.20	ii.20				ii.20					
169		ii.26	ii.26									
170		ii.27	ii.27				ii.27					
171		ii.41	ii.41				ii.41					
172		ii.45	ii.45				ii.45					
173		ii.61	ii.61				ii.61					
174		1.8	1.8				1.8					
175		1.10	1.10				1.10					
176		1.15	1.15				1.15					
177		2.29	2.29				2.29					

	o	G	Y	Vienna 44	j	Venice IV.16	D	J	W	H	L	B
178	see above 121-9	2.32	2.32				2.32					
179		i.79	i.79				i.79					
180		2.23	2.23									
181		2.24	2.24				2.24					
182		2.27	2.27				2.27					
183		2.36	2.36				2.36					
184		2.37	2.37				2.37					
185		2.38	2.38				2.38					
186		2.56	2.56				2.56					
187		2.59	2.59									
188		2.67	2.67				2.67					
189		2.72	2.72				2.72					
190		2.73	2.73									
191		i.136	i.136				i.136					
192		ii.36	ii.36				ii.36					
193		ii.25	ii.25				ii.25					
194		3.49					3.49					
195		2.168										
196		2.130										
197	see above 130-1	i.1										
198		i.21										
199		i.36										

	o	G	Y	Vienna 44	j	Venice IV.16	D	J	W	H	L	B
200		i.53										
201		i.58										
202		i.60										
203		i.130										
204		Cass 55										
205		Cass 66										
206		Cass 67										
207		Cass 80										
208		app.45U										
209		app.46U										
210		i.139										
211		i.142										
212	see	2.74										
213	above	2.79										
214	107-19	2.84										
215		Cass 24										
216		Cass 27										
217		Cass 39										
218		Cass 49										
219		Cass 50										
220		Cass 53										
221		ii.65										
222		ii.70										

	o	G	Y	Vienna 44	j	Venice IV.16	D	J	W	H	L	B
223		ii.71										
224		ii.75										
225	see	ii.76										
226	above	1.12										
227	96-105	1.17										
228		1.22										
229		2.2										
230		i.16										
231		i.17										
232		2.13										
233		i.26										
234		i.27										
235			fresh hand in Y?[148]									
236			i praef.									
237			i.1									
238			i.2									
239			i.3									
240			i.4									
241			i.5									

[148] Information from Carl-Gustaf Lindqvist.

	o	G	Y	Vienna 44	j	Venice IV.16	D	J	W	H	L	B
242			i.6									
243			i.7									
244			i.8									
245			i.9									
246			i.10									
247			i.11									
248			i.12									
249			i.13									
250			i.14									
251			i.15									
252			i.16									
253			i.17									
254			i.18									
255			i.19									
256			i.20									
257			i.21									
258			i.22									
259			i.23									
260			i.24									
261			i.25									
262			i.26									
263			i.27									
264			i.28									

	o	G	Y	Vienna 44	j	Venice IV.16	D	J	W	H	L	B
265			i.29									
266			i.30				i.30					
267			i.31									
268			i.32									
269			i.33				i.33					
270			i.34									
271			i.35									
272			i.36				i.36					
273			i.37				i.37					
274			i.38									
275			i.39									
276			i.40				i.40					
277			i.43				i.43					
278			i.44				i.44					
279			i.45				i.45					
280			i.46				i.46					
281			i.47				i.47					
282			i.48				i.48					
283			i.49				i.49					
284			i.50				i.50					
285			i.51									
286			i.52				i.52					
287			i.53				i.53					

	o	G	Y	Vienna 44	j	Venice IV.16	D	J	W	H	L	B
288			i.55				i.55					
289			i.58									
290			i.59				i.59					
291			i.60				i.60					
292			i.61									
293			i.64				i.64					
294			i.65									
295			i.66				i.66					
296			i.67				i.67					
297			i.68									
298			i.69									
299			i.70				i.70					
300			i.71				i.71					
301			i.72				i.72					
302			i.73				i.73					
303			i.74									
304			i.75				i.75					
305			i.76				i.76					
306			i.77				i.77					
307			i.78				i.78					
308			i.79				i.79					
309			i.80				i.80					
310			i.81				i.81					

	o	G	Y	Vienna 44	j	Venice IV.16	D	J	W	H	L	B
311			i.82				i.82					
312			i.83				i.83					
313			i.84									

The order in H is disturbed in the MS further than is shown here; for ease of comparison, in the case of this MS only we have rearranged the sequence to that of the group of which it forms part. The actual sequence is: i.115, i.117, i.123, i.127, 3.47, ii.25, ii.26, 3.46, ii.68, i.138, ii.13, i.10, i.23, i.31, i.34, i.35, i.26, i.27, i.28, i.29, i.2, i.12, i.38, i.39, i.41, i.42, i.54, i.56, i.63, i.65, i.68, i.74, i.89, i.97, i.102, i.128, ii.22. This arrangement might in part be due to considerations of subject matter (e.g. 3.47 might have been placed immediately after i.127 Ideler because both mention oil; but why then put 3.46 in the place previously occupied by 3.47, rather than just putting it next after 3.47?). But mechanical dislocation must surely have played a part too.

It should be added that the arguments in the following discussion differ in kind; those derived from the omission of whole problems relate to the entire table, while those based on readings within specific problems are drawn only from the (relatively small) part of the material in the table which is included in the books which we are editing here. In what follows, references such as "i.10-i.68" are to be understood in terms of the sequence of problems *in the table*. Figures in [brackets] refer to line numbers in the table.

However, the sequence at [1-20], *before* D starts following the BHJLW group, includes, after the items from book i Ideler, our 2.59. This occurs in GY [187] and in o [129], and D could have taken this problem from that branch of the tradition along with the problems at [72]ff.; but there seems no obvious reason for it to be taken out of sequence in this way. Shortly afterwards [80-81] D skips i.127-8, presumably as already included from the BHJLW group; similarly with (for example) ii.26 [168], and 2.59 (above) [186]. But sometimes D repeats problems it already has (i.102 at [57] and [149], ii.25 at [69] and [192]) or skips ones it does not (4.73 at [189]); keeping track of what had already been included and what had not was clearly difficult. After 3.49 D resumes from i.30 [265] in sequence, filling in gaps left by what it already has. D shows considerable freedom in altering the text. Usener 1859 viii remarks that D and E are "to be avoided" (*a quibus erat abstinendum*).

GY are clearly closely related as far as ii.25 [192]. o is also closely related but with a disturbed order. After ii.25, Y starts again from the preface to book 1 [235], at first including all problems whether included before or not, but later skipping some already included (i.41-2 [275-6], i.54 [286-7], i.56-7 [287-8], i.62-3 [291-2]) but sometimes repeating (i.61 at [49] and [291], i.79 at [178] and [307]).

2.2.1. DGYjo

D in lines [77-193] of the table cannot derive from: Y, which omits 3.49 (unless D switched to another source for this problem only); Venice, App. gr. IV.16, which omits ii.7-3.49 [143-193]; j, which has none of the items included in this section of D; o, which omits 2.56.4-end and 3.49; Vienna, med. gr. 44, which omits i.9 [30] and all subsequent items. D and o share the omission of i.127-8 between 126 and 133 [80-81], but D has these texts just before [61-62] from its other source. D agrees with other MSS against GY, 2.73.5 (ἐξατμίζει), but that would require explanation on any view of the DGYo group.

G cannot derive from: D, for chronological reasons, because D does not include Us. app. 45 and 46, and because of readings in 2.56.8, 2.72.2, 3.48.1,3, 3.49.2,10-13; Y, for chronological reasons, because Y does not include Us. app. 45 and 46 or 3.49, and because of Y's omission at 2.89.4-5; Venice, App. gr. IV.16, for chronological reasons, and because Venice, App. gr. IV.16 omits i.10 [26], i.8 [29], i.9 [30], i.3 [33], i.31 [37] , i.54 [45], 156-7 [46-47], i.118 [77], i.122 [78], i.128 [81], i.132 [83], i.134 [84]; j, for chronological reasons, and because j omits i.8-i.9 [29-30], i.3 [33], i.23 [34], i.35 [39], i.54 [45], i.57 [47], i.62-3 [50-51], i.118 [77], i.122 [78], i.126-7 [79-80], i.132 [83], i.134 [84]; o, which omits i.9 [30], i.54 [45], i.41 [141], ii.17 [145], ii.19 [146], 2.56.4-end, 3.49; Vienna, med. gr. 44, for chronological reasons and because Vienna, med. gr. 44 omits i.9 [30] and all subsequent items.

Y cannot derive from: D, for chronological reasons, and cf. readings at 2.56.8, 2.72.2, 3.48.3; Venice, App. gr. IV.16, for chronological reasons, and because Venice, App. gr. IV.16 omits i.10 [26], i.8 [29], i.9 [30], i.3 [33], i.31 [37] , i.54 [45], 156-7 [46-47], i.62-3 [50-51], i.118 [77], i.122 [78], i.128 [81], i.132 [83], i.134 [84]; G, 2.98.3; j, for chronological reasons, and because j omits i.8-i.9 [29-30], i.3 [33], i.23 [34], i.35 [39], i.54 [45], i.57 [47], i.118 [77], i.122 [78], i.132 [83], i.134 [84]; o, which omits i.9 [30], i.54 [45], i.41 [141], ii.17 [145], ii.19 [146], and 2.56.4-end; Vienna, med. gr. 44, for chronological reasons and because Vienna, med. gr. 44 omits i.9 [30] and all subsequent items. At 2.103.3, where AP and [Ar.] have *καταφέρεται*, but the other MSS of our problems *καταφαίνεται*, G has *καταφαίνεται* but Y *καταφέρεται*; see above, §2.1.1.

j cannot derive from: D, for j clearly follows the sequence in G and includes items (i.7 [28], i.28 [31], i.5 [32], i.24 [35], i.25 [36]) which D omits at this point though including them either earlier from the BHJLW sequence (i.7 [7], i.28 [23], i.5 [5]) or later from the GYo sequence (i.24-5 [72-73]); Y, for j does not share the omission with Y at 2.89.4-5, though it *does* share its reading with Y alone at 2.89.2.; Venice, App. gr. IV.16, for chronological reasons, and because Venice, App. gr. IV.16 omits i.10 [26], i.31 [37], i.56 [46], i.62-3 [50-1], i.128 [81]; o, which omits i.24-5 [35-36] and has i.31 in a different part of the sequence ([137] not [37]); Vienna, med. gr. 44, which omits i.9 [30] and all subsequent items.

o cannot derive from: D, for chronological reasons, because D omits Us. app. 45, 46, and because of readings at 2.56.8, 3.48.2,3; Y, for chronological reasons, and because Y omits Us. app. 45, 46; Venice, App. gr. IV.16, for chronological reasons, and because Venice, App. gr. IV.16 omits i.10 [26], i.8 [29], i.3 [33], i.56-7 [46-7], i.63 [51], i.118 [77], i.122 [78], i.132 [83], i.134 [84], i.128 [136]; j, for chronological reasons, and because j omits i.8 [29], i.3 [33], i.23 [34], i.57 [47], i.118 [77], i.122 [78], i.132 [93], i.134 [84]; Vienna, med. gr. 44: for chronological reasons, and because Vienna, med. gr. 44 omits i.9 [30] and all subsequent items.

DGY (and e) correct error against o and all other MSS, 2.56.3 (j does not include this problem). Yj agree against G and all other MSS in error at 2.89.2. DY agree against G, 1.15.6; against Go, 2.59.5. At 2.38.1 D agrees with Y after correction and with no other MS. DY agree against other MSS at 2.56.2. 2.59.2. DY share minor omissions against G at 1.10.2, 2.27.2. GY share the same variant text of 2.103 against D; but we know from elsewhere that D expands and adapts the text (cf. 1.15.4), and the version in D is further removed from the standard text than is that in GY (n.b. especially *ἢ καὶ διὰ τὸ εἶναι ψυχρά* in D). Yo agree at 2.36.2 and share omission against G at 3.48.1. GY agree, 2.98.1.

Y was previously owned by Ulrich Fugger; Lehmann 1960, 455. G is dated to the 12th century by Omont 1886-98, pt.1 282.

2.2.2. BDHJLW

B cannot derive from: D, which omits i.58 [48]; H, which omits i.51 [44], i.58 [48], ii.16 [65]; W, which omits i.10-i.68 [26-53].

D cannot derive from: BL, because of readings in 3.46.1,2; H, which omits i.51 [44], ii.16 [65], and cf. reading in 3.46.1; W, which omits i.10-i.68 [26-53], and cf. reading in 3.46.3.

H cannot derive from: BD, on chronological grounds; W, which omits i.10-i.68 [26-53], and cf. reading in 3.46.3.

J cannot derive from: BL, on chronological grounds, and cf. readings in 3.46.1,2; D, on chronological grounds, and because D omits i.58 [48]; H, on chronological grounds, because H omits i.51 [44], i.58 [48], ii.16 [65], and because of reading in 3.46.1; W, which omits i.10-i.68 [26-53], and cf. reading in 3.46.3.

L cannot derive from: D, which omits i.58 [48]; H, which omits i.51 [44], i.58 [48], ii.16 [65]; W, which omits i.10-i.68 [26-53].

W cannot derive from: BL, on chronological grounds, and cf. readings in 3.46.1,2; D, on chronological grounds; H, which omits ii.16 [65], and cf. reading in 3.46.1.

Because the problems (3.46-47) in the present edition that are included in HJW, in BL's "book 2", or in the part of D that derives from this group are not included in any of the other MSS studied here, we are not in a position to comment on the relation of this group of MSS either to the MSS with the standard sequence or to the source of the DGYjo group. That is a question that must be determined by the editor of books i-ii Ideler; in the stemma we have taken the safest option and linked J to α. We know that B and L are siblings (or nephew and uncle) in their book 1 (above); it is likely that they are similarly related here too. On the basis of content and of readings in the problems included here there is no reason why DLW cannot all derive from J. It cannot be determined from these problems whether H derives from J or from λ. Similarly, we are not in a position to comment on the affinities of D in its third section ([266-312 of the table]). For the date of W see Hunger 1969, 89; it was purchased in Constantinople by Augerius von Busbeek, ibid. 90.

2.3. MSS not containing our text

For completeness, we note the following:

Wartelle 1963 (no. 1958) reports cod. Vat. Palat. gr. 400 as containing at fols. 174v-178 extracts from Aristotle's physical problems "*De canitie*", on grey or white hair. In fact the text which extends from 174v12 to 177v3 and is headed *From Aristotle's [writings] on nature: on grey hair* ('Εκ τῶν ἀριστοτέλους φυσικῶν· Περὶ πολιᾶς), with no reference to *problems*, is a version, with some abbreviated paraphrasing and some omissions, of Aristotle, *GA* 5.4 784a23-786b5. It is followed on the remainder of 177v by a number of short items on diverse topics which we have been able to identify, with the aid of the *Thesaurus Linguae Graecae*, as taken from various places in Aristotle *HA, DA, De sensu, De somno* and *GA*.

Paris, ancien gr. 1943 has a subscription stating that it contains at 43v-54 "Cassii iatrosophistae et Aristotelis quaestiones medicae de animalibus et quadrupedibus" (cf. Wartelle 1963, no.1429). In fact the texts are only from Cassius. (We are grateful to Antonio Garzya for confirming this: cf. his 1994, 184, and Garzya and Masullo 2004, 17, noting that this is an indication of the various groups of problems being treated as a single body of material.)

MS Berlin, cod gr. fol.39, mentioned in connection with Alexander by Diels 1906, 10 as containing "various excerpts", is described by Studemund and Cohn 1890, MS no. 275, as "Grosse Pappschachtel enthaltend die Sammlunger C. Weigel's zu Griechischen Medicinern, 42 Stücke in 15 Mappen, welche zusammen den beiden vorherigen HSS 1873 angekauft werden." Item 16 is described as "Ein heft mit Abschriften und Collationen kleinerer Stücke des ... Alexander von Aphrodisias ..." [among others], and for details reference is made to *Catalogue des livres anciens et rares en vente chez T.O. Weigel*, 6ème partie, Leipzig 1873. We have not been able to consult a copy of the latter or to ascertain what the Alexander material might be. The "MS" itself was taken from Berlin to Krakow, probably after World War II (Garzya and Masullo 2004, 20; Jürgen Wiesner, private communication), but Garzya was unable to locate it (Garzya and Masullo, loc. cit.) and notes that the Cassius material it contains, at least, dates from the 18th century.

Although the known Latin version by Peter of Abano includes only books i-ii Ideler (Cranz 1960, 127), attributed by him to Alexander, references in his *Conciliator* suggest that he also translated books 1 and 2, which he found attributed to Aristotle (Olivieri 1988, 194-199; we are grateful to Pieter De Leemans for drawing this to our attention). The attribution, together with his apparently having read αὐτὴν in 1.16.4 (q.v.) suggests that the MS he used was akin to MRe or to m. At first sight Peter's *sortiti sunt <u>ab</u> astrorum impressione* looks more akin to ἀπὸ τῆς *ἀστρῴας* of BLNOSbi than to the *ὑπὸ τῆς ἀστρῴας* of MRPTemw and Psellus, but can be explained by the change from a passive verb with *ὑπὸ* to a deponent.

3. Principles adopted in editing and apparatus

We have not restored readings from parallel texts in the [Aristotle] *Problems* automatically; rather, where the apparent reading of the archetype of our MSS is possible, we have printed it (e.g. 2.40.1,2,5, 2.45.11, 2.51.8-9, 2.83.3, 2.87.2, 2.99.3, 2.100.4). Where the sense of the apparently reconstructed archetype is not possible, we follow the [Aristotle] *Problems* (e.g. 2.52.2, 2.97.3). In general we have followed the MSS, or the majority of them, more closely than Bussemaker and (particularly) Usener did: e.g. 1.3.9, 1.15.5, 1.17.6, 2.10.2, 2.13.4, 2.42.1,3, 2.75.7, 2.78.6, 2.161.3, 2.178.3. We have not followed Usener's fantasies at 1.17.24 and 2.39.8 or his unnecessary alteration at 2.78.5. We have not however followed the MSS everywhere; thus at 2.42.10 Wegehaupt's *τούτω*, and at 2.78.13 l^{2}'s *μύκητες*, the latter independently conjectured by Usener, clearly give the best sense.

Bussemaker and Usener frequently print the definite article where our MSS, or most of them, do not have it (e.g. 1.17.26, 1.19.1, 2.9.1, 2.55.1, 2.62.3, 2.75.3, 2.81.1,4,7, 2.91.3, 2.109.1). We have not followed suit. Where our tentative stemma suggests that a particular reading is found in more branches of the tradition we have generally preferred it to one found in fewer (e.g. 2.37.2; 2.53.1; 2.101.2, where *ἐφικνεῖται* is the more appropriate word in any case; 2.105.2 *ναυτίασις*; 2.127.4; 2.142.3; 2.149.2,4; 2.190.2); but not always, for sometimes the reading of one branch is superior in sense to that of several (e.g. KOSbi at 2.90.5). At 1.17.33 we have adopted the reading found in Psellus as superior to that of all our MSS and best explaining the variants. We have, with some hesitation, reserved angle brackets in the text for supplements conjectured by the modern editors and other scholars; many of the readings found in only one or two late MSS (notably K) are clearly conjectures not based on any earlier MSS tradition, but because of the difficulty of drawing a hard-and-fast line between conjectural readings, especially where they are found in several MSS, and transmitted ones we have thought it best not to introduce angle brackets in any such cases.

The apparatus is in principle negative: for each problem it may be assumed that all the MSS cited at the start of the individual problem have the reading printed in our text except where otherwise indicated. However, where the adopted reading is that of only a few MSS, or where it is particularly significant for the relations between the MSS, we have indicated which MSS have it, rather than requiring the reader to work this out by elimination. We have indicated where the choices of previous editors, Buss(emaker) and Us(ener) differ from ours; references to Bussemaker may be either to his printed text or to the more or less definite suggestions for emendations included in his introductory *variae lectiones,* and we have generally not noted cases where his text diverges from the text we have adopted but the *variae lectiones* indicate agreement with it. We have in principle disregarded variations and errors in spelling and accentuation which are significant neither for the reconstruction of the MS transmission nor for the interpretation of the text. "BC^{ac}D" (for example) = B, C before correction, D.

The consensus of all those MSS which contain a given problem is indicated by "codd.", and that of all of them not otherwise cited by "cett." Exceptions to these rules are (i) Psellus' paraphrase (*p*), which we have cited only where its readings are particularly significant for choosing between MSS readings; absence of a reference to *p* should not therefore be taken to mean it agrees in wording with our text; (ii) Gaza's translation (*g*) and the Latin version (*r*) which we have treated in a similar way because it is not always possible to infer which of the variant readings in the Greek the Latin represents, and because in the case of *r* many of the variations appear to be modifications and abbreviations rather than evidence for an earlier and better state of the Greek text; for *g* see above, §2.1.17; (iii) Valla's translation (*v*), which derives from an extant MS (K) and which we have therefore cited only where its renderings are significant for interpreting the Greek. As always, the Greek text implied by a Latin translation may not be that which the translator actually read in the MS he used, but may rather reflect either a conjectured reading in the Greek text or simply an assumption as to the correct sense. Since we have noted the readings of *gprv* only where they are relevant to the reconstruction of the Greek text on which these versions were based, those who wish to study any of these in their own right should not rely on our apparatus for this purpose. The readings we give for *r* are those of the earliest, eighth-century MS, Bamberg, Staatsbibl., cod. med. 1 itself, as reported by Stoll 1992, 68-75 (cf. Rose 1863, 666-676), as opposed to the emendations in the later, 12th-century MSS which Rose reports in addition.

4. Acknowledgements

Sophia Kapetanaki undertook the bulk of the work in an earlier project to catalogue the MSS of the *scripta minora* attributed to Alexander, and has performed by far the greater part of the collation of the Greek MSS of these *Problemata* and of the construction of the apparatus. Bob Sharples drafted the introduction, translation and notes, and undertook the comparisons with the Latin versions (*grv*), Psellus and the Bekker *Problems*. Research on the earlier cataloguing project was generously supported by grants from the British Academy and the A.G. Leventis Foundation, and that specifically on the *Problemata* by a grant from the Leverhulme Trust. We are also grateful to the many libraries that have provided prints or microfilms of the various MSS, and for discussion and advice to Bruce Barker-Benfield, Dr A.T. Bouwman, Charles Burnett, Myles Burnyeat, Ioana Ciuca, John Dudley, Pat Easterling, Philip van der Eijk, Lou Filius, Klaus-Dietrich Fischer, Antonio Garzya, Alan Griffiths, Simon Hornblower, Jason Koenig, Wolfgang Kullmann, Inna Kupreeva, Pieter De Leemans, Carl-Gustaf Lindqvist, Naomi van Loo, John Monfasani, Vivian Nutton, Yannis Papadoyonnakis, Marlein van Raalte, Hans Ramminger, Georges Sidéris, Ineke Sluiter, Harold Tarrant, Manuela Tecusan, Stephen White and Jürgen Wiesner. The Thesaurus Linguae Graecae database has given incalculable help in searches for parallels and for points of word-usage.

SK
RWS
June 2006

5. Bibliography

P. Accattino and P.L. Donini, *Alessandro* di *Afrodisia: L'anima*. Rome and Bari, Laterza, 1996.

D. Amand, *Fatalisme et liberté dans l'antiquité grecque*, Louvain 1945, reprinted as E. Amand de Mendieta, *Fatalisme et liberté dans l'antiquité grecque*, Amsterdam 1973.

S. Amigues, *Théophraste: Recherches sur les plantes*, Paris, Budé, 4 volumes, 1988-, in progress.

G. de Andrés, *Catálogo de los códices griegos de la Real Biblioteca de El Escorial*, vol.2, Madrid, El Escorial, 1965

—— *Catálogo de los codices griegos de la Biblioteca Nacional*, Madrid, Ministerio de Cultura, 1987

G. Bastiniani and D.N. Sedley, eds., 'Anonymus, *Commentarius in Platonis* Theaetetum', in *Corpus dei papiri filosofici greci e latini*, Parte 3, *Commentari*, Florence: Olschki, 1995, 227-562.

A. Blair, 'The *Problemata* as a natural philosophical genre', in A. Grafton and N. Siriaisi, *Natural particulars: nature and the disciplines in Renaissance Europe*, Cambridge, Mass.: MIT Press, 1999, 171-204.

Bodleian Library: *Greek manuscripts in the Bodleian Library: an exhibition held in connection with the 13th International Congress of Byzantine Studies*, Oxford, Bodleian Library, 1966.

T. Brennan, 'The text of Anaxagoras fragment DK 59B22', *AJPh* 116 (1995) 533-7.

I. Bruns, *Supplementum Aristotelicum* 2.2, Berlin: Reimer, 1892.

C. Burnett, ed., *Adelard of Bath, conversations with his nephew*, Cambridge: Cambridge University Press, 1998.

U.C. Bussemaker (ed.), *Aristoteles, Opera Omnia*, vol.IV. 1, Paris: Didot, 1857; reprinted Hildesheim: Olms, 1973.

P. Canart, 'Scribes grecs de la Renaissance', *Scriptorium* 17 (1963) 56-82.

V. Capocci, *Bibliotheca Apostolica Vaticana: Codices Barberiniani graeci* 1, *Codices 1-163*, Vatican: Vatican Library, 1958.

V. Caston, 'Epiphenomenalisms: Ancient and Modern', *Philosophical Review* 106 (1997) 309-63.

A. Chaniotis, 'Epigraphic evidence for the philosopher Alexander of Aphrodisias', *Bulletin of the Institute of Classical Studies* 48 (2004) 79-81.

H. Cherniss, *Plutarch: Moralia* vol.13.1 (Loeb Classical Library), London and Cambridge, Mass.: Heinemann/Harvard University Press, 1976.

F.E. Cranz, 'Alexander of Aphrodisias', in P.O. Kristeller, ed., *Catalogus Translationum et Commentariorum,* 1, Washington D.C., 1960, 77-135.

H. Daiber, review of F.E. Peters, *Aristoteles Arabus*, *Gnomon* 42 (1970) 538-547.

J. Demianczuk, ed., *Supplementum comicum*, Krakow 1912.

R. Devreesse, *Catalogue des manuscrits grecs de la Bibliothèque nationale, pt. II. Le fonds Coislin*, Paris, Imprimerie nationale, 1945.

H. Diels, *Die Handschriften der antiken Aerzte*, Abh. Preuss. Akad. Wiss., phil.-hist. Kl., 1905, 3; 1906, 1; 1907, 2; reprinted Leipzig, Zentral-antiquariat, 1970.

F.R. Dietz, ed., *Commentarii in Hippocratis aphorismos* (Scholia in Hippocratem et Galenum, vol.2), Königsberg: Bornträger, 1834.

J.M. Duffy, *Michaelis Pselli philosophica minora* I, Teubner 1992.

P. van der Eijk, *Diocles of Carystus*, Leiden: Brill, 2000-2001.

S. Fazzo, 'Alessandro di Afrodisia e Tolomeo: Aristotelismo e astrologia fra il II e il III secolo d.C.', *Rivista di storia di filologia* 4 (1988) 627-649.

S. Fazzo and M. Zonta, *Alessandro d'Afrodisia, Sulla Provvidenza.* Milan: Rizzoli, 1998.

L.S. Filius, 'The theory of vision in the *Problemata Physica.* A comparison between the Greek and the Arabic versions', in G. Endress and R. Kruk, eds., *The ancient tradition in Christian and Islamic Hellenism. Studies ... dedicated to H.J. Drossaart Lulofs*, Leiden: Research School CNWS, 1997, 77-83.

—— *The Problemata Physica attributed to Aristotle: the Arabic version of Hunain ibn Ishaq and the Hebrew version of Moses ibn Tibbon*, Leiden: Brill, 1999.

H. Flashar, 1962a. 'Beiträge zur spätantiken Hippokratesdeutung,' *Hermes* 90 (1962) 402-418.

—— 1962b. *Aristoteles: Problemata Physika*, Berlin 1962 (^{4}1991) (Aristoteles Werke in Deutscher Ubersetzung, vol.19).

E.S. Forster, *The Works of Aristotle: Problemata*, Oxford: Clarendon Press, 1927.

W.W. Fortenbaugh, R.W. Sharples and M. Sollenberger, *Theophrastus on Sweat, Dizziness and Fatigue*, Leiden: Brill, 2003.

A. Garzya, 'La tradition manuscrite des *Problemata* de Cassius le iatrosophiste', in A. Garzya and J. Jouanna, eds., *Storia e ecdotica dei testi medici greci, Atti del II Convegno Internazionale, Parigi 24-26 maggio 1994*, Naples: M. d'Auria, 1994, 181-9.

—— 'Ancora sulla tradizione manoscritta dei *Problemi* di Cassius iatrosofista', in K.-D. Fischer, D. Nickel, P. Potter, eds., *Text and Tradition: Studies in Ancient Medicine and its Transmission presented to Jutta Kollesch*, Leiden: Brill, 1998, 85-89.

A. Garzya and R. Masullo, *I Problemi di Cassio Iatrosofista*, Naples: Accademia Pontaniana, 2004 (Quaderni dell'Accademia Pontaniana, 38).

Theodore Gaza, translation of Alexander, *Problems*, in Aristotelis omnia quae extant opera (et) Averrois Cordubensis in ea opera omnes commentarii, Venice: Giunta, 1562, repr. Frankfurt am Main: Minerva, 1962, vol.7 169r-204r.

C. Genequand, 'Quelques aspects de l'idée de la nature, d'Aristote à al-Ghazālī', *Revue de Théologie et de Philosophie* 116 (1984) 105-129.

—— *Alexander of Aphrodisias On the Cosmos*, Leiden: Brill, 2001.

C. Graux, *Essai sur les origines du fonds grec de l'Escurial: épisode de l'histoire de la renaissance des lettres en Espagne*, Paris, F. Vieweg, 1880.

C. Hubert, rev. H. Drexler, *Plutarchi Moralia*, vol.6.1, Lepizig: Teubner 1959.

H. Hunger, *Katalog der griechischen Handschriften der Österreichischen Nationalbibliothek. T.2, Codices juridici; Codices medici*, Vienna, G. Prachner, 1969.

I. Hutter, *Corpus der byzantinischen Miniaturenhandschriften*, Bd.5.1, *Oxford college libraries: Textband* (Stuttgart: Hiersemann, 1997).

J.L. Ideler, *Physici et medici graeci minores*, Berlin 1841-2; reprinted Amsterdam: Hakkert, 1963.

C. Jacob, 'Questions sur les questions: archéologie d'une pratique intellectuelle et d'une forme discursive', in A. Volgers and C. Zamagni, eds., *Erotapokriseis: Early Christian question-and-answer literature in context*, Leuven: Peeters, 2004, 25-54.

J. Jouanna, 'Présentation d'un nouveau corpus de *problemata médicaux et physiques:* les *Problèmes Hippocratiques*', in U. Criscuolo and R. Maisano, eds, *Synodia, studia humanitatis Antonio Garzya ... dicata*, Naples: M. D'Auria, 1997, 511-539.

E. Kalinka, 'De codice Mutinensi 145', in *Innsbrucker Festgruß von der philosophischen Fakultät dargebracht der 50. Versammlung deutscher Philologen und Schulmänner in Graz*, Innsbruck 1909, 167-201.

S. Kapetanaki and R.W. Sharples, 'A glossary attributed to Alexander of Aphrodisias', *BICS* 44 (2000) 103-143.

J.F. Kindstrand, 'Sostratus-Hercules-Agathion: the rise of a legend', *Annales Soc. Lit. Human. Uppsal.* 1979-80, 51-52.

J. Kraye, 'The legacy of ancient philosophy', in D.N. Sedley, ed., *The Cambridge Companion to Greek and Roman Philosophy*, Cambridge: Cambridge University Press, 2003, 323-52.

B. Lawn, *The Salernitan questions: an introduction to the history of Medieval and Renaissance problem literature*, Oxford: Clarendon Press, 1963.

—— *The Prose Salernitan questions*, London: Oxford University Press for the British Academy, 1979.

H.D.P. Lee, *Aristotle: Meteorologica*, London/Cambridge, Mass., Heinemann/ Harvard University Press (Loeb Classical Library), 1952.

P.J.G. Lehmann, *Eine Geschichte der alten Fuggerbibliotheken*, vol.2, Tübingen, Mohr, 1960.

H. Linke, *Quaestiones de Macrobii Saturnaliorum fontibus*, Bratislava 1880.

P. Louis, Les manuscrits parisiens des *Problèmes* d'Aristote', *Revue de Philologie* 60 (1986) 197-204.

—— ed., *Aristote: Problèmes*, Paris: Les Belles Lettres, t.1, sections I à X, 1991; t.2, sections XI à XXVII, 1993; t.3, sections XXVIII-XXXVIII, Index, 1994.

W. Luppe, 'Der Kommentar Pap. Oxy. 2744', *Archiv für Papyrusforschung* 20 (1970) 29-42.

—— 'Nochmals ps.-Aristoteles *Problemata inedita* 2.153 in POxy 2744', *ZPE* 36 (1979) 55-6.

G. Marenghi, 'La tradizione manoscritta dei Problemata physica aristotelici', Bollettino del Comitato per la Preparazione della Edizione Nazionale dei Classici Greci e Latini 9 (1961) 47-57.

—— *[Aristotele]: Profumi e miasmi*, Naples: Arte Tipografica, 1991 (Università degli studi di Salerno, Quaderni del Dipartimento di scienze dell'antichità no.10).

G. Mercati, *Per la cronologia della vita e degli scritti di Niccolò Perotti*, Rome: Biblioteca Apostolica Vaticana, 1925 (Studi e testi 44).

C. Meredith-Owens, 'A tenth century Arabic miscellany', *British Museum Quarterly* 20 (1955-56) 33-34.

K.A. de Meyier, *Codices Vossiani graeci et miscellanei*, Leiden, Universiteitsbibliotheek, 1955.

E. Miller, *Catalogue des manuscrits grecs de la bibliothèque de l'Escurial*, Paris, Imprimerie nationale, 1848.

E. Mioni, *Bibliothecae Divi Marci Venetiarum codices Graeci manuscripti. Thesaurus antiquus*, Rome, Istituto poligrafico e zecca dello Stato, vol.1 1981, vol.2 1985.

P. Moraux, *Der Aristotelismus bei den Griechen,* vol.2 (Peripatoi, 6), Berlin: De Gruyter, 1984.

P. Moraux et al., *Aristoteles Graecus*, vol.1 (Peripatoi 8), Berlin: De Gruyter, 1976.

S.N. Mouraviev, 'The Moving Posset Once Again: Heraclitus fr. B 125 in context', *Classical Quarterly* 46 (1996) 34-43.

E. Müntz and P. Fabre, *La Bibliothèque du Vatican au xve siècle d'après des documents inédits*, Paris: E. Thorin, 1887.

L. Olivieri, *Pietro d'Abano e il Pensiero Neolatino*, Padua: Antenore, 1988.

H. Omont, *Inventaire sommaire des manuscrits grecs de la Bibliothèque nationale*, Paris, A. Picard, 1886-98: pt. 1, Ancien fonds grec: théologie, 1886; pt. 2, Ancien fonds grec: droit, histoire, sciences.,1888; pt.3, Ancien fonds grec: belles-lettres. Manuscrits grecs de Coislin. Manuscrits du supplément grec, etc. 1888-98.

T. S. Pattie and S. McKendrick, *Summary catalogue of Greek manuscripts in the British Library*, vol.I, London, British Library, 1999.

L. Pearson and F.H. Sandbach, *Plutarch: Moralia*, vol.9, Loeb Classical Library, 1965.

A.L. Peck, *Aristotle: Generation of Animals*, London/Cambridge, Mass., Heinemann/Harvard University Press (Loeb Classical Library), 1942.

I.N. Pontikos, *Anonymi miscellanea philosophica, a miscellany in the tradition of Michael Psellos (Codex Baroccianus graecus 131),* Athens: Academy of Athens, 1992 (Corpus philosophorum medii aevi, Philosophi Byzantini 6).

C. Prantl, 'Über die Probleme des Aristotles', *Abh. Königl. bayerischen Akad. der Wiss.* 6 (1851) 339-377.

—— review of Bussemaker, *Gelehrte Anzeigen der Königlichen bayerischen Akad. der Wiss.* 25 (1858) 201-207.

A. Preus, 'Drugs and Psychic States in Theophrastus' *Historia Plantarum* 9.8-20', in W.W. Fortenbaugh and R.W. Sharples, eds., *Theophrastean Studies*, New Brunswick: Transaction, 1988 (Rutgers University Studies in Classical Humanities, 3) 76-99.

J.M. Riddle, *Disocorides On Pharmacy and Medicine*, Austin: University of Texas Press, 1985.

E. Richter, *De Aristotelis Problematibus*, diss. Bonn, 1885.

G. Rochefort, 'Une anthologie grecque du 11e siècle, le Parisinus suppl. gr. 690', *Scriptorium* 4 (1950) 3-17.

V. Rose, *Aristoteles Pseudepigraphus*, Leipzig: Teubner, 1863.

A. Roselli, 'Ps.-Aristotele *Problemata Inedita* 2.153 in POxy 2744', *ZPE* 33 (1979) 9-12.

C.-E. Ruelle, 'Alexandre d'Aphrodise et le prétendu Alexandre d'Alexandrie', *Revue des Études grecques* 5 (1892) 103-107.

—— ed., *Problemata Aristotelis*. Leipzig: Teubner, 1922.

F.H. Sandbach, 'Plutarch and Aristotle', *Illinois Classical Studies* 7 (1982) 207-232.

—— *Aristotle and the Stoics*, Cambridge Philological Society suppl. vol.10, 1985.

D. Sedley, '*Metaphysics* Λ 10', in M. Frede and D. Charles, eds., *Aristotle's* Metaphysics Lambda, Oxford: Clarendon Press, 2000, 327-350.

R.W. Sharples, 'Theophrastus on tastes and smells', In W.W. Fortenbaugh et al. (eds.), *Theophrastus of Eresos: on his life and work*. New Brunswick: Transaction, 1985. (Rutgers Studies in Classical Humanities, 2) 183-204.

—— 'Alexander of Aphrodisias: Scholasticism and Innovation', in H. Temporini and W. Haase (eds.), *Aufstieg und Niedergang der römischen Welt*, vol.II.36.1. Berlin: De Gruyter, 1987, 1176-1243, at 1198-9.

—— *Alexander of Aphrodisias,* Ethical Problems, London: Duckworth, 1990.

—— *Alexander of Aphrodisias,* Quaestiones *2.16-3.15,* London: Duckworth, 1994.

—— *Theophrastus of Eresus: Sources for his Life, Writings, Thought and Influence, Commentary volume 5, Sources on Biology*, Leiden: Brill, 1995.

—— 1998a, 'Alexander and pseudo-Alexanders of Aphrodisias: *Scripta minima. Questions* and *Problems, makeweights* and prospects', in: Wolfgang Kullmann, Jochen Althoff, Markus Asper (eds.), *Gattungen wissenschaftlicher Literatur in der Antike* 383-408.(Scriptoralia, 95). Tübingen: Gunter Narr Verlag, 1998.

—— 1998b, *Theophrastus of Eresus: Sources for his Life, Writings, Thought and Influence, Commentary volume 3.1, Sources on Physics*, Leiden: Brill, 1998.

—— 'Schriften und Problemkomplexe zur Ethik', in P. Moraux, *Der Aristotelismus bei den Griechen,* vol.3 *Alexander von Aphrodisias,* 513-616, ed. J. Wiesner; Berlin, De Gruyter, 2001.

—— 'Aristotelian Theology after Aristotle', in D. Frede and A. Laks, eds., *Traditions of Theology*, Leiden: Brill, 2002, 1-40.

—— 'Theophrastus, *On Dizziness*', in W.W. Fortenbaugh, R.W. Sharples and M. Sollenberger, *Theophrastus on Sweat, Dizziness and Fatigue*, Leiden: Brill, 2003.

——'Evidence for Theophrastus *On Hair*, *On Secretion*, *On Wine and Olive-Oil?*', *Bulletin of the Institute of Classical Studies* 47 (2004) 141-151.

—— (forthcoming), 'Natural Philosophy in the Peripatos after Strato', forthcoming in W.W. Fortenbaugh & S.A. White (eds), *Lyco of Troas and Hieronymus of Rhodes* (Rutgers University Studies in Classical Humanities, 12).

I. Sluiter, 'Two Problems in Ancient Medical Commentaries', *CQ* 44 (1994) 270-275 (at 270-273).

H. von Staden, *Herophilus : the art of medicine in early Alexandria*, Cambridge: Cambridge University Press, 1989.

P. Steinmetz, *Die Physik des Theophrast*, Bad Homburg: Max Gehlen, 1964.

U. Stoll, *Das "Lorscher Arzneibuch": ein medizinisches Kompendium des 8. Jahrhunderts (Codex Bambergensis medicinalis 1),* Stuttgart: F. Steiner, 1992 (Sudhoffs Archiv, Beihefte, 28)

W. Studemund and L. Cohn, *Verzeichniss der griechischen handschriften der Königlichen bibliothek zu Berlin*, Berlin: A. Asher, 1890.

F. Susemihl, review of Richter 1885, *Wochenschrift für klassische Philologie* 1885, 1481ff.

P. Tassinari, *Ps.-Alessandro di Afrodisia. Trattato sulla Febbre*, Alessandria: Edizioni dell'Orso, 1994.

P. Thillet, *Alexandre d'Aphrodise: Traité du Destin*, Paris: Les Belles Lettres, 1984.

D'Arcy W. Thompson, *A Glossary of Greek Fishes*, London: Oxford University Press, 1947 (St. Andrews University Publications, 45).

R.B. Todd, *Alexander of Aphrodisias on Stoic Physics*, Leiden: Brill, 1976.

M. Ullmann, *Die Medizin im Islam*, Leiden: Brill, 1970 (Handbuch der Orientalistik, ed. B. Spuler, Abt. 1, Der Nahe und der Mittlere Osten, Ergänzungsband 6, Abschnitt 1)

H. Usener, *Alexandri Aphrodisiensis quae feruntur Problematorum libri 3 et 4*, Berlin 1859.

G. Valla, translation of Alexander of Aphrodisias *Problems*, in *Problemata Aristotelis cu(m) dpulici [sic] translatio(n)e, antiqua. vz. [et] nouas. Theodori gaze: cu(m) expo(sitio)ne Petri Aponi ... Problemata Alexa(n)dri aphrodisei. Problemata Plutarchi.* Venice: B. Locatellus, 1501.

G. Verbeke, *L'évolution de la doctrine du* pneuma *du stoïcisme à S. Augustin*, Paris: de Brouwer and Louvain: Institut supérieur de philosophie, 1945.

M. Vogel and V. Gardthausen, *Die griechischen Schreiber des Mittelalters und der Renaissance*, Hildesheim, G. Olms, 1966 (Zentralblatt für Bibliothekswesen, Beiheft 33; originally published Leipzig 1909).

A. Wartelle, *Inventaire des manuscrits grecs d'Aristote et de ses commentateurs*. Paris: Les belles lettres, 1963.

H. Wegehaupt, 'Zur Überlieferung der Problemata des sogenannten Alexanders von Aphrodisias', *Berliner Philologische Wochenschrift* 35 (1915) 95-96.

—— 'Zur Überlieferung der pseudo-Aristotelischen προβλήματα ἀνέκδοτα', *Philologus* 75 (1918) 469-473.

L.G. Westerink., ed., *Michael Psellus: De omnifaria doctrina*, Nijmegen: Centrale Drukkerij, 1948.

J. Wiesner, review of D. Harlfinger, *Specimina griechischer Kopisten der Renaissance*, 1, *Gnomon* 85 (1978) 481-484.

—— and U. Victor, 'Griechische Schreiber der Renaissance', *Rivista di studi bizantini e neoellenici* n.s. 8-9 (1971-2) 51-66

N.G. Wilson, 'A Byzantine Miscellany: MS Barocci 131 revisited', *Jahrb. Oest. Byzant.* 27 (1978) 157-179.

S. Zervos, 'ΑΕΤΙΟΥ ΑΜΙΔΗΝΟΥ ΛΟΓΟΣ ΕΝΑΤΟΣ', *ΑΘΗΝΑ* 23 (1911) 265-392.

6. SIGLA AND STEMMA

A Paris, BN, ancien gr. 2047A [14th] (A Buss./Us.). (34r-35v, 1.1-12, incomplete at beginning and end; 36r-51r, 2.66-192, incomplete at beginning; latter part of 2.113-start of 2.117 on leaf misplaced between i.52 and i.58, now 27rv; 76v, 81rv, 63r-64v, 3.1-45)

B Paris, BN, ancien gr. 1884 [1503] (B Buss./Us.) (9r-13v, 1.4-2.89, sel.; 17r, 3.46-7)

C Paris, BN, ancien gr. 2047 [15th] (C Buss./Us.) (7v-9r, 1.9-2.66, sel.)

D Paris, BN, ancien gr. 2381 [15th or 16th?] (D Buss./Us.) (93r-96v, sel., see above, Introduction §2.2).

E Paris, BN, ancien gr. 1883 [14th] (E Buss./Us.) (13v-14r, 1.5-21, sel.)

F Florence, Laurentianus gr. 71.33 [14th] (178r-186v, 2.89-192 and 1 prologue)

G Paris, BN, ancien gr. 1268 [12th] (295v-304v, sel.; see above; Introduction §2.2)

H Paris, BN, ancien gr. 1630 [14th] (181r-183r, sel.; see above, Introduction §2.2)

J Paris, BN, Suppl. gr. 690 [1075-1085] (25r-28r, sel.; see above, Introduction §2.2)

K Modena, MS gr. Alpha V.7.17 (II.E.12/gr.145) [15th] (60r-70r, 2.60-192)

L London, BL, Harleian 6295 [second half of 15th] (10v-16r, 1.4-2.89, sel.; 16r-20r, sel.; see above, Introduction §2.2)

M Madrid, BN, MS gr, 4616 (LXXXIV:N.84/34) [second half of 15th] (M Buss./Us.) (23r-64v, 1-2)

N Oxford, New College MS 233 [1450-1472] (N Us.) (56v-64r, 1 prol.-2.22).

O Oxford, Corpus Christi MS gr. 113 [c.1490] (O Us.) (85v-98r, 1-2)

P Venice, Marc. gr. 521 (316) [mid 13th] (P Us.) (89v-97r, 1-2)

R Venice, Marc. gr. 257 (622) [beginning of 14th] (R Us.) (35v-49v, 1-2)

S Venice, Marc. gr. 259 (982) [mid 14th] (S Us.) (29r-45v, 1-2)

T Modena, MS gr. Alpha V.6.12 (III.G.6/gr.210) [16th] (T Us.) (343v-371r, 1-2)

V Vienna, NB, med. gr. 49 [16th] (V Us.) (1r-2r, 1 prol.)

W Vienna, NB, med. gr. 37 [first half of 14th] (67rv-70r and 95rv, sel.; see above, Introduction §2.2)

X Vatican, Palat. gr. 199 [13th] (27r, 1.5-21, sel.)

Y Vatican, Palat. gr. 237 [14th] (153r-171r, sel., see above, Introduction §2.2).
b Bologna, MS gr, XXI.3635 (15) [1325-1334] (51v-79r, 1-2)
c Cambridge, UL, Ms gr. Mm 1.17 (2278) [15th] (87rv, 3.5-21, sel.)
e Escorial, MS gr. Upsilon I.9 (248) [mid 16th] (55r-76r, 1-2)
i Paris, BN, Coislin. gr. 332 [end of 15th] (51v-81v, 1-2)
j Paris, BN, ancien gr. 2505 [15th] (1r-3v, sel. including 2.89; see above, Introduction §2.2)
k Paris, BN, ancien gr. 2314 [14th-15th] (96r-98r, 2.1-2.41, incomplete at end)
l Leiden, Voss. Misc. 16 [mid-15th] (18r-25v, 2.23-2.82)
m Modena, MS gr. Alpha P.5.20 (II.D.10/gr.109) [14th-15th] (24v-31r, 1 prologue - 2.68; 78r-79v, 2.68 and 2.175-192)
o Oxford, Bodleian, Barocc. 131 [1250-1275] (62v-64v, sel.: see above, Introduction §2.2)
w Vatican, Barb. gr. 147 [early 16th] (33r-43v, 1)

p Psellus in o, 441v-446v (ed. Duffy 1992): see above, Introduction §1.2 and n.62.
r *Problemata Bambergensia* (ed. Stoll 1992, 68-75; cf. Rose 1863, 666-676)

𝔓 POxy 2744

g Gaza (G Buss./Us.) [1453] (Gaza 1562, 199v-204r, 1.1-2.38)
v Valla, transl. of 1501 (V Buss./Us.) (Valla 1501, 274r-288v; see above, Introduction §2.1.14)

Buss. Bussemaker 1857 (291-322, 1-3)
Us. Usener 1859 (1-2)

For the Bekker *Problems* we have used the Budé edition of Louis (1991-4) and have adopted his sigla. "Ap." = the lost MS used by Apostolius; cf. Louis vol.1, 1991, xliii-xlv.

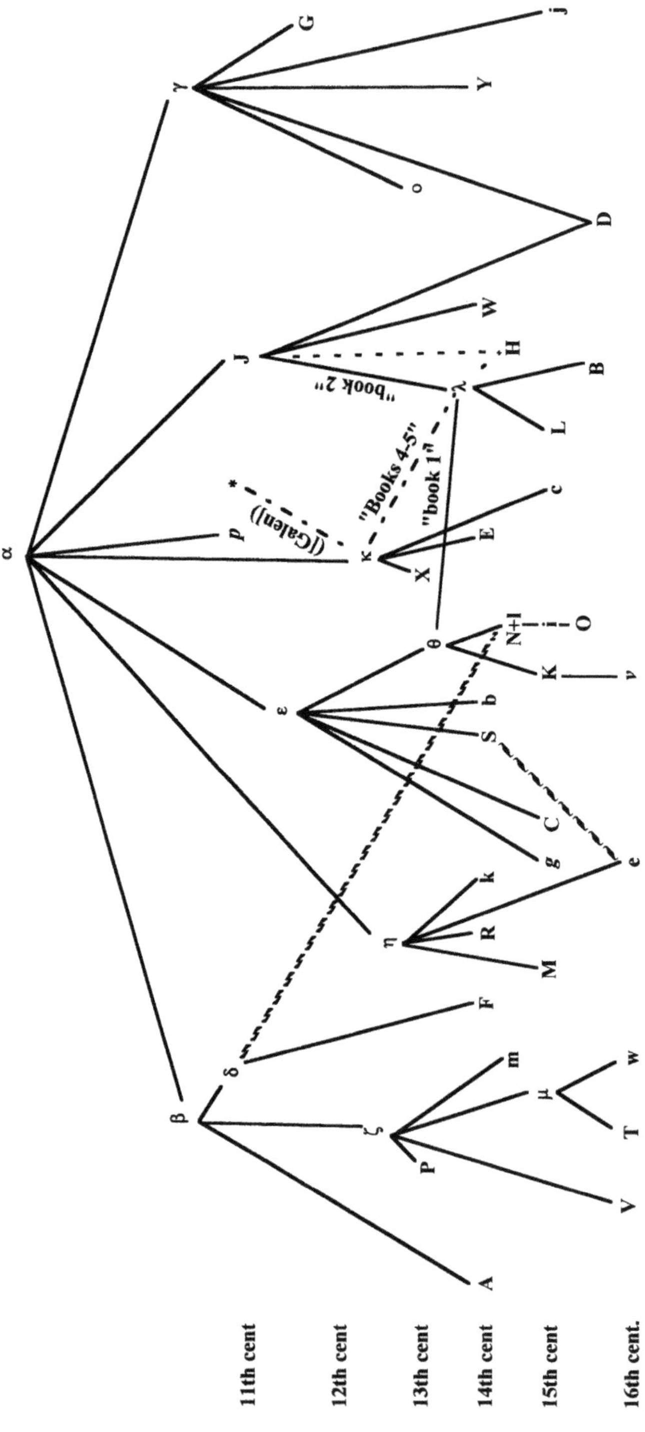
α
β
γ
δ
ε
ζ
η
θ
κ
λ
μ
G
Y
j
o
D
W
J
H
B
λ
L
"book 2"
"Books 4-5"
"book 1"
([Galen])
*
p
c
E
X
N+l
O
K
v
b
S
C
g
e
k
R
M
F
m
w
T
P
V
A
11th cent
12th cent
13th cent
14th cent
15th cent
16th cent.

Pseudo-Aristoteles (Pseudo-Alexander),

Supplementa Problematorum

φυσικῶν ἀπορημάτων καὶ ἰατρικῶν προβλημάτων ἐκλογαί

Οἱ μὲν πλεῖστοι τῶν παλαιοτέρων ἰατρῶν σποράδην ἐξηῦρόν τινα τῆς ἰατρικῆς, ἐκ κληδόνων ἢ τριόδων συνάγοντες καὶ ἀπὸ τύχης ἢ μαντείας ἢ κλήσεως δαιμόνων ἢ ἄλλῳ τινὶ τρόπῳ. Ἱπποκράτης δὲ ὁ τῶν Ἀσκληπιαδῶν ταύτην ὡς ἂν εἴποι τις πλαζομένην συνάξας καὶ τελείως ὑφάνας πλήρη καὶ ἀρτίαν εἰργάσατο κεφαλὴν ἐπιθείς· καὶ οὐκ ἄν τις ἁμάρτοι λέγων, ὡς ὁ προνοητικὸς Θεὸς ἐλεήσας τὸ ἀνθρώπινον γένος

Tit. *ἀρχὴ τοῦ τρίτου λόγου ἀλεξάνδρου ἀφροδισέως φυσικῶν ἀπορημάτων καὶ ἰατρικῶν προβλημάτων ἐκλογαί* (*ἐκλογαί* in marg. c) EXc: *ἀλεξάνδρου ἀφροδισιέως φυσικῶν ἀπορημάτων καὶ ἰατρικῶν προβλημάτων τόμος* Γ N: *ἀριστοτέλους περὶ ζώων ἰδιότητος ἐπιγράφεται· προοίμιον* (et ad initium prooemii in margine *ἔοικεν ἀριστοτέλει*), ante primum problema *ἀριστοτελικῶν* (in *ἀριστοτέλους* corr. sup. lin.) *ἰατρικῶν προβλημάτων καὶ φυσικῶν ἐκλογαί τὸ* Γ M: *ἀριστοτέλους περὶ ζῴων ἰδιότητος* Γ PRTm (in fine huius libri R): *τόμος* Γ[ος]. *ἀριστοτέλους περὶ ζώων ἰδιότητος* w: *ἀριστοτέλους ἰατρικῶν προβλημάτων καὶ φυσικῶν ἀπορημάτων ἐκλογαὶ τό(μος)* Γ in fine huius libri Tmw: *ἀλεξάνδρου βιβλίον τρίτον* F: *τοῦ αὐτοῦ* (sc. *ἀλεξάνδρου ἀφροδισιέως*) Sb: *τμῆμα ἐκ τοῦ δευτέρου βιβλίου τῶν ἀλεξάνδρου προβλημάτων· ἀριστοτέλους περὶ ζώων ἰδιότητος* e: *ἀλεξάνδρου προβλημάτων βιβλίον γ*[ον] e[mg]: *ἀριστοτέλους φυσικὰ ἐρωτήματα* k: hunc librum et praecedentem in unum coniungunt O et i

Prooemium (FMNOPRSTVbeim*p*w Us. Flashar (1962a, 404-405); 1-3, 6-8, 14-15 interpretatus est Petrus Aponensis, *Conciliator* ap. Olivieri 1988, 194-196; cf. Introd. 2.3)

3 *κλήσεων* Flashar || *δαιμόνων*] *δαιμονίων* m || *ἄλλῳ τινὶ τρόπῳ* Us.: *aliis quibusdam modis* Petrus: *ἄλλων τινῶν τρόπων* codd. 3-4 *ὁ τῶν ἀσκληπιαδῶν ἡγέμων p* (ex coniectura?) 4 *εἴπη* N[pc]OSi || *συναγαγὼν* NOSbe[mg]i 5 *καὶ*[2] om. Oi

NATURAL PUZZLES AND MEDICAL PROBLEMS: SELECTIONS[149]

Most of the earlier doctors discovered certain parts of medicine in an unsystematic way, collecting them from chance remarks and encounters[150] and from chance or prophecy or invocation of divine spirits or in some other way. Hippocrates, from the Asclepiadae,[151] gathered [medicine] together when it was as one might say wandering in perplexity, and having woven it into a finished [fabric] made it complete and fully developed by placing the head upon it.[152] One would not be wrong to say that god in his providence, having taken pity on the human race which was being destroyed by a

---

149 The title appears as follows in the MSS: *Beginning of book 3 of the natural puzzles and medical problems of Alexander of Aphrodisias, selections* (EXc); *Alexander of Aphrodisias, natural puzzles and medical problems, book 3* (N); *By Aristotle, entitled On the peculiar characteristic of animals* M, which also has *Aristotelian medical problems and natural puzzles, selections, book 3* before problem 1.1; *Aristotle on the peculiar characteristic of animals, book 3* (PRTmw; R has the title at the end of the book, and Tmw add there *Aristotle's medical problems and natural puzzles, selections, book 3*); *Alexander, book 3* (F); *By the same author* (i.e. Alexander) (Sb); *Section from the second book of Alexander's Problems: Aristotle on the peculiar characteristic of animals* (e); *Alexander, Problems book 3* (e, in the margin); *Aristotle's natural questions* (k). MSS O and i treat this book not as a separate book but as a continuation of book ii Ideler. See the Introduction, §1.4. – The preface was edited from MS V, where it appears as an unattributed preface to Hippocrates, *Aphorisms*, by Dietz 1834, 244.16-245.29. Flashar 1962a argues that it belongs to our collection; though there is some exaggeration in describing the latter as "including the whole of medicine in a bare account". Sluiter 1994 however points to parallels with Galen's commentary on the first *Aphorism* (17.2 345-356 Kühn). See also Usener 1859, xi-xii; above, Introduction, at nn.8, 79.

150 *Klêdon* also has the specific sense of an oracle derived from a chance utterance, and *triodos* (literally: road-junction) may imply more than just chance encounters, as it has a particular association with fortune-tellers and the like.

151 Literally "the descendants of Asclepius", but in fact the title of a medical guild. Psellus has "Hippocrates, the leader of the Asclepiadae", possibly a conjectural addition.

152 A more natural metaphor in Greek than in English (we might speak of the keystone of an arch); cf. Plato, *Gorgias* 505b, *Timaeus* 39d, *Philebus* 66d. Nevertheless, one may feel that our author has mixed his metaphors somewhat.

ἀλλεπαλλήλοις νόσοις ἀπολλύμενον αὐτὴν τὴν φύσιν σαρκώσας Ἱπποκράτην κατήγαγε πρὸς ἀρτίαν ταύτης παράδοσιν· ἴσως γὰρ καὶ τοῦτο αἰνίττεται κατὰ τὸ προοίμιον τῶν Ἀφορισμῶν λέγων ὡς "ἐπειδὴ κατὰ τὴν πεῖραν ἡ ἰατρικὴ σχεδὸν ἀκατάληπτός ἐστιν (οὔτε γὰρ ὅτε βουλόμεθα, τοῖς πάθεσι τῶν ἀνθρώπων ἐντυγχάνουσιν οἱ ἰατροί, τύχῃ γὰρ καὶ τῷ σπανίῳ τῆς γενέσεως δουλεύει ταῦτα), ἔτι γε μὴν καὶ ἐπικίνδυνος τῷ ἐν σώματι ῥευστῷ μὲν διὰ τὴν ὕλην καὶ ἀβεβαίῳ κεκτημένῳ δὲ καὶ θείαν δύναμιν ψυχικὴν γυμνάζεσθαι τὴν ἰατρικὴν τέχνην καὶ οὐκ ἀψύχῳ καὶ ἀτίμῳ καθάπερ τὰς ἄλλας τέχνας, πρὸς δὲ τούτοις καὶ τὰ πάθη ὑπὸ πολλῶν αἰτίων γεννᾶσθαί τε καὶ αὐξάνεσθαι καὶ διὰ τοῦτο δυσχεραίνειν τὴν πεῖραν ἐν τῷ διακρίνειν τὸ ποιητικὸν αἴτιον — φέρε τῷ λόγῳ ψιλῷ χωρὶς ὑποκειμένης ὕλης ἀσωμάτως πᾶσαν περιλαβὼν ἐν ὀλίγῳ χρόνῳ διδάξω καὶ διὰ τοῦτο καὶ ἐπιστημονικὸν λόγον ἔχειν ποιήσω, καὶ τὰς αἰτίας πάσας ὑποθήσομαι σὺν ταῖς διαγνώσεσι πρὸς τὸ σὲ λοιπὸν τῇ πείρᾳ γυμνάζειν τὸν λόγον, ὅτε δ᾽ ἄν σοι κατὰ τύχην περιπέσῃ τι πάθος ἐφαρμόζειν τὸν λόγον καὶ γυμνάζειν καὶ ἀληθῆ τοῦτον εὑρίσκειν."

τὸ δὲ μέγιστον τοῦ ἀνδρός, ὅτι οἱ παρ᾽ αὐτοῦ λεγόμενοι Ἀφορισμοὶ οὐχ ἁρμόζουσι μόνῃ ἰατρικῇ ἀλλὰ καὶ κοινῶς παντὶ τῷ βίῳ· νόμοι γάρ εἰσι καθολικοὶ θεσπίζοντες καὶ κανονίζοντες τὰ γινόμενα· ὅταν λέγῃ "ἐν τοῖσι γυμναστικοῖσιν αἱ ἐπ᾽ ἄκρον εὐεξίαι σφαλεραί"· καὶ "πᾶν τὸ πολὺ τῇ φύσει πολέμιον"· καὶ "ὕπνος ἀγρυπνίη, μᾶλλον γιγνόμενα τοῦ μετρίου κακόν", καθολικὸν δίδωσιν ἡμῖν κανόνα καὶ νόμον ἐπὶ παντὸς πράγματος, ὡς πᾶν

7 *αὐτὴν* om. V 7-39 *Ἱπποκράτην- παραπλησίων* om. F 7-8 *Hipocras ... quem ... tanquam nunquam peccantem* Petrus *ἀρτίαν*] *αἰτίαν* T || *ταύτην* P 9 *τὸ* om. T 10 *ἡ* MV Us.: om. cett. 11 *ἰατροὶ* ex *ἰατρικοὶ* corr. O 12 *τῇ γενέσει* P || *ἐπικίνδυνον* NOSbe^{mg}i Flashar 13 *τῷ*] *τὸ* NOSbi 15 *καὶ*^{2} om. w 16 *γενέσθαι τὲ* NOSbe^{mg}i || *αὐξάνεσθαι* ex *αὔξεσθαι* corr. T || *δυσχεραίνει* NOi 20 *προς τῶ σε* TVw || *τῇ πείρᾳ λοιπὸν* NOSbe^{pc}i (*λοιπὸν* add. sup. lin. i) 21 *ὅτε*] *ὅταν* M: *τότε* Us. || *σοι*] *σε* m || *περιπέσοι* NSbe^{mg}i: *περιπέσει* O || *πάθοι* N^{ac}Sb: *πάθος* vel *πάθοις* N^{pc}: *πάθοις* Oe^{pc}i 22 *ἐφαρμόζειν* Us.: *ἐφαρμόζων* codd. || *γυμνάζειν* Us.: *γυμνάζων* codd. || *εὑρίσκειν* Us.: *εὑρίσκων* codd. 24 *μόνον* MPV Us. 25 *ὡς ὅταν* V Us. 25-6 *ἐν τοῖσι γυμναστικοῖσιν* om. NOi 26 *καὶ πᾶν* Us. ex Hippocrate, *p*: *καὶ πάλιν* codd. 27 *ὕπνος σοι* M: *ὕπνος καὶ* Pm 28 *διδοὺς* m || *νόμον* ex *νόμων*

succession of diseases, having made nature herself incarnate sent down Hippocrates to impart her[153] adequately. Perhaps he hints even at this in the prologue of the *Aphorisms*,[154] when he says "Medicine can hardly be grasped by experience; for we doctors do not encounter people's afflictions at the time we wish to, for these things are subject to chance and to the infrequency of their occurrence. Moreover it involves risks, because the art of medicine is practised on a body which is on the one subject to flux, because of matter, and unstable, but which on the other hand possesses a divine power in the soul, and [is] not, like the other crafts, [practised] on [a body] which is lifeless and of small value. In addition to this afflictions are produced and intensified by many causes, and for this reason practice finds it difficult to decide the cause that produced them. Look, [for this reason] by including the whole [of medicine] in a bare account, without the underlying matter and not involving body, I will teach it in a short time, and on account of this I will make [you] possess an account based on understanding, and I will set before you all the causes together with their diagnoses, so that you can for the rest exercise the account in practice, and whenever you encounter any affliction by chance you can fit the account to it and exercise it and find that it is true."

The greatness of the man [is apparent] from the fact that the *Aphorisms* he uttered are appropriate not only to medicine but in general to the whole of life: for they are universal laws foretelling and regulating the things which happen; when he says "in gymnastics extreme fitness is precarious" [1.3] and "all abundance is an enemy to nature" [2.51] and "[both] sleep [and] sleeplessness are bad[155] in excess" [7.72], he gives us a universal rule and

---

153 I.e. Nature: or perhaps (so Flashar) "it" (medicine). Sluiter 1994 suggests taking "nature herself" as "the very nature [of medicine]". See also above, n.136.

154 Not in our text of the *Aphorisms*. Flashar regards it as a paraphrase of *Aphorism* 1.1, "Life is short, the art long, the right moment hard to catch, experiment precarious, judgement difficult." For the structure of the sentence see Sluiter 1994; we have simplified it in the translation. *Asômatôs*, "not involving body", belongs to philosophical, rather than medical, vocabulary.

155 Galen, *In Hippoc. Aphorism.* 7.73 18.1 189.10-13 Kühn, observes that "some copies" have this aphorism earlier, with the reading "bad", as here, which he prefers to the reading "disease". Most of our MSS of the *Aphorisms* in fact have "bad" in 7.72 itself; Galen must have found "disease" in his text there, as in our

μέτρον ἄριστον· καὶ πάλιν "τὰ μὴ καθαρὰ σώματα ὁκόσῳ ἂν θρέψῃς μᾶλλον βλάψεις", ὃ δύναται καὶ περὶ ψυχὴν μεταφέρεσθαι· οὕτω γὰρ ὁ θεῖος Πλάτων ἔφη "τὰς μὴ καθαρὰς ψυχὰς ὅσῳ ἂν παιδεύσῃς, μᾶλλον βλάψεις, τῆς γὰρ ἀσωμάτου ψυχῆς ἀσώματος τροφὴ παίδευσις." καὶ ἐπὶ ἄλλων δὲ ἀφορισμῶν τὸ εἰρημένον κρίνων εὑρήσεις, ἔχουσι δὲ καὶ τὸ διωρισμένον καὶ τὸ συνεχές, ὃ καὶ αὐτὸ τελείας ἀρετῆς ἐστι σύμβολον· ὡς γὰρ ἐπὶ τῶν γραμμικῶν, καὶ ἐπ' αὐτῶν εἰ οὕτως ἔτυχεν ὁ πέμπτος τὸν τέταρτον σαφηνίζει καὶ ὁ τέταρτος τὸν τρίτον καὶ ὁ τρίτος τὸν δεύτερον καὶ ὁ ἐφεξῆς ὡσαύτως· καὶ προβαίνων σαφῶς τοῦτο ποιεῖται τὴν τάξιν φυλάττων τοῦ λόγου καὶ τὸ ἀκόλουθον τῆς αἰτίας ἐπὶ ἡλικιῶν ἢ ὡρῶν καὶ παθῶν καὶ τῶν παραπλησίων.

I.I. *Διὰ τί τῶν ἐν γάλακτι τρεφομένων ἀλεκτρυόνων οἱ ὄρχεις εὐμεγέθεις καὶ εὔπεπτοι γίνονται; ὅτι τὸ αἷμα ἔχει ἐν αὐτῷ ὀρρόν, ὅνπερ ἕλκει ὁ νεφρὸς κατ' ἔλλειψιν τοῦ ὑγροῦ οὖρον ὄντα· ἔχει δὲ καὶ θορόν, ὅνπερ οἱ ὄρχεις ἕλκοντες τρέφονται, ὃς παχύτερός ἐστιν οὔρου, λεπτότερος δ' αἵματος, ὃ*

---

corr. O 29 *ὁκόσῳ*] *ὅσω* MNORSbeim 30 *περὶ*] *ἐπὶ* V Us. 31 *ἔφη* om. NOi || 32 *βλάψῃς* NOim*p* || *ἀσωμάτου*] *σωμάτου* O 33 *τὸ*] *τῶ* w || *κρίνω* N^{ac}O 34 *διορισμένον* e || *αὐτοτελείας* w || *τελείας* ex *λείας* corr. O 35 *γραμματικῶν* Flashar 37 *ὁ* om. V 39 *καὶ*^{1}] *ἢ* PTm || *καὶ*^{2}] *ἢ* V

1.1 (AMNOPRSTbeim*p*w Buss. Us.; *g* 2.78)

1-12 *Διὰ τί ... οὖν* amissa in A 1 *ἐν τῷ γάλακτι* NOSbe^{mg}i Buss. 2 *εὔπετοι* O || *αὐτῷ* w Buss. Us.: *ἑαυτῶ* m: *αὐτῶ* cett. || *ὀρὸν* Tw || *ὅνπερ* Buss. Us.: *ὅνπερ* R: *ὅπερ* cett. 3 *κατέλειψιν* T: *καὶ ἔλλειψιν* w || *κατ' – ὄντα* om. *g* || *ὑγροῦ οὖρον*] *οὔρου* tantum Mm Buss. || *ὄντα* Us.: *ὄν* NOPRSTbeim (add. supra lin. m): *ὂν* w: om.

law for every thing, that measure is best [in] everything; and again, [when he says] "the more you nourish bodies that are not pure, the more you will harm them" [2.10], which can also be transferred to the soul. For this is what the divine Plato[156] said, "the more you educate impure souls the more you will harm them, for the incorporeal soul is nourished by [something] incorporeal, [namely] education."[157] And in the case of other aphorisms too, if you judge what [I] have said, you will find [it is so]. They are both discrete and continuous,[158] and this too is a token of complete excellence; for as in the case of geometry,[159] so with them too, if it so chance, the fifth clarifies the fourth and the fourth the third and the third the second, and so on. And as he proceeds he makes this clear by keeping to the order of the account and the sequence of cause[s] in respect of periods of life or seasons and afflictions and the like.

1.1. Why are the testicles of cockerels fed on milk large and easy to digest?[160] Because blood has in it a watery serum, which the kidneys attract through lack of moisture, and this is urine. It also [has in it] semen, which the testicles attract and are nourished by; this is thicker than the serum but

Aphorisms MSS D'YO'. See Littré 4 603.

156 Rose 1863, 219 argued that the similar reference to "the divine Plato" at i.87 Ideler showed Neoplatonist influence. But Alexander himself uses the expression in the genuine part of his commentary on the *Metaphysics*, at 18.3.

157 Not in fact in Plato.

158 I.e., they can be read both individually and as a connected whole.

159 So Dietz. But later theorems depend on earlier ones, not vice versa.

160 For the alleged fact cf. Galen, *Alim. fac.* 3.20.6 (*CMG* 5.4.2 359.3-4 Helmreich), specifying that it is on the whey of the milk that the cockerels are fed; Oribasius, *Coll. med.* 3.17.3 (*CMG* 6.1.1 79.30-31 Raeder), *Synopsis ad Eusthathium* 4.16.4 (*CMG* 6.3.3 133.12-13 Raeder), *Ad Eunapium* 1.34.4 (*CMG* 6.3 340.12 Raeder); the testicles of cockerels fed in this way are also referred to by Galen, *Meth. med.* 7.6 (10 481.16-17 Kühn: coupled with rock fish, cf. 1.6 below), 12.6 (10 848.14K; cf. n.182 below); Aetius Amidenus, *Iatrica* 2.133 (*CMG* 8.1 201.14-15 Olivieri). For the testicles of cockerels as a foodstuff, without reference to their being fed on milk cf. also the following, all of which also mention rock fish: Galen, *Meth. med.* 8.5 (10 575.17 Kühn), *Comp. med. secundum locos* 2.1 (12 .518.1 Kühn); Aëtius Amidenus, *Iatrica* 5.68 (*CMG* 8.2 40.25 Olivieri), 5.80.22 (*CMG* 8.2 58.12 Oliveri).

καὶ σχῆμά ἐστιν αὐτοῦ· ἀλλ' ἕως μὲν ἡ φύσις πολλῆς δεῖται τροφῆς, ὡς ἐπὶ παιδίων, καὶ τὸν θορὸν εἰς τὸ πᾶν ἀναλίσκει σῶμα· ὅτε δὲ οὐ χρείαν ἔχει πολλῆς, τοῦτον περὶ τὸ ἡβάσκειν περὶ τοὺς ὄρχεις συνάγει τῶν τε θηλειῶν καὶ τῶν ἀρρένων, ὡς ἔστιν ἰδεῖν τότε τοὺς ὄρχεις εὐρυνομένους καὶ τρίχας φυομένας περὶ τὴν ἥβην· ἔπειτα δὲ μετὰ τὴν σύλληψιν τοῦ θήλεος ἡ φύσις χρείαν ἔχουσα πολλοῦ αἵματος πρός τε τροφὴν τῷ βρέφει καὶ πρὸς γάλακτος γένεσιν, τοῦτον τὸν θορὸν ἀπὸ τῶν ὄρχεων λαμβάνουσα, πάλιν ἀναμίγνυσι τῷ αἵματι καὶ οὕτως ἀπεργάζεται τὸ γάλα· οἱ οὖν ὄρχεις τῶν ἀλεκτρυόνων πλέον ἕλκουσιν ἐκ τοῦ γάλακτος τοῦ βρωθέντος ὑπὸ τοῦ ἀλεκτρυόνος διὰ τὴν ὁμοιότητα καὶ ὅτι τὸ οἰκεῖον ἑαυτῶν, τὸν θορὸν λέγω, πάλιν εὑρίσκουσιν ἐν τῷ γάλακτι, καὶ οὕτω γίνονται εὐτραφεῖς οἱ ὄρχεις καὶ εὔπεπτοι καὶ τρόφιμοι.

1.2. Διὰ τί, εἰ τὸ δριμὺ ἀπὸ πολλῆς γεννᾶται θερμότητος, τὸ δριμείας ποιότητος μετέχον μέλι ἐπαινοῦσιν, εἴ γε καὶ μετρίως αὐτὸ βούλονται εἶναι θερμόν; φημὶ ὅτι οὐκ ἀφ' ἑαυτοῦ ἐστι δριμύ, ἀλλ' εὔκρατον, ἡ δὲ γλῶσσα τῇ οἰκείᾳ θερμότητι θερμαίνουσα αὐτὸ πλέον ἐπὶ τὸ δριμὺ μετάγει· πᾶν γὰρ γλυκὺ θερμαινόμενον πρῶτον μὲν δριμὺ γίνεται, ὕστερον δὲ ὑπερθερμαινόμενον ἄγαν πικρὸν ἀποτελεῖται. τὸ δὲ σάκχαρ παρὰ τοῖς Ἰνδοῖς οὕτω λεγόμενον μέλιτός ἐστι πῆξις, τοῦ ἡλίου τὴν ἐν τῷ ἀέρι δρόσον πηγνύοντος ἐπὶ τὸ γλυκύ, ὥσπερ καὶ ἐν τῷ ὄρει τῷ Λιβάνῳ καλουμένῳ γίγνεται τοιοῦτον· ἔστι δὲ ὅμοιον χόνδρῳ ἅλατος, λευκόν, εὔθρυπτον,

---

M Buss. 4 *ὃς*] *ὡς* Oi 6 *παιδίων* ex *παιδίον* corr. e || *ὅτε* Us.: *ὅταν* cett. 7 *ἔχῃ* MRTempw^{ac} Buss. || *πολλὴν* NOi: || *τὸ* om. Oi, add. supra lin. N || *ἡβάσκειν*] *διδάσκειν* m: om. *p* || *συνάγει*] *συνέχει* m 7-8 *τῶν τε θηλειῶν καὶ τῶν ἀρρένων* om. *g* 8 *ἀρρέων* O 9 *ἔπειτα* Us.: *ἐπειδὴ* codd. Buss.: *πάλιν p* || *δὲ*] *δὴ* m: om. *p* 10 *ἔχουσα*] *ἔχει* NOSbe^{mg}i: *ἔχοι* coni. Buss. || *χρείαν ἔχουσα*] *δεομένη p* 11 *θορὸν*] *ὀρρὸν* NOSbe^{mg}i || *πάλιν* om. *p* Buss. 12 *οὖν*] *γοῦν* m 13 *πλέων* A 13-14 *πλέον – ἀλεκτρυόνος* add. in. marg. m || *τοῦ βρωθέντος – καὶ ὅτι* om. *g* || *τῶν ἀλεκτρυόνων* M 14 *τὸ* om. m || *ἑαυτὸν* A: *ἑαυτοῦ* P 15 *πάλιν* Tw Buss. Us.: om. cett. || *εὑρίσκουσα* P

1.2 (AMNOPRSTbeim*p*w Buss. Us.; *g* 2.79; 4-6 Petrus ap. Olivieri 1988, 198-199) 1 *γεννάται ἀποπολλῆς* A || *τὸ δὲ δριμείας* AMPRTmew Buss. 2 *μετέχον ποιότητος* MNORSbei || *μετέχων* A || *αὐτὸ* A || *βούλονται*] *existimemus g* (?) 3 *οὐ καθ' ἑαυτοῦ* A: *οὐ καθ' ἑαυτὸ* m 4 *αὐτὸ* A || *πλέων* A: *πλεῖον* m 4-13 *πᾶν – ἐπιτιθέμενον* om., et reliqua paucissimis verbis excerpsit *p* 5 *πρῶτον*] *πρότων* A 5-6 *ὑπερθερμαινόμενον* ex *ὑπερμαινόμενον* corr. O 6 *σάκχαρ* NO^{pc}ig Buss.: *σάχαρ* AMPRem: *σάγχαρ* O^{ac}STw: *σάκχαρον* Us. || *τοῖς* om. NOSbi 7 *δρόσσων* A: *rores g* 8 *πηγνύοντες* N: *πηγνύονες* Oi || *ἐν τῶ ὄρει*] *ἐν τω ὄρι τὸ γλυκυ τῶ λιβάνω* A || *λιβάνω*] *βάνω* O || *καλουμένῳ* om. MNORSbegi 9 *ἔθριπτον* A 10

thinner than the blood, [seeing that] it is its form.[161] While nature needs much nourishment, as in the case of children, it uses up the semen too [as nourishment for] the whole body. But when it does not need much, at the time of puberty, it collects this in the region of the testicles of both females[162] and males, as one can see the testicles enlarging and hair growing in the genital region. Then, after conception, the nature of the female, needing much blood for nourishment of the embryo and for the production of milk, takes this seed from the ovaries, mixes it with the blood again, and so produces the milk. The testicles of cockerels, then, attract more of the milk which the cockerel feeds on, because of the similarity and because they find what is appropriate to them, I mean the seed, in the milk,[163] and in this way the testicles become well-nourished and easy to digest and nourishing.

1.2. Why, if pungency comes from great heat, do they praise honey which has a pungent quality, if they also want it to possess moderate heat? I say that of itself it is not pungent but well-blended, but the tongue, warming it by its own heat, changes it to be more pungent; for everything that is sweet, when it is heated, first becomes pungent, and then, when it is heated to excess, bitter. What is called *sakkharon* (cane-sugar) in India is congealed honey, when the sun sets the dew in the air[164] and makes it sweet, as also happens in this sort of way on the mountain called Libanus. It is like granules of salt, white, easily crushed, and sweet. It too has a cleansing

---

161 Literally "shape". The Greek could mean either that semen is the form of the blood or vice versa; but Aristotle regards semen as concocted from blood, i.e. as more perfect than it (*GA* 1.19 726b7ff., cf. 4.1 766b19ff., and below, n.249).

162 I.e. ovaries – the existence of which was unknown to Aristotle, their discovery being one of the achievements of Hellenistic anatomy: cf. Herophilus, T61 von Staden 1989 (184-6), along with von Staden 1989, 167-8, 230, 232-3. Above, Introduction, n.121.

163 Our author apparently thinks that the milk of one species contains seed appropriate to another. For the analogy between white of egg and milk cf. Aristotle, *GA* 3.2 752b24 (pointing out that the analogy is false, for it is the yolk that is nourishment for the embryo), and more generally Brennan 1995. But the cockerels are presumably being fed on cows' or goats' milk, not on white of egg.

164 It was believed honey was formed made by the bees from the same dew when it settled on the leaves of plants. Cf. Aristotle, *HA* 5.22 553b29; Theophrastus **435** FHS&G.

γλυκύ· ἔστι δὲ ῥυπτικῆς καὶ αὐτὸ δυνάμεως ὁμοίως τῷ μέλιτι τῷ μετέχειν ἰχωροειδοῦς τινος ῥύψεως, ὅθεν καὶ ἑψόμενον καὶ τοῦτο μεταβάλλον, οὐκέτι μὲν σμήχει ὡς τὸ ἄνεφθον, τρέφει δὲ μᾶλλον ἐσθιόμενον ἢ ἔξωθεν ἐπιτιθέμενον· ἔστι δὲ καὶ ἑτέρα διαφορὰ μέλιτος χαλβανόριον λεγόμενον, τῶν μελισσῶν βοσκομένων ἐν ἐκείνῳ τῷ ὄρει τὰ ἄνθη, ἐν ᾧ καὶ χαλβάνη γίνεται καὶ τοῦ μέλιτος μεταλαμβάνοντος τῆς ποιότητος ὅπερ δύναται πρὸς ἀμβλυωπίαν ἐγχριόμενον ποιεῖν, θερμῆς μᾶλλον δυνάμεως ὄν· ὁμοίως καὶ τὸ θύμων ὄζον ποιεῖ πρὸς ψυχρὰν γαστέρα· καὶ τὸ Ποντικὸν *ἀψινθίου πικρὸν ὄν, καὶ πρὸς τὴν διαφορὰν τῆς ποιότητος καὶ ἡ δύναμις αὐτοῦ· τὸ δ'* Αἰγύπτιόν *ἐστιν ὑγρὸν καὶ ὑδατῶδες τῷ πολλῆς μετέχειν ὑγρότητος τὰ ἄνθη ἐπὶ πολὺ ποτιζόμενα τῷ ποταμῷ· τὸ δὲ πάντων ἄριστον τὸ νησιωτικὸν καὶ τῶν παραθαλαττίων ὀρῶν καὶ τόπων, ὡς ἀπέριττον καὶ ξηρότερον, καὶ τὸ* Ὑμήττιον *μάλιστα ἐν τῷ* Ὑμηττῷ *ὄρει τῆς* Ἀττικῆς *ἔχοντι ἄνθη βοτανῶν καὶ δένδρων εὐώδη καὶ καθαρὰ καὶ εὔχροα καὶ λεπτομερῆ.*

1.3. *Διὰ τί τὸ γάλα ψυχρόν ἐστι καὶ τοιοῦτον ὂν εὐαλλοίωτον καὶ τρόφιμον ὑπάρχει; λέγω ὅτι καὶ καθὸ ἐξ αἵματος μετεβλήθη, κατὰ τοῦτο πάλιν ἐπὶ*

αὐτῷ A || *ὁμοίως*] *ὁμοίος* sic N^{pc}: om. APRTmw: add. in. marg. e || *τὸ μέλιτη* sic A 10-11 *τῷ μετέχειν – ῥύψεως* om. g 11 *ἰχωροειδοῦς* PT^{ac} Buss. Us.: *ἰχοροειδοῦς* T^{pc}: *εἰχωροίδοῦς* A: *ἰχυροειδοῦς* w: *ἰσχυροειδοῦς* cett. || *ἑψομένου* M || *μεταβαλλῶν* A 12 *ἔσωθεν* m 12-13 *ἢ ἔξωθεν ἐπιτιθέμενον* om. *g* 13 *διάφορα* m: *διαφθορᾶ* A || *χαλβανώριον* A 14 *τῶν μελισσῶν* ex *τὸν μελισσον* corr. T: *τῶν μελισῶν* A 14-15 *ἐν ἐκείνῳ – ποιότητος*] *quem galbana vitiat* tantum *g* 14 *ἐν ᾧ*] *ἐννοῖς* A || *καὶ* add. sup. lin. O || *χαλβάνοις* A 15 *λαμβάνοντος* Buss. || *τῆς*] *τοῖς* A || *τῆς ποιότητος* add. in marg. O 15-16 *ὅπερ δύναται – δυνάμεως ὄν*] *prodest hoc illitum ad oculorum claritatem g* 16 *ἐχριόμενον* MO^{ac} || *ὂν*] *ὃν* A: *ὢν* T 17 *θύμων*] *θύμος* m: *θυμίον* e || *ὄζον*] *ὄζωον* A: *ὄζων* i^{ac}: *ὄξος* Mem || *γαστέραν* A || *ἀψινθίου* Us.: *ἀψίθιον* NOi: *ἀψίνθιον* cett.: del. Buss. 17-18 *καὶ τὸ Ποντικὸν – αὐτοῦ* om. *g* 18 *ὃν* A || *δύναμεις* A || *αὐτοῦ* A 19 *τῷ πολλῆς μετέχειν* in *τὸ πολλῆς μετέχον* mutavit m 19-20 *τε ἄνθυ* A 21 *ὀρῶν* $MPR^{pc}Te^{ac}m$ Buss. Us.: *ὁρῶν* Aw: *ὅρων* $NOR^{ac}Sbe^{pc}i$: *ὀρῶν καὶ* om. *g* || *ὡς*] *ὁς* A || *ἀπέριττον* Buss. Us.: *ἀπερίττον* A: *ἀπερίττων* cett. 22 *ξηρότερον* P, Buss. Us.: *ξηροτέρον* sic A: *ξηροτέρων* cett. || *τὸ* om. NOTi || *μάλιστα* post *ἀττικῆς* transtulit Buss. || *ἐν … ἀττικῆς* om. A 23 *ἔχοντα* A || *βοτανῶν καὶ δένδρων* om. A || *καὶ καθαρὰ καὶ εὔχροα*] *καὶ καθαρὰ καὶ εὔχρεια* m: om. A 23-24 *ἔχοντι – λεπτομερῆ*] *propter florum singularem praestantiam g*

1.3 (AMNOPRSTbeim*p*w Buss. Us.; *g* 2.80)

1-2 *ἐστι – ὑπάρχει*] *ὂν θερμαίνει p* 1 *καὶ τοιοῦτον ὂν* om. A || *καὶ*[2] om. A 2 *λέγω*

quality, like honey, because it shares in a certain cleansing like serum in nature, for which reason when this is boiled it too changes and no longer cleans as it does when unboiled, but rather nourishes, when it is eaten or applied externally. There is also another variety of honey called "all-heal mountain [honey]", when the bees feed on the flowers in the mountain where all-heal (*Ferula galbanifera* Boiss. & Buhse) grows and the honey acquires their quality, which can be effective when it is used as an ointment for dimness of sight, since it has a rather hot power. [Honey] which has a fragrance of thyme[165] acts similarly on a cold stomach, [as does] Pontic [honey] from wormwood, which is bitter; and the power of [honey] corresponds to the difference in quality. Egyptian [honey] is fluid and watery because the flowers [there] contain much moisture as they are irrigated to a considerable extent by the river. Best of all is [honey] from the islands and from mountains and regions that are adjacent to the sea, because it is[166] plain and drier, especially Hymettus [honey] from Mount Hymettus in Attica, which has flowers of herbs and trees which are fragrant and pure and beautiful in colour and rare in texture.

1.3. Why is milk cold and [though] it is like this, easily changed and nourishing? I say that inasmuch as it has changed from blood,[167] in this respect milk changes more easily back to this again, and is virtually blood

[165] Galen, *De antidotis* 1.4 (14 27.4-6 Kühn) and Oribasius, *Coll. med.* 2.62.3 (*CMG* 6.1.1 60.22-24 Raeder) say that the best honey has the pungency of thyme but *not* the fragrance or much of the taste.

[166] So MS A, followed by Bussemaker and Usener; the majority of the MSS have "they [the places] are". MS T has "they are plain and it is drier". MS A is careless about case-endings (which is the point at issue here in the Greek) but its reading seems to give a better sense.

[167] Milk is concocted blood: Aristotle, *GA* 4.8 777a7 ff.

τοῦτο ῥᾷον μετέρχεται [τὸ αἷμα] καὶ σχεδὸν αἷμά ἐστι λελευκωμένον. ὁμιλοῦν δὲ τῷ ἀέρι πλέον ψύχεται· βδαλλόμενον δ' ἔτι ποσῶς θερμόν ἐστι, μὴ ὁμιλοῦν τῷ ἀέρι καὶ σῴζει τὴν οἰκείαν δύναμιν, καθάπερ καὶ τὸ σπέρμα· λευκὸν δὲ καὶ ψυχρὸν ποσῶς ἐστιν, ὅτι ὑπ' ἀδένων λευκῶν καὶ ἀναίμων καὶ ἧττον θερμῶν τῶν ἐν τοῖς μαστοῖς μεταβάλλεται. οὐ δεῖ δὲ αὐτὸ πυτίᾳ πηγνύειν· κακοστόμαχος γὰρ καὶ δυσανάδοτος· ἀλλὰ μᾶλλον ὀπῷ συκῆς ἢ ἐχίνῳ ὄρνιθος πλύνας ἢ κνήκῳ. ἡ μέντοι πτισάνη εὐαλλοίωτος μέν, ἀλλ' ἧττον τρόφιμος· ὁ δὲ χόνδρος τοῦ ἄλικος τρόφιμος μέν, δυσαλλοίωτος δέ, ὅθεν καὶ ὀξυμέλιτι ἐσθίεται τριβείς.

1.4. Διὰ τί τὰ ἐν πνεύμονι ἕλκη ἀνίατα; πρῶτον μὲν ὅτι πᾶν τὸ δεόμενον ἰάσεως ἠρεμίας δεῖται· εἶπε γὰρ Ἱπποκράτης ὡς "ποδὸς ἄκος ἀκινησίη"· ἡ γὰρ κίνησις σύρροιαν ποιεῖ χυμῶν περιττῶν, οὗτοι δὲ φλεγμονάς· ὁ δὲ πνεύμων ἀεικίνητος διὰ τὴν ἀνάγκην τῆς ἀναπνοῆς· ἄλλως τε τὸ ἕλκος ἀεὶ ἐπίπαγον ποιοῦν δεῖται ἀνακαθάρσεως· οὐ δύναται δέ, εἰ μὴ μόνον [ἢ] βηχί, ἀνακαθαίρεσθαι, ἡ δὲ βὴξ σύντονος οὖσα κίνησις σπαράσσει τὸ ἕλκος καὶ ἀνευρύνει· ἄλλως τε ὑγρὸν καὶ θερμὸν τὸ μόριον καὶ εὔσηπτον, καὶ βρόγχιον

---

om. A*p* Buss.: *censendum g* || καὶ om. ANOPi Buss. 2-3 πάλιν ἐπὶ τοῦτο om. A Buss. 3 τὸ αἷμα delevimus: εἰς αἷμα A Buss.: τὸ γάλα Us. || λευκαινόμενον ὑπὸ φύσεως Buss.: λευκενόμενον ὑπὸ φύσεως A 4-6 ὁμιλοῦν – τὸ σπέρμα post μεταβάλλεται (7), 4-5 ἔτι – δύναμιν omissis, collocavit *g* 4 δὲ] δὴ Us. || πλέον om. T || βαλλόμενον MN^{ac}ROeim || δ' ἔτι m, Buss. Us.: δέ τι ANOi : δέ τοι MPRSTbew 5 καθάπερ] ὃς A: ὡς Buss. 6 δὲ om. A || ποσός A || ἀδένων ex ἀδέων corr. O 7 ἀναίμων] ἀναίμων A: ἀνέμων R^{ac}T || καὶ ἧττον θερμῶν om. *g* || θερμῶν] θερμὸν A: θερμὸν ὂν *p* || τῶν] ὥντων A: ὄντων Buss.: om. *p* || ἐν A: om. *p* || μασθοῖς R^{pc}Sbew: μασθεῖς R^{ac} 8 αὑτὸ A || πυτία P^{pc}Tw Buss. Us.: πητία AP^{ac}: πιτύα cett. || κακοστόμαχον *p* || κακοστόμαχος – δυσανάδοτος om. *g* 9 ὀπῶ AR^{ac}Tew || πλύνας] πλήνας Re: πληνοις M: πλύναντα coni. Buss.: πλῦναι Us.: πλύνειν *p*: om. *g* || κνήκῳ Buss. Us.: κνίκω codd. || ἡ μέντοι πτισάνη] *sic enim caseum concoqui* g 10-11 μέν – τριβείς om. *g* 10 δὲ] γε b^{ac} || ἄλικος T: ἀλικοῦ w: ἅλατος P: γάλακτος NOi 11 ὅθεν τὲ καὶ m

1.4 (ABLMNOPRSTbeim*p*w Buss. Us.; *g* 2.81)

1 ἐν τῶ Buss. || πνεύμωνι AL || ἕλκει A 2 εἶπε γὰρ ἱπποκράτης] εἶπε δὲ καὶ ἱπποκράτης m: om. A Buss. || ὡς ποδὸς ἄκος] ὡς ποδὸς σάκης B: ὅς καὶ το ποδὸς ἄκος A: ὡς καὶ τὸ· ποδὸς ἄκος Buss. 3 οὗτοι] οὕτοι T: οὗτο m 4 καὶ τὸ ἕλκος BL 4-5 τὸ ἕλκος – ἀνακαθάρσεως] *purulenta pulmonis* tantum *g* 5 ἐπίπαγον ποιοῦν Buss. Us.: ἐπὶ παγοποιοῦ APRTw: ἐπιπαγοποιοῦ Me: ἐπίπαγοποι sic m: ἐπὶ ποιοῦ Sbe^{mg}: ἐποιοῦ NOi: ἐμποιοῦ BL || δεῖται Us.: οὗ δεῖται Buss.: οὐ δέεται A: οὐ δέεται cett. || εἰ μὴ] εἰμὶ A || μόνον] μόνη Us.: om. A, Buss. || ἢ om. A Buss. Us.: exhib. codd. (ἡ w) || βηχὶ ex βραχὺ in. marg. corr. e 7 ἀνευρύνει] ἀναβρύει B: ἀναβρύνει L || ἄλλος A || post θερμὸν etiam *laxum* exhibet *g* || ἄσηπτον m ||

that has been turned white. In the presence of air it becomes colder; when it is still being milked [from the animal] it is to a certain extent hotter, not being in the presence of air, and it preserves its own proper power, as seed does too. It is white and cold to a certain extent, because it is changed [from being blood] by glands in the breasts which are white and bloodless and less hot. It should not be set with rennet, for that is bad for the stomach and hard to assimilate, but rather it should be [set] with fig-juice or [by] washing a bird's gizzard [in it][168] or safflower. Barley-gruel is easily assimilated, but less nourishing;[169] rice-wheat gruel is nourishing but hard to assimilate, and for this reason it is eaten ground up with vinegar and honey.

1.4.[170] Why are wounds in the lung incurable? First because everything that needs healing needs rest; for Hippocrates said that "rest cures feet".[171] Movement causes a flowing together of excessive humours, and these cause inflammation: but the lung is always in movement because of the necessity of respiration.[172] Moreover the wound, which constantly forms a scab, needs to be cleansed; but it cannot be, except by coughing, and coughing, being a violent movement, tears the wound and enlarges it.[173] Moreover the [lung] is moist and

168 A bird's gizzard is listed among things which set milk also by *Geoponica* 18.19.2, which specifies "a domestic fowl", i.e. a chicken, which "bird" on its own can mean in Greek, and doubtless does here.

169 Cf. [Ar.] *Probl.* 1.37. [Cross-references to this collection are largely taken from Bussemaker and Rose: following their example, = indicates a text which is identical or nearly so, "cf." a text addressing the same problem, often with a different solution.] The connection of thought is unclear; *khondros*, the word used here for rice gruel, could be made with milk (Aetius Amidenus, *Iatr.* 9.42.237-8 = p.385.26-7 in Zervos 1911), but *ptisane*, barley-gruel, is contrasted with milk at Galen, *Meth. Med.* 7.6 (10 481.10 Kühn) = Aetius Amidenus, *Iatr.* 9.34.78-81 (= p.357.9-12 in Zervos 1911) = Oribasius, *Synopsis ad Eustathium* 5.53.9 (*CMG* 6.3 181.5-7 Raeder). It seems therefore as if reference to the digestibility of milk may have prompted the addition of a strictly irrelevant comment on the digestibility of other foods for invalids.

170 An expanded version of the second part of [Alex.], *Probl.* ii.18 Ideler, where the reference is explicitly to consumptives.

171 Not in our text of Hippocrates, but also cited by [Stephanus] in *Scholia in Hippocratis De fracturis*, 55.11 ed. D. Irmer 1977.

172 So Galen, *Meth. med.* 5.8 (10 338.4ff. Kühn)

173 Galen, *Meth. med.* 5.11 (10 360.1ff. Kühn) = Aetius Amidenus, *Iatr.* 8.75 *(CMG* 8.2 540.26-541.1 Oliveri).

τοῦ πνεύμονος ἀναβρωθὲν οὐ δύναται γενέσθαι, εἰ μὴ ὑπὸ μόνης φύσεως πάλιν ἐκ σπέρματος.

1.5. Διὰ τί οἱ φρενιτικοὶ ἐν μὲν τῷ πάθει ἰσχυροί, ἀπαλλαγέντες δὲ ἀσθενεῖς; ὅτι ξηρὰ δυσκρασία κατέχει τὸν ἐγκέφαλον καὶ τὰ νεῦρα, αὕτη δὲ τόνον δίδωσι τοῖς νεύροις πρὸς ἐνέργειαν· ὕστερον μέντοι ἀπαλλαγέντες αἰσθάνονται λοιπὸν τοῦ κόπου, τῆς διακριτικῆς διανοίας καθαρᾶς ἐπεισηκούσης τοῖς πάθεσι, καὶ ἐνυγρανθέντα δὲ τὰ νεῦρα μαλακὰ γίνονται καὶ ἀσθενῆ.

1.6. Διὰ τί οἱ πετραῖοι ἰχθύες εὔπεπτοι; ἐπειδὴ πᾶν τὸ πρὸς ταῖς πέτραις ὕδωρ καὶ γαλήνης οὔσης κινεῖται· σκληρὸν γὰρ καὶ μαλακὸν ἐναντιότητα

βρόγχιον] *βροχὺ* ex *βραχὺ* corr. m: [*τὸ*] *βρόγχιον* Buss. 8 *πνεύμονος* APSTe^{mg}mw Buss. Us.: *πνεύματος* cett. || *ἀναβρωθὲν* om. *g* || *οὐ* A || *ὑπὸ* add. sup. lin. T: *ἀπὸ* m || *ὑπὸ μόνης*] *ὑποκειμένης* BL

1.5 (ABELMNOPRSTXbceim*p*w Buss. Us.; *g* 2.82)

1 *φρενητικοὶ* BELMSTXce^{pc}m*p*w: *φρενοτικοὶ* e^{ac} || *τω πάθη* A 2 *διότι* EXc || *αὐτῇ* E: *αὐτὴ* Am: *αὐτὴ* PRTce 3 *δίδοσι* E || *πρὸς ἐνέργειαν ὕστερον* om. *g* 3 post *ἀπαλλαγέντες* add. *καὶ τοῦ ἐγκεφάλου καθαροῦ ἐφεστηκῶτος· καὶ ὑγραθέντα τὰ νεῦρα τοῦ σώματος σὺν τοῦ ἐγκεφάλου* E 4 *κόπου*] *κόσμου* EXc 4-5 *ἐπεισηκούσης* Us.: *ἐπιστοιχούσης* Buss.: *ἐφεστηκούσης* Xc: *ἐφεστηκοῦσης* sic E: *ἐπιστηκοῦσης* sic A: *ἐπιστηκούσης* cett. || *ἐπεισηκούσης τοῖς πάθεσι*] *est* tantum *g* 5 *γίνεται* Buss. 5-6 *καὶ ἐνυγρανθέντα – ἀσθενῆ*] *καὶ οὕτως* (om. *g*) *μαλακὰ γίνονται τὰ νεῦρα καὶ ἀσθενῆ* E*g* 6 *καὶ* om. O

1.6 (AEMNOPRSTXbceim*p*w Buss. Us.; g 2.83)

1 *ἰχθῦαις* sic A: *ἰχύες* O 2 *οὔσης* bis X: om. NOi 3 *πρὸς ἄλληλα* om. Buss. || *οὐ*

warm and putrefies easily, and when the bronchial tube of the lung has been eroded it cannot be [re]generated, except by nature alone from seed again.

1.5. Why are those suffering from inflammation of the brain strong when suffering the affliction, but weak when it leaves them? Because a dry imbalance takes hold of the brain and the sinews, and this gives the sinews tautness for activity; subsequently when it leaves [the patients][174] they are now aware of their weariness, their discerning reason applying itself to their afflictions in a pure condition, and the sinews, having been moistened, become soft and weak.

1.6. Why are fish that live among rocks easily digested?[175] Since all the water in the region of rocks is in movement even when there is calm; for

---

[174] MS E adds here: "and the brain applies itself in a pure condition, and the sinews of the body are moistened along with the brain".

[175] The claim is commonplace: explicitly at Galen, *Alim. fac.* 3.27.1-2 (*CMG* 5.4.2 366.20-367.3 Helmreich), *Meth. med.* 8.2 (10 548.13 Kühn), *Pro puero epileptico* 5 (11 373.4 Kühn), *In Hippocr. aphorism*. 2.18 (17.2 489.12 Kühn); Anon., *De alimentis* 75 (2 279.28 Ideler 1841-2); Aëtius Amidenus, *Iatr.* 2.254 (*CMG* 8.1 246.10-11 Oliveri); Paulus Aegineta, *Epitomae* 1.90 (*CMG* 9.1 64.3-4 Heiberg). Rock-dwelling fish are also mentioned as part of a dietary recommendation at (e.g.) Aretaeus, *De cur. acut. morb.* 1.1.9 (*CMG* 2 93.21-2 Hude); Galen, *San. tuend.* 4.4.82 (*CMG* 5.4.2 115.17 Koch), 5.4.13 (*CMG* 5.4.2 143.35 Koch), *Meth. med.* 8.2 (10 537.9 Kühn), *Ad Glauconem* 1.10 (11 35.1 Kühn: see below, n.182), *in Hippocr. de vict. acut.* 1.26 (*CMG* 5.9.1 147.5 Helmreich), sometimes coupled with the testicles of cockerels (above, 1.1), as at Galen, *Meth. med.* 7.6 (10 .481.15 Kühn), 8.5 (10 575.17 Kühn), *Comp. med. secundum locos* 2.1 (12 518.1-2 Kühn); Aëtius Amidenus, *Iatrica* 5.68 (*CMG* 8.2 40.25 Oliveri), 5.80 (*CMG* 8.2 58.12 Oliveri). (However, [Alex.], *Probl.* i.52 Ideler says that rock-dwelling fish are *hard* to digest, and id. ii.17 is concerned with why those who can digest beef cannot digest rock fish.)

What is more variable is the explanation given. That given here, that the fish are in constant movement, appears more briefly at Stephanus, *In priorem Galeni libr. ad Glauconem* 119 (164.15-17 Dickson); but the reverse, that rock fish have lighter flesh because they do not travel far, is given by Hippocrates, *De diaet.* 2.48.1 (*CMG* 1.2.4 170.5-7 Joly). More often reference is made to the quality of the water: Galen, *De rebus boni malique suci* 9 (*CMG* 5.4.2 417.9-13 Helmreich); Oribasius, *Coll. med.* 3.15.4 (*CMG* 6.1.1 76.23-5 Raeder), *Synops. ad Eustath.* 4.14.4 (*CMG* 6.3 130.10-13 Raeder); Joannes Actuarius, *De spir. anim.* 2.7.8 (1 368.5-7 Ideler). Galen, *Alim. fac.* 3.29 (*CMG* 5.4.2 368.18-20 Helmreich) refers to the diet of the fish.

ἔχοντα πρὸς ἄλληλα οὐ δύναται ἡσυχάζειν ἀλλήλοις ὁμιλοῦντα, τὸ δὲ κινούμενον λεπτότερον γίνεται τοῦ ἠρεμοῦντος· καὶ οἱ ἰχθύες οὖν τῷ ὕδατι συγκινούμενοι ἀπέριττοι γίνονται· τῷ αὐτῷ τρόπῳ καὶ αἱ πτέρυγες συνεχῶς κινούμεναι εὔπεπτοί εἰσιν.

1.7. Διὰ τί ἐπὶ τῶν συνεχῶν πυρετῶν ἐν ταῖς παρακμαῖς οὐ λούομεν αὐτούς; ὅτι οἱ παρὰ φύσιν χυμοὶ ἄπεπτοί εἰσιν ἔτι καὶ παχεῖς καὶ ἀντεχόμενοι τῶν σωμάτων, οἱ δὲ κατὰ φύσιν λεπτοὶ καὶ εὐδιάρρυτοι· τὸ θερμὸν οὖν τοῦ λουτροῦ ἑλκύσει τῇ βίᾳ προτέρους ὡς λεπτοὺς τοὺς κατὰ φύσιν ἐν ἀρχῇ τοῦ νοσήματος, ἐὰν καὶ ἐν ἀπεψίᾳ λούσωμεν.

1.8. Διὰ τί ἐπὶ τῶν ἑκτικῶν ὕδωρ παρέχουσιν οἱ ἰατροὶ πρὸ τῆς τροφῆς βραχύ; ἵνα τὰ κατεξηραμμένα στερεὰ μόρια προϋγράνωσι, καὶ ἐπιτήδεια γένωνται πρὸς φύσιν στερεᾶς τροφῆς· οὕτω γὰρ καὶ οἱ κονιαταὶ προσραίνουσι τὸν τοῖχον ὕδατι καὶ οὕτω χρίουσι τὴν κονίαν.

1.9. Διὰ τί τὰ γλυκέα αὔξει σπλῆνα καὶ ἧπαρ; ὅτι τὸ μὲν ἧπαρ τῇ ὁμοιότητι τῆς κράσεως ἕλκει ῥᾷον τὰ γλυκέα· θερμὰ γὰρ καὶ ὑγρὰ ἀμφότερα· ὁ δὲ σπλὴν ἕλκων τὸ τρυγῶδες τοῦ αἵματος ἐκ τοῦ ἥπατος συνεφέλκεται καὶ αἷμα πλεῖον τῷ εἶναι πλῆθος, τοῦτο δὲ γίγνεται μᾶλλον,

---

A || δύνανται Em || ὁμιλοῦντα] ὁμοιοῦντα NOi 4 κινουμένοις Oi || γίνεται post ἠρεμοῦντος w || οἱ om. EXc || ἐν τῷ ὕδατι MRe 5 αὐτῶ A || τῶ αὐτῶ λόγῳ καὶ τρόπω EXc || πτέρυγες ex πτέρυχες corr. S 6 συνεχῶς κινούμεναι om., *quam reliquas corporis partes experiamur* add. *g* || κινούμενοι A Buss.

1.7 (ABLMNOPRSTbeim*p*w Buss. Us.; g 2.84)

1 ἐπὶ – πυρετῶν om. BL || οὐ A 2 αὐτοῦς A || ⟨ὅ⟩τι e || ἔτι] ἔτοι A 3 ἀντέχονται Buss. 4 ἑλκήσει B: ἐλκήσει T: ἕλκυσει sic A: ἕλκης L: *trahat g* || τῇ βίᾳ] τῇ μίᾳ BL: om. *g* || πρότερον e: πρὸς πόρους BL 5 νοσήματος ex σήματος corr. A || καὶ om. M || λούσομεν BL

1.8 (ABDGLMNOPRSTYbeim*p*w Buss. Us.; *g* 2.85; *v* 5.63)

1 ἐκτικῶν BGLNOPTYei || παρέχουσιν οἱ ἰατροὶ] παρέχου οἱ ἰατροὶ G: παρέχουσι Y: δίδοται D*p* || 2 ἵνα τὰ] ἵνα κατὰ O || ἐπιτήδεια ex ἐπιτήδειος corr. e: ἐπιτήδειαι Us.: ἐπιτίδια A 2-3 καὶ ἐπιτήδεια γένωνται] ὥστε γενέσθαι ἐπιτήδεια D 3 γένωνται] γένονται AB || πρὸς φύσιν om. BL: εἰς πρόσφυσιν vel πρὸς πρόσφυσιν coni. Buss.: πρὸς ἀποδοχὴν τῆς D: πρὸς δοχὴν *p*: *ad agglutinationem agnationemque g*: *ad cibi solidi excipiendam naturam v* || post φύσιν add. προσφοραὶ Us. || καὶ om. m || οἱ κονίαν ποιοῦνται D 4 προσραίνουσι] πρὸς ραίνουσι ALR: προσρένουσι B: προβρέχουσι DGY: προειραίνουσι corr. in προηγραίνουσι e: πρότερον ῥαίνουσι *p*: *aspergere soliti sunt g*: *madefaciunt v* || κονίαν] ἀκονίαν w: τιτάνην GY: τίτανον D: *calcem gv*

1.9 (ABCELMNOPRSTXbceim*p*w Buss. Us.; *g* 3.86)

1 τὰ om. BL || τὸν σπλῆνα R^{pc}e || τὸ ἧπαρ^{1} MR^{pc}e || ὅτι MRe Us.: διότι EXc: om. cett. 2 ἕλκη B 3 ἕλκον BL 4 σὺν ἐφέλκεται E: *quoque elicit g*:

hard and soft, which are opposites, cannot be at rest in the presence of each other, and what is in motion becomes rarer than what is at rest. The fish too, being moved with the water, are purified.[176] In the same way wings,[177] which are constantly in movement, are easily digested.

1.7. Why, in the case of continuous fevers, do we not wash [the patients] when the fever is abating? Because the unnatural humours are still undigested and are thick and cling to the body, while natural ones are thin and flow through [the pores] easily. So the warmth of the bath forcibly draws out first those which are natural, at the beginning of the disease, because they are rarer, if we wash them even when they have not digested.

1.8. Why, in the case of hectic fevers, do doctors give water shortly before food? So that they may moisten in advance the solid parts which have been dried, and so that these may become suitable for the nature of solid nourishment.[178] For in this way plasterers, too, moisten the wall with water and then apply the plaster.

1.9. Why do sweet things cause the liver and spleen to grow? Because the liver, through the similarity of its blending, attracts sweet things more easily; for both are warm and moist. The spleen, attracting the dregs of the blood from the liver, attracts a larger quantity of blood along with it because there is an abundance, and this happens more, when it has its attractive [faculty]

---

176 Literally: become plain.

177 The term could also mean "birds", but this usage is rare and poetic., and Galen, *Meth. med.* 7.6 (10 481.16-17 Kühn) and Aëtius Amidenus, *Iatrica* 5.80 (*CMG* 8.2 58.13) make it clear that the reference is to the wings of birds. Wings are coupled with rock-dwelling fish and cockerel's testicles (above 1.1) in these passages and also at Galen, *Comp. med. secundum locos* 2.1 (12 518.2 Kühn).

178 Usener's emendation would give: "and so that there may be <administerings> of solid food which are suitable for nature." For the connection of hectic (continuous) fevers with undue heating of the *solid* parts of the body cf. (e.g.) Galen, *Diff. febr.* 1.9 (7 304.7, 313.8 Kühn), *Inaeq. intemp.* 5 (7 743.11 Kühn), *Meth. med.* 9.1 (10 600.9 Kühn), 10.10 (10 718.12 Kühn), *In Hippocr. progn.* 2.60 (*CMG* 5.9.2 312.7, 313.9 Heeg), Philoponus, *in DA* 293.17, and Alexander of Tralles, *Febr.* 1 (1 291.19-20 Puschmann). [Alexander of Aphrodisias], *Febr.* 16.6 (14.14-19 Tassinari) draws a parallel with the solid walls of an ancient bath growing hot first; cf. ibid. 19.3, 15.26-16.3 Tassinari; Theophilus Protospatharius, *De febr. diff.* 8 (17.20ff. Sicurus) uses the analogy of adding cold water to a heated but empty metal vessel.

ὅταν ἐρρωμένην ἔχῃ τὴν ἑλκτικήν· ἀπὸ γὰρ τοῦ ἐπισπᾶν καὶ σπλὴν εἴρηται.

1.10. *Διὰ τί οἱ τὸν κρόταφον πληγέντες σφοδρῶς ἀπόλλυνται; ὅτι ὁ μῦς οὗτος, ἐξ αὐτοῦ τοῦ ἐγκεφάλου ἐκπέφυκεν, ὅθεν καὶ πλησίον αὐτοῦ ὑπάρχει· συμπάσχων οὖν ὁ ἐγκέφαλος καὶ συναλγῶν καὶ συναπολαύων τῆς πληγῆς, συστέλλεται, σὺν αὐτῷ δὲ καὶ τὸ ψυχικὸν πνεῦμα πιλοῦται καὶ παχύνεται· καὶ οὕτως ἡ ψυχὴ μὴ εὐποροῦσα τοῦ ἐπιτηδείου ὀργάνου ἀφίπταται ἐκ τοῦ σώματος· δεσμοῦ γὰρ φθορὰ λύσιν τῶν συμπλακέντων ἀποτελεῖ.*

1.11. *Διὰ τί οἱ ἐμβρόντητοι ἀπόλλυνται ἢ νωθοῦνται καὶ ἀνόητοι γίνονται; τῷ αὐτῷ πάλιν λόγῳ· ὁ γὰρ ἄμετρος κτύπος τῆς βροντῆς, διὰ τῆς ἀκοῆς εἰσιὼν πρὸς ἐγκέφαλον πλήττει πάλιν σφοδρῶς τὸν ἐγκέφαλον καὶ τὸ ψυχικὸν πνεῦμα, καὶ τοῦτο πάλιν συστέλλεται καὶ παχύνεται καὶ ἀνόητον ψυχῇ γίνεται πρὸς ἐνέργειαν· εἰ δ' ἀμέτρως παχυνθείη, καὶ χωρισμὸν ποιεῖται τῆς ψυχῆς καὶ τοῦ σώματος, καὶ διὰ τοῦτο θάνατον.*

1.12. *Διὰ τί τοῦ ᾠοῦ τὸ μὲν λεκιθῶδες εὔπεπτόν ἐστι, τὸ δὲ λευκὸν*

συνεφρίκεται Oi || *πλέον* EXc || *τῷ*] *τὸ* B || *μᾶλλον* om. BLNOi et fort. *g* 5 *ὅταν*] *ὅτε* Buss. (in var. lect.) || *ἐρωμένην* BC: *ἐρρομένην* E || *ἔχει* BCELTw Buss. (in var. lect.): *ἔχει* A || *ἐπισπᾶν*] *ἐσπᾶν* BL || *σπλὴν*] *πλὴν* O^{ac}i 5-6 *σπλὴν εἴρηται*] *λέγεται σπλὴν* C 6 *εἴρηται* A: *ἥρηται* T

1.10 (ABDEGLMNOPRSTXYbceim*p*w Buss. Us.; *g* 2.87)

1 *πληγέντες* ex *πληγέντις* corr. e || *σφοδρῶς εὐθὺς* D || *διότι* EXc 2 *τοῦ* add. sup. lin. O || *καὶ* om. DY || *πλησίων* E || *αὐτοῦ²*] *αὑτοῦ* A 3 *οὖν* bis T || *συμπάσχων οὖν ὁ ἐγκέφαλος*] *ὁ γοῦν ἐγκέφαλος συμπάσχων* P || *σὺν ἀλγῶν* E: *συναλγῶν αὐτῷ* D || *συναπολαύων*] *συναλάβων* T^{ac}: *σὺν ἀπολάβων* E: *συναπολαμβάνει* Y: *συνεπιλαμβάνεται* D 4 *καὶ συστέλλεται* D || *αυτῷ* sic Aw || *ψυχικὸν*] *φυσικὸν* M 4-5 *καὶ παχύνεται* om. D 5 *παχύνεται*] BDLNOPST^{pc}bc^{mg}e^{mg}im^{ac} Us.: *πλατύνεται* AEGMRT^{ac}XYcem^{pc}w Buss. *οὗτως* E: *οὗτος* A || *ψυχῇ* AE || *εὐποροῦσα*] *ἀπορροῦσα* Oi 6 *σώματος ὡς γνωστικὸν* M || *δεσμοῦ*] *δεσμοῦται* w || *λύσις* M 7 *ἀποτελεῖ*] *ἐργάζεται* D

1.11 (AMNOPRSTbeim*p*w Buss. Us.; *g* 2.88)

1 *οἱ* Am^{ac}*p* Buss. Us.: om. cett. || *ἐκβρόντητοι* e || *ἀπόλλυται* T^{ac} || *νωθοῦνται* nos: *νωδοῦνται* codd.: om. *p* 2 *αὐτῷ* A || *κτύπος* ex *κτύλος* corr. M || *βροντῆς*] *aeris g* 3 *εἰσιὼν*] *εἴσεισι* NOSbe^{mg}i || *πρὸς*] *εἰς* Buss. *πλήττει πάλιν σφοδρῶς τὸν ἐγκέφαλον* Aw Buss. Us., add m in marg.: om. cett. 5 *ἀνόητον* Us.: *ineptus g* (ex coniectura?): *ἀνόμοιον* cett. 6 *καὶ διὰ τοῦ σώματος* m^{pc}

1.12 (AEGMNOPRSTXbceimo*p*w Buss. Us.; *g* 2.89)

1 *διὰ τί* om. Go: *διατοῦ* S || *λεκυθῶδες* MP*p*: *λυκιθῶδες* X || *τοῦ ᾠοῦ τὸ μὲν λεκιθῶδες*] *τὸ γὰρ τοῦ ᾠοῦ τὸ λεκιθῶδες* Go 2 *ὅτι*] *διότι* EXc || *δὲ* om. O || *ὑγρὸν*

strengthened; for it is from "attracting" (*epispan*) that it has been called "spleen".[179]

1.10. Why do those who have been struck hard on the temples die? Because this muscle grows out from the brain itself, and therefore is close to it. So the brain, being affected together with it and feeling pain with it and being influenced by the blow together with it, contracts, and along with it the psychic *pneuma* is compressed and thickened, and so the soul, not being provided with its proper instrument, departs from the body; for the destruction of a bond effects the release of what was intertwined.

1.11. Why do those who are struck by thunderbolts die, or become dull[180] and stupid? For the same reason again: for the immoderate blow of the thunder, entering through their hearing to the brain, again strikes the brain and psychic *pneuma* vehemently, and this again contracts and is thickened and becomes useless[181] for the soul's activity. If it is immoderately thickened, it also brings about a separation of soul and body, and thus death.

1.12. Why is the yolk-coloured [part] of an egg easy to digest, the white

---

179 So Meletius, *De nat. hom.* 20 (*Anecdota Oxoniensia* 3 103.13-14 Cramer) and Leo, *De natura hominum synops.* 68 (*CMG* 10.4 52.7-8 Renehan), using the same term as here for "dregs"; also Orion, *Etymologicum* s.v. *splên* (143.27-8 Sturz), *Etymologicum Magnum* s.v (724.17-18 Gaisford) and ps.-Zonaras, *Lexicon* s.v. (1663.16-17 Tittmann) "Spleen" and "attract" are used in close proximity, without any explicit allusion to the etymology, by Galen, *Meth. med.* 11.16 (10 796.16-17 Kühn).

180 Reading *nôthountai.* Both this and the MSS *nôdountai* are otherwise unattested: but if *nôdountai* meant anything it would appear to mean "have no teeth".

181 Following Gaza's *ineptus*, which gives a good sense but is presumably just a guess. Usener's *anoêton* ("foolish, stupid") is uncovincing; the sense that results is extremely awkward. But the MSS *anhomoion* ("dissimilar") is awkward too, and could be seen as implying a grossly materialistic view of soul (something of which there is indeed evidence elsewhere in these texts, notably in i.26 Ideler; above, Introduction, at n.115).

δύσπεπτον; ὅτι τὸ μὲν χρυσοειδὲς θερμόν ἐστι, τὸ δὲ λευκὸν ὑγρὸν καὶ ψυχρόν. γίνωσκε γὰρ ὡς ἔστιν ἐκ τεσσάρων στοιχείων· τὸ μὲν γὰρ ὀστρακῶδες ἔοικε τῇ γῇ ξηρὸν καὶ ψυχρὸν ὑπάρχον, τὸ δὲ ὑγρὸν τῷ ὕδατι ὑγρὸν καὶ ψυχρὸν τυγχάνον· τὸ δ᾽ ἐν αὐτῷ πνεῦμα τῷ ἀέρι ὑγρὸν καὶ θερμόν, τὸ δὲ λεκιθῶδες τῷ πυρὶ θερμὸν καὶ μετρίως ξηρόν· μετέχει δὲ καὶ τῆς χροιᾶς τοῦ πυρός· καὶ εἴ μοι συγχωροίης εἰπεῖν, ἔοικεν εἶναι μίμημα κόσμου· καὶ αὐτὸ γὰρ σφαιροειδές, ἐκ στοιχείων τεσσάρων, μετέχον καὶ δυνάμεως ζωτικῆς.

1.13. Διὰ τί τῶν κολικῶν ἡ κόπρος βολβιτώδης; διότι ὡς ἐπὶ τὸ πλεῖστον ἐκ φλέγματος ἔχει τὴν γένεσιν τὸ πάθος, τοῦτο δὲ τῷ πλήθει σηπόμενον πνεύματα τίκτει, ἅτινα τὴν κόπρον ὑπενδυόμενα καὶ διογκοῦντα ταύτην καὶ φυσῶντα ὡς τὴν τῶν βοῶν ποιεῖ· ὅθεν πολλάκις καὶ ἐπινήχεται ἐν ὕδατι διὰ τὴν κουφότητα τοῦ πνεύματος· καὶ ἡ τῶν βοῶν δὲ ἐμπνευματοῦται κόπρος διὰ τὸ πολλῇ τροφῇ κεχρῆσθαι ἀχυρώδει καὶ τῷ πλήθει πνεύματα πλεῖστα τίκτειν.

ἐστιν EXc 2-3 ὑγρὸν καὶ ψυχρὸν] ψυχρὸν καὶ ὑγρὸν EXc 3 γίνωσκε – στοιχείων] ἐκ τεσσάρων στοιχείων γάρ ἐστι Go: γίνωσκε δὲ ὅτι (ὅτοι sic X) τὸ τοιοῦτον, καὶ αὐτὸ ἐκ τεσσάρων στοιχείων ἐστί EXc: ἴσθι οὖν ὅτι ἐκ τεσσάρων στοιχείων ἐστὶν A Buss. || οἷον τὸ μὲν γὰρ EXc 4 ψυχρὸν καὶ ξηρὸν A Buss. || ξηρὸν καὶ ψυχρὸν ὑπάρχον] ψυχρὸν καὶ ξηρὸν (χρὴ ξηρὸν X) ὑπάρχον τῇ συστάσει (ἔχων τὴν σύστασιν E) καθάπερ καὶ ἡ γῆ· φύσει ψυχρὰ καὶ ξηρὰ εἴη EXc || ὑπάρχον] ὑπάρχων ANR^{ac}: ὑπάρχει Oi 5 ὑγρὸν καὶ ψυχρὸν] ψυχρὸν καὶ ὑγρὸν EXc || τυγχάνον] τυγχάνων AE: ὑπάρχον P || τὸ δ᾽] τῶ δ᾽ Ao || ὑγρὸν om. Go 5-6 καὶ θερμὸν om. AGMPRTow, add. in marg. e 6 λεκιθῶδες] λεκιθῶδαις sic A: λεκυθῶδες MP: λευκῶδες o || μετρίως om. Go || ξηρὸν ὄν A: ξηρὸν ὄν Buss. || δὲ add. sup. lin. O 6-7 καὶ τὸ χροιᾶς EXc 6-9 μετέχει – ζωτικῆς] με ... tantum A, ceteris amissis 7 μοι om. Go || συγχωροίης] συγχωροίεις E: συγχωρήσεις Gow: συγχωρήσης T || μίμημα ex μίμηκα corr. e 8 ἐκ τεσσάρων στοιχείων G: ἐκ τεσσάρων o || μετέχων EGX^{ac}cw

1.13 (BLMNOPRSTbeim*p*w Buss. Us.; *g* 2.90)

1 κωλικῶν B Buss.: κολλικῶν T || βολβιτώδης BLNOSbe^{mg}im^{ac} Us. Buss.: βολβιθώδης MPRem^{pc}w: βολβηθώδης T 2 ἔχη Ni || σηπόμενα M || 3 πνεύματα] πνεῦμα BLNOSbgim^{ac} || ὑπενδυόμενα] πρὸς ὑπενδυόμενα BL: προσυπενδυόμενα Buss.: ἐπενδυόμενα M^{pc}: ὑπεκδυόμενα M^{ac} 4 φυσσῶντα NOi || βοῶν ex βῶν corr. O 5 ἐν om. BL Buss. || ἐν ὕδατι trans. post τοῦ πνεύματος P || διὰ τὴν – πνεύματος om. *g* || τὴν τοῦ κουφότητα S 6 δὲ om. B || ἐμπνευματοῦται] διαπνευματοῦται B Buss.: διεμπευματοῦται L: ἐμπνευτοῦται O || ἡ κόπρος P 7 ἀχυρώδη BLN^{ac}Tm^{ac}: ἀχυρῶν e^{ac}

hard to digest? Because the gold-coloured [part] is hot, the white moist and cold.[182] You should know that [the egg] is made of the four elements: the shell is like earth, being dry[183] and cold, the moist part like water, being moist and cold, the *pneuma* in it is like air, being moist and hot, and the yolk like fire, being hot and quite dry. It also shares in the colour of fire. And, if you will allow me to say it, [the egg] seems to resemble the universe: for it too is spherical, made of the four elements, and shares in the life-giving faculty.[184]

1.13. Why does the excrement of those with colic resemble cow-dung? Because for the most part the affliction has its origin in phlegm, and this, when it putrefies because of its abundance, generates gases, which enter the excrement and cause it to swell and inflate it and make it like that of cattle. For this reason it often also floats on water because of the lightness of the gas. The excrement of cattle is full of gas because the consume much nourishment containing bran, and because of its quantity they produce a very great amount of gas.

---

182 A similar explanation at Stephanus, *In priorem Galeni librum ad Glauconem* 123 (168.2-4 Dickson), on Galen, *ad Glauconem* 1.10 (11 35.6-7 Kühn); Galen gives the fact but not the reason. For the white being hard to digest cf. also Galen, *Meth. med.* 12.6 (10 848.14 Kühn: immediately followed by a reference to the testicles of cockerels fed on milk; cf. 1.1 above).

183 For "being dry and cold" MSS EX have "and having a constitution like that of earth; for by nature it would be cold and dry".

184 The idea that the world as a whole possesses a life force is Stoic or Platonic, rather than Aristotelian. The egg was familiar as a symbol for the *three* parts of Stoic philosophy (logic, physics, ethics): cf. Diogenes Laertius 7.40, Sextus *M* 7.18.

1.14. Διὰ τί ἐπὶ τῶν τὴν μήνιγγα τὴν παχεῖαν τοῦ ἐγκεφάλου τετρωμένων μύκητες ἐξίασι διὰ τῆς διαιρέσεως; ὅτι ἀεὶ τοῦ ἐγκεφάλου ἀτμοὺς γεννῶντος καὶ διαφοροῦντος αὐτοὺς διὰ τῶν ῥαφῶν τοῦ κρανίου, νῦν εὑρίσκοντες ἀναπέτειαν πλείστην τῆς διαιρέσεως δι᾽ αὐτῆς ἐξίασιν ἀθρόως· καὶ ὁμιλοῦντες ψυχρῷ τῷ ὀστέῳ πήγνυνται καὶ τῇ ἁπαλότητι ἐξαπλοῦνται καὶ λαμβάνουσι σχῆμα μυκήτων, ἄνωθεν μὲν ὄντες πλατεῖς, κάτωθεν δὲ στενοί· οὕτω γὰρ καὶ καπνὸς εὑρίσκων πολλὰς ὀπὰς καὶ θυρίδα μίαν, ῥᾶον καὶ ἀθρόως διὰ τῆς θυρίδος ἔξεισιν ἤπερ διὰ τῶν ὀπῶν.

1.15. Διὰ τί τὸ φλέγμα παχὺ ὂν καὶ ψυχρὸν καὶ διὰ τοῦτο κατωφερές, πολλάκις ἀπὸ τῶν ποδῶν ἄνεισιν ἐπὶ τὴν κεφαλὴν ἢ ἐπὶ τοὺς ὀφθαλμοὺς τοῦτο; φημὶ ὅτι οὐ τὸ φλέγμα καθ᾽ ἑαυτὸ ἄνεισιν, ἀλλ᾽ ἡ ἀποκριτικὴ δύναμίς ἐστιν ἑνὸς ἑκάστου μορίου ἀποδιώκουσα τοῦτο ἐπὶ τὸ ἄνω, ἄχρις ἂν εἰς τὸ ἀσθενέστερον πάντων τῶν μορίων ἀποδιωχθὲν ἐκεῖ ἔλθῃ, κἀκεῖνο δι᾽ ἀσθένειαν μὴ δυνηθῇ αὐτὸ ἀποσείσασθαι, ἀλλὰ μείνῃ ὑπ᾽ αὐτοῦ βαρούμενον καὶ ῥευματιζόμενον· οὕτω γὰρ ἐάν τις ἄνθρακα πυρὸς

1.14 (BLMNOPRSTbeim*p*w Buss. Us.; *g* 2.91)
1 ἐπὶ τῶν τὴν μήνιγγα] ἐπὶ τὸν μήνιγγα Oi: ἐπὶ τῷ τὴν μήνιγγα N^{ac}: ἐπὶ τὴν μήνιγγα m: τὸν μήνιγγα BL ‖ τῶν add. sup. lin. P ‖ τὴν[1] add. sup. lin. N ‖ μήνιγκα PT ‖ τὴν παχεῖαν τοῦ ἐγκεφάλου] τὴν παχείαν ἐγκεφάλου BL: τοῦ ἐγκεφάλου τὴν παχεῖαν P*p*: παχεῖαν τοῦ ἐγκεφάλου M 2 τετρωμένην BLT^{ac} ‖ μύκητες] μήνιτες BL ‖ διαιρέσεως ex αιρέσεως corr. O 3 γενῶντος BL ‖ κρανίου ex κραου corr. O 4 ἀναπέτειαν BLPT Buss. Us.: ἀναπετίαν MNORSbeim: ἀναπέτλεαν sic w, qui ει compend. velut λ legisse videtur 5 πήγνυται m ‖ καὶ – ἐξαπλοῦνται om. *g* (sed cf. infra) ‖ ἁπαλότητι PTw Us.: ἀπαλότητι m: ἁπλότητι BMNORSbei Buss.: ἀπλότητι LM ‖ ἐξαπλοῦνται] ἐξαπνοῦνται NOi 6 ἄνωθεν μὲν ὄντες πλατεῖς om. BL 7 post στενοί haec habet *g*: *videlicet loci ratione perficiantur* 7 εὑρίσκον L ‖ ὀπὰς BLw 7-8 οὕτω – ὀπῶν om. *g*: καὶ θυρίδα – διὰ τῆς om. BL 8 θυρίδας BL ‖ ἤπερ] ἥπερ Rew: εἴπερ BLTm ‖ ὀπῶν w
1.15 (BDGLMNOPRSTYbeim*p*w Buss. Us.; *g* 2.92)
1 φθλέγμα O ‖ καὶ ψυχρὸν om. Buss. ‖ διὰ τοῦτο om. D 2 ἀπὸ om. BL Buss.: ἐπὶ Y ‖ πολλάκις ἀπὸ τῶν ποδῶν] οὐκ ἐπὶ τοὺς πόδας ἀεὶ κάτεισι ἀλλ᾽ D ‖ ἢ ἐπὶ] καὶ ἐπὶ D 2-3 ἢ – τοῦτο om. *g* 3 τοῦτο om. DGY Buss. ‖ φημὶ ὅτι om. DGY: ὅτι om. O ‖ οὐ om. B ‖ τὸ φλέγμα om. BL ‖ καθ᾽ ἑαυτῶ BL: καθ᾽ αὐτὸ D 4 ἐστὶν add. sup. lin. m ‖ ἡ ἑνὸς Buss. ‖ μωρίου G ‖ τοῦτο om. Buss. ‖ ἑνὸς ἑκάστου – τὸ ἄνω] ἡ ἀπὸ τούτου τοῦ μορίου πρὸς ἐκεῖνο διώκουσα D 5 εἰς] πρὸς D ‖ ἀσθενέστερον] ἀσθενέστατον M Buss. (coni.) Us. ‖ πάντων τῶν μορίων] μέρος D ‖ ἀποδιδιωχθὲν O ‖ ἀποδιωχθὲν ἐκεῖ ἔλθῃ] καταντήσει D ‖ ἐκεῖ ἔλθῃ] ἐκεῖθεν BL: καταντήσῃ D: ἔλθῃ Buss. 6 κἀκεῖνο οὖν D ‖ δυνηθῇ Us. Buss.: δυνηθὲν DY: δυνηθείη cett. ‖ αὐτὸ om. BL Buss. ‖ ἀποσείσασθαι] ἀποσήσασθαι Tw: ἀπόσησθαι G: ἀπώσασθαι τοῦτο D: ἀπώσασθαι Y ‖ ἀλλὰ om. D ‖ μείνῃ] μεῖνον e^{ac}: μένει D 6-7 ὑπ᾽ αὐτοῦ] ἐπὶ τὸ

1.14. Why, in the case of those who are wounded in the thick membrane of the brain, do fleshy excrescences [literally: "mushrooms"] emerge through the split? Because the brain continuously generates vapours and distributes them through the sutures of the skull; now, finding considerable [scope for] expansion through the split they exit through it in bulk, and being in the presence of the bone, which is cold, they are congealed, and spread out through their softness, and take on the shape of mushrooms being broad above and narrow below. For in this way smoke too, finding many chimneys and one window, exits more easily in bulk through the window than through the chimneys.

1.15. Why does phlegm, which is thick and cold and for this reason travels downwards, often rise up from the feet to the head or eyes? I say that the phlegm does not rise up of itself, but it is the expulsive power of each single part [of the body] which pursues it upward, until it is pursued to the weakest of all the parts and arrives there, and this through weakness cannot shake it off, but continues being weighed down and caused to undergo flux by it. For in this way, when many people are present, one might throw a live coal to

πολλῶν παρόντων ῥίψειέ τινι, κἀκεῖνος ἑτέρῳ, καὶ ἄλλος ἄλλῳ τῇ ἰσχύι αὐτὸν ἀπορρίπτοι, ἄχρις ἂν εἰς τὸν ἀσθενέστατον ἐλθὼν ἀπομείνῃ.

1.16. *Διὰ τί ἐν λοιμικῇ καταστάσει πολλοὶ ἀπόλλυνται κατὰ πάθος ἓν ὡς ἐπὶ τὸ πλεῖστον καὶ κατὰ τὴν αὐτὴν ἡλικίαν, ὃ καὶ δοκεῖ ὥσπερ ἀνατρέπειν τὴν ἀπὸ τῶν ἀστέρων εἱμαρμένην; φημὶ ὅτι ὅσοι ἐκληρώθησαν ὑπὸ τῆς ἀστρῴας κινήσεως κατὰ τὴν αὐτῶν ἀπότεξιν ὀλιγοχρόνιοι, οὗτοι θᾶττον ἐλέγχονται καὶ ἁλίσκονται ὑπὸ τοῦ λοιμοῦ κινούμενοι· οὕτω γὰρ καὶ οἰκία παλαιὰ καὶ σαθρὰ θᾶττον ὑπ' ἀνέμου σφοδροῦ καταπίπτει ἤπερ ἡ ἰσχυρὰ καὶ νέα· τῷ δὲ αὐτῷ πάθει διά τινα τοιαύτην ἀστρῴαν ἀπόρροιαν λοιμικήν.*

1.17. *Διὰ τί ὁ πολὺς δῆμος ἐπὶ τῶν κραιπαλῶν φησιν οἶνον οἴνῳ διαλύεσθαι τὴν ἐξ αὐτοῦ κεφαλαλγίαν; φημὶ τοῦτο ἰδιωτικὸν μέν, ἀληθὲς δὲ τυγχάνειν· τὴν γὰρ ἀπὸ τῆς πολλῆς θερμότητος τοῦ οἴνου γενομένην ἔκλυσιν τῶν μορίων ἕτερος οἶνος εἰσιὼν καὶ ἀποδιώκων αὐτὴν ῥώννυσι ταῦτα καὶ παρασκευάζει διώκειν καὶ ἀποκρούεσθαι τὰ συναχθέντα ὑπ' αὐτοῦ πνεύματα καὶ βαροῦντα τὴν κεφαλὴν καὶ τὸ σῶμα· καὶ ὁ μάλιστα κεκερασμένος ἐκλύει τὴν τοῦ προτέρου θερμότητα δι' ἣν καὶ ἡ ἔκλυσις·*

BL 6-9 *ἀλλὰ – ἀπομείνῃ* om. *g* 7 *ἄν τις* MY Us. 8 *ῥίψειέ τινι*] *ῥίψειε τινῶν* P: *ῥίψῃ ἔν τινι* DGY: *ῥίψι τινὶ* L: *ῥίψιν τινὶ* B: *ῥίψῃ τινὶ* Buss. || *κακεῖνο* e || *ἐν ἑτέρῳ* D || *ἐν ἄλλῳ* D || *ἰσχύει* BLSbm^{ac}w 9 *αὐτῶν* BLNOi || *ἀπορρίπτοι* M Us.: *ἀπορίπτει* BL: *ἀπορρίψουσι ἕκαστος* D: *ἀπορρίψουσιν* Buss.: *ἀπορρίπτει* cett. || *ἄχρις ἂν*] *ἕως* BL || *τὸν*] *τὸ* BDLN^{ac}m || *ἀσθενέστατον* DGTYm^{pc}w Buss. Us.: *ἀσθενέστερον* BLMNORSbeim^{ac}: *ἔλαττον* P || *ἐλθὼν ἀπομείνῃ*] *μέρος ἔλθῃ* BL

1.16 (BLMNOPRSTbeim*p*w Buss. Us.; *g* 2.93; 3-7 Petrus, Olivieri 1988, 197-198) 1 *καταστάσι* e || *ἓν*] *ἐν* O 2 *καὶ*^{1} om. BLNOSbi Buss. || *καὶ κατὰ τὴν αὐτὴν ἡλικίαν*] *et <u>in</u>aequales g* || *ὃ g* Us.: om. cett. Buss. || *ὥσπερ*] *ὅτι* BL 3 *ἀνατρέπειν*] *ἀνατρέχει* BL 4 *ὑπὸ*] *ἀπὸ* BLNOSbi Buss. || *ἀστρόας* BLT: *ἀθροᾶς p* || *αὐτῶν*] *αὐτὴν* MRem*p* Petrus: om. PTw Us. || *ἀπόταξιν* LMRe 5 *θᾶττον* om. *g* || *λιμοῦ* S 7 *ἤπερ* LOw: *εἴπερ* BL || *τῷ δὲ αὐτῷ πάθει* om. Buss. 7-8 *τῷ δὲ – λοιμικὴν*] *sic etiam homines, qui ex fato vitam sortiti sunt breviorem, expeditius ab aeris vitio moventur, et capiuntur; malum autem commune propterea est, quoniam aer, qui syderum pestilenti defluvio vitiatus est, idem continet omnes, atque ab omnibus trahitur g* 8 *ἀστρόαν* T || post *λοιμικὴν* lacunam posuit Us., Gazae versione collata

1.17 (B[G]LMNOPRSTbeim[o]*p*w Buss. Us.; *g* 2.94)
1-44 *Διὰ τί – φέρεται* om. Go 1 *φησιν* post *διὰ τί* collocavit P || *ἐπὶ*] *ὑπὸ* P || *οἴνῳ* add. sup. lin. B 2 *κεφαλαλγίαν*] *εὐκρασίαν p* ("immo *ἀκρασίαν*, nisi euphemismo utitur", Duffy) 3 *τυγχάνον* BLNOST^{ac}bi || *τοῦ οἴνου* om. Buss. 4 *ἕλκυσιν* T || *τῶν μορίων* om. Buss. || *καὶ* om. B 5 *παρασκευάζειν* BLMw || *διώκει* BLM 6 *καὶ*^{1} om. Buss. || *καὶ τὸ σῶμα* om. *g* (sed cf. infra ad 9) || *καὶ μάλιστα ὁ*

someone, and he to another, and one person might use his strength to throw it to another, until it arrives at the weakest and stays there.

1.16. Why, in conditions (*katastasis*) of plague, do many very frequently die from a single affliction and of the same age, which seems to disprove fate from the stars? I say that as many as were allotted a short life by the movement of the stars at their birth are more swiftly tested and caught when affected by the plague; for in this way an old and rotten house collapses from a strong wind more swiftly than one that is strong and new. By the same effect, on account of some such plague-bringing effluence from the stars.[185]

1.17. Why does the general public say of drinking bouts that wine is dissolved by wine, [i.e.] the resultant headache? I say that this is laymen's talk,[186] but still true. The great heat of the wine produces relaxation of the [bodily] parts; more wine, entering in and expelling this [relaxation], strengthens them and equips them to chase away and beat off the vapours that have been brought together by it and weigh down the head and the body. It is the wine most mixed [with water] that removes the heat of the previous [wine] which brought about the relaxation. For a small quantity of

---

185 Usener marks a lacuna here. For the last sentence Gaza has: "so too human beings, who have been allotted a shorter life by fate, are affected more swiftly by the corruption in the air, and are seized by it; but the evil is shared, because the same air, which is corrupted by the plague-bringing effluence of the stars, surrounds everyone and is breathed by everyone." The objection that if astrology is true people born at different times should not suffer a similar fate is commonplace; cf., with Amand 1973, 53-54, Cicero, *De div.* 2.97, Favorinus ap. Gellius, *Noct. Att.* 14.1.27, and Sextus, *M* 5.91-3. It is not entirely clear whether the response intended here is (a) that people's birth horoscopes do indeed make them more or less susceptible, but a single cause of plague can be sufficient to affect those who are less susceptible as well as those who are more so, or (b) that many different horoscopes can indicate short life (and many others, by implication, long life) and that the plague discriminates between these two general groups, other differences in individual horoscopes being irrelevant as far as the plague is concerned. (b) allows horoscopes more influence in the outcome where plague is concerned than does (a); it is also perhaps the more natural reading of the text.

The argument is made less clear than it might be by the attribution of the plague *itself* to the influence of the stars, in a different way. See above, Introduction, at nn.112-114.

186 As with our "hair of the dog".

ὀλίγος μὲν γὰρ οἶνος πινόμενος ὑγείαν ποιεῖ, σύμμετρος δὲ εὐφρασίαν, ἄμετρος δὲ μέθην καὶ παραφροσύνην, καθὼς καὶ Θεόγνις ὁ ποιητὴς εἶπε περὶ τοῦ συμμέτρου οἴνου φάσκων· "οἶνος πινόμενος πουλὺς κακός· ἢν δέ τις αὐτὸν πίνῃ ἐπισταμένως, οὐ κακός, ἀλλ' ἀγαθός".

ὅθεν οὐκ ἀσκόπως μυθεύουσι τῷ Διονύσῳ βάκχην ἕπεσθαι, διὰ τὴν ἐκ τοῦ οἴνου χορείαν. Σατύρους δὲ διὰ τὴν εὐκινησίαν· Λύδην δὲ διὰ τό τινας ἐξ αὐτοῦ ἐκλύεσθαι· πάρδον δὲ διὰ τὴν ποικίλην φαντασίαν τῶν οἰνωμένων· διάφορον γὰρ ἕκαστος καὶ ποικίλον ἀναδέχεται λογισμὸν ὑπὸ τοῦ οἴνου κινούμενος· πολύστικτος γὰρ καὶ ἡ δορὰ τοῦ ζῴου· μίαν δὲ βάκχην μανεῖσαν καὶ δράσασαν φόνον, τὴν Ἀγαυὴν μητέρα τοῦ Πενθέως, πολλοὶ γὰρ ὑπερμεθυσθέντες ἐτράπησαν εἰς φόνον, γυμνὸν δὲ τὸν Διόνυσον διὰ τὴν ἀπὸ τοῦ οἴνου γινομένην φανέρωσιν τῶν πινόντων τοῦ λογισμοῦ· ποθῆσαι δὲ Ἀφροδίτην καὶ Ἀριάδνην, τοῖς γὰρ οἰνωθεῖσι παρέπεται πλείστη γυναικῶν ἐπιθυμία· ἔχειν δὲ καὶ φαλακρόν τινα διὰ τὸ μάλιστα

---

g Buss. Us. 7 *κεκερασμένος*] *κερασμένος* Oi: *κεκραμένος* Buss. || *πρωτέρου* B || *δι' ἣν καὶ ἡ ἔκλυσις* om. *g* || *ἔκλυσις*] *ἔκκλησις* BL 8 *μὲν* add. sup. lin. O || *γὰρ* add. sup. lin. m || *ὁ οἶνος* BL || *εὐφρασίαν*] *εὐκρασίαν* BL 9 *μέθην καὶ παραφροσύνην*] *et menti obest et corpori g* || *παραφροσύνη* O^{ac} || *καθὼς*] *καθὰ* P || *Θεόγνις* Buss. Us.: *θεαγένης* codd. || *φάσκων*] *πάσχων* T 10 *πουλὺς* Theognidis codd. plerique Buss. Us.: *πολὺς* codd. huius operis || *κακὸς* codd. huius operis, Theognis ap. Artemidorum 1.66, Clemens et Stobaeus: *κακὸν* Theognidis codd. || *ἢν*] *ἂν* BL 11 *πίνῃ* B Theogn. Buss. Us.: *πίει* T: *ποιεῖ* m^{ac}: *πίῃ* cett. || *κακὸς*^{2} ex *κακῶς* corr. i || *οὐ κακός, ἀλλ' ἀγαθὸς* codd. huius operis, Theognis ap. Artemidorum et Stob.: *οὐ κακόν, ἀλλ' ἀγαθὸν* Theognidis codd. 12 *οὐκ* add. sup. lin. N: om. BLOSbi: exhib. *g* || *μηθεύουσι* m^{ac}: *μεθύουσι* BL || *βάκχην*] *Liberum patrem*, i.e. *βάκχον*, *g* 13 *οἴνου* add. sup. lin. O || *χωρείαν* i^{ac}: *χωρίαν* BL || *σατύρας* MR 14 *ἐκλύεσθαι ἐξ αὐτοῦ* BL Buss. || *αὐτῶν* M || *πάρδον*] *πάροδον* Be^{ac} || *ποικίλον* Tw 15 *οἰνομένων* w || *γὰρ*] *δὲ* BL || *ἕκαστον* BNOPi || *καὶ ποικίλων* O: *καὶ ποικίλον ἕκαστος* Se^{pc}: *ποικίλον* B 17 *μανεῖσα* BL || *δράσα* BL || *εἰς τὴν* M || *ἀγενὴν* BL: *ἀγανὴν* M: *ἀγαυνὴν* Se^{pc} 18 *ὑπερμεθισθέντες* T: *ὑπερμαθέντα* BL 19 *διὰ* om. BL || *τοῦ*^{2} om. BL 20 *ποθῆσαι*] *ποθῆναι* e^{ac}: *ποθεῖσα* BL 21 *γυναικῶν*] *inebrietatis g* (?) || *ἔχει* BLw || *δὲ*] *τε* T || *καὶ*] *τι* BL 22 *ἐρημοῦν*] *ἠρεμοῦν* BL || *βλάπτει* BL ||

wine when drunk produces health, a moderate amount cheerfulness, too much drunkenness and madness, as Theognis[187] the poet said when he spoke about a moderate [amount of] wine:

> Much wine is bad when drunk: but if someone drinks it
> Wisely, it is not bad, but good.[188]

And for this reason it is not without consideration that they say in the myths[189] that a Bacchant accompanies Dionysus because wine leads to dancing, the Satyrs because [it leads to] ease of movement; a Lydian female[190] because some are relaxed (released, *ekluesthai*) by it; a leopard because of the variegated imagination of those who are drunk (for each of them takes on a different and variegated [mode of] reasoning[191] when affected by the wine; and the skin of the animal too is variegated); a single Bacchant who went mad and committed murder, Agave the mother of Pentheus (for many when excessively drunk have turned to murder). [They speak of] Dionysus as naked because of the revelation of the thought of those who drink that is brought about by wine. [They say that] he fell in love with Aphrodite[192] and Ariadne (for those who are drunk are characterised by very great desire for women). He has a certain bald person[193] [as companion]

---

187 So Bussemaker and Usener: "Theagenes" codd.

188 Theognis 509-510; cf. 211-212. Also cited (D. Young, *Theognis*, Leipzig: Teubner, 1961, p.33 ad loc.) by Galen, *Quod animi mores* 3 40.20 Mueller; Artemidorus, *Onir.* 1.66; Clement, *Strom.* 6.2.11.5; Stobaeus, *Ecl.* 3.18.11.

189 The following account combines allegorical interpretation and etymology.

190 Alan Griffiths comments that the text seems to give "a first cluster of four generic followers (Bacchant, satyrs, Lydian woman, panther) followed by remarks on named characters (Agave, Dionysus, Aphrodite and Ariadne, Silenos and Pan)", so that 'Lydian woman' has no more specific reference here than 'Bacchant'. For the association between Dionysus and Lydia he compares e.g. Euripides, *Bacchae* 55. It is however odd, as he notes, that of the first group of four only "Satyrs" is in the plural.

191 *logismos*. The use of this term in connection with imagination is remarkable.

192 For the love of Aphrodite and Dionysus cf. Pausanias 9.31.2 and Schol. on Apollonius Rhodius 1.932: R. Graves, *The Greek Myths*, Harmondsworth: Penguin, 1955, 71 n.3.

193 Silenus. As Alan Griffiths suggests, our author's omission of the name is probably learned affectation.

τὸν ἐγκέφαλον ἐρημοῦν καὶ βλάπτειν καὶ μαραίνειν τὸν πολὺν οἶνον, ὅθεν καὶ Μάρωνα προσηγόρευσαν· ἔχειν δὲ καὶ Πᾶνα, τουτέστιν αὐτὸ τὸ πᾶν εἶναι τοῦ οἴνου τὴν δύναμιν, ἢ πάλιν διὰ τὴν διάφορον χροιὰν καὶ δύναμιν τοῦ οἴνου. καὶ γὰρ οὗτος πολυσχημάτιστος καὶ πολυχρώματος· ὑπὸ πυρὸς δὲ κεραυνωθέντα καὶ ἐν μηρῷ βληθέντα, διότι πολλάκις οἶνος ἡλιαζόμενος τελειοῦται τῇ κράσει καὶ τῇ δυνάμει τοῖς κεράμοις κρυπτόμενος· τέσσαρας δὲ γυναῖκας εἶναι ἀδελφὰς διὰ τὸ τέσσαρας τροπὰς ἔχειν καὶ μεταβολὰς τὸν οἶνον.

ἄκουε δὲ καὶ περὶ τοῦ Πανός, ὅτι βούλονται αὐτὸν εἶναι τὸ πᾶν τοῦ κόσμου· δύο μὲν γὰρ κέρατα περιτιθέασι τῇ κεφαλῇ διὰ τὴν διπλῆν φύσιν τοῦ παντός· τὸ γὰρ πᾶν ἄρρεν καὶ θῆλυ τῶν κατὰ μέρος καθέστηκεν, ὥσπερ καὶ ἥλιος καὶ σελήνη, ὁ μὲν γάρ ἐστι προετικὸς ἀντὶ σπέρματος τῶν ἰδίων φώτων πρὸς τὴν σελήνην, ἡ δὲ δεκτικὴ τούτων ὑπάρχει καθάπερ καὶ τὸ θῆλυ γένος, ἔτι γε μὴν ζωὴ καὶ θάνατος, γένεσις καὶ φθορά, ὡς δηλοῖ σελήνης τὰ φῶτα αὐξανόμενα καὶ μειούμενα καὶ ἡ σφαῖρα κινουμένη καθ' ἑκάστην ἀπὸ ἀνατολῶν ἐπὶ δυσμάς, οἱ δὲ ἑπτὰ ἀστέρες ἐναντίαν ὁδὸν ἀπὸ δυσμῶν ἐπὶ ἀνατολάς, ὡς δείκνυσι τοῦτο πλέον τῇ ὄψει πάντων τῶν κινουμένων ἀστέρων· ἔχει δὲ καὶ τὸ Πᾶν σῶμα διφυές, καὶ τὸ μὲν ἄνω ἀνθρωποειδές, τὸ δὲ κάτω προβατοειδές, διότι καὶ τοῦ παντὸς τὸ μὲν ἄνω μέρος περιέχει λογικὰς δυνάμεις, τὸ δὲ κάτω καὶ ἀλόγους· μὴ ἠρεμεῖν δὲ τὴν στάσιν διὰ τὴν ἀεικινησίαν τῆς φορᾶς, καὶ τὴν ῥευστὴν

μαραίνει BL: μωραίνειν MR^{ac}e^{ac} Buss. || τὸν πολὺν οἶνον] ὁ πολὺς οἶνος BL 23 Μάρωνα] μάρων L || ἔχειν add. sup. lin. O: ἔχει BL || πάνα sic BLNOPRSb^{ac}ei || τὸ add. sup. lin. O 24 εἶναι] διὰ τὸ πάντα κρατεῖν Us. || τὴν om. P || διὰ τὴν add. in marg. O: ἐπὶ τὴν BL 25 καὶ γὰρ – πολυχρώματος om. *g* || πολυσχημάτιστος] πολυχρημάτιστος ἐστὶ B: πολυσχημάτιστος ἐστὶ L Buss. 26 ἐν add. sup. lin. R || ἐν τῷ μηρῷ M Buss: ἐν μηνῶ B: ἐν μικρῶ L || βληθέντα] βλαθέντα O: ἐμβληθέντα PTmw || διότι ex διότις corr. e || ὁ οἶνος M Buss. Us. 28 ἀδελφὰς] ὑδαρὰς BL 29 ἔχει BL || καὶ μεταβολὰς om. BL Buss. 30-44 ἄκουε δὲ – φέρεται om. *g* 30 ἄκουε δὲ καὶ περὶ τοῦ πανὸς om. T || δὲ] δὴ L || βούλεται BLm 32 τὸ κατὰ μέρος BL Buss. 33 καὶ[1] om. P || προετικὸς *p*: προεκτικὸς MPRTe^{ac}mw: παρεκτικὸς BLNOSbe^{pc}i Buss. Us. 34 φώτων om. M: φωτῶν Buss. || τῇ σελήνῃ M || δὲ add. sup. lin. R || τούτου P: om. Buss. 36 αὐξόμενα PTmw 37 καθεκάστην ex καθέστην corr. O 38 ἐναντία T || πάντων om. M 38-39 πάντων τῇ ὄψει Buss. 39 ἔχειν Us. || πᾶν om. BLNOTi Buss. Us. || διφυὲς] διαφυὲς NOei || καὶ[2] om. B Buss. Us. 40 τὸ δὲ κάτω προβατοειδὲς om. T 41 λογικοὺς Buss. 42 δὲ ex διὰ corr. m || τὴν στάσιν ex τῇ στάσει corr. B: τῇ στάσει Buss. || ἀεικινησίαν ex ἀεικίνησαν corr. T || φορᾶς MTm^{ac}w Buss. Us.: φθορᾶς BLNOPi:

because much wine greatly lays waste and harms and withers (*marainein*) the brain, for which reason they also called him Maron.[194] He also has Pan as a companion, that is: the power of wine is universal,[195] or again because of the differing colour and power of wine (for [Pan] too has many shapes and many colours). [They say that] Dionysus was struck by a thunderbolt and placed in a thigh[196] because often wine which has been warmed by the sun is perfected in its blending and power while concealed in a jar. [They say that] four women are his sisters because wine undergoes four changes and transformations.

Hear also concerning Pan, that they intend him to be the whole (*pan*) of the world.[197] For they place two horns on his head because of the double nature of the whole. For the whole has been established as male and female in its parts, like the sun and the moon (for the former projects its own light to the moon in place of seed, the latter, receiving this, is like the female race), and moreover [the whole has been established as including] life and death, generation and destruction, as the light of the moon shows by waxing and waning, and the sphere [of the fixed stars] moving every [day] from east to west, and the seven planets [moving] in the opposite way from west to east, as he shows more by the sight of all the planets.[198] Pan's body is of two natures, like a man above and like a goat below, because of the universe (*pan*) too the upper part[199] contains the rational powers and the lower also irrational ones.[200] [They say that] he does not remain still in his position because of the continuous movement of locomotion and the flowing nature

---

194 Maron was in fact the priest of Apollo who gave Odysseus the wine which he had with him in the Cyclops' cave (*Odyssey* 9.197); his name was used figuratively for wine. Hesiod (fr. 238 Merkelbach-West) made Maron Dionysus' great-grandson.

195 Literally: "is the universe itself". Usener's supplement "that is: everything, because the power of wine prevails over all things" has no basis in the MSS.

196 Dionysus was rescued by Zeus from the thunderbolt which killed his mother Semele, and kept in Zeus's thigh as a substitute womb.

197 Flashar, 1962a 408 and n.1 compares Cornutus, *ND* 27, p.49.5 Lang, and Orphic *Hymn* 11.1 Quandt.

198 Literally "the moving stars". But the point of this clause is unclear.

199 I.e. the heavens, as opposed to the lower part around the earth at the centre.

200 This division of the universe at first sounds Platonic, but the heavenly bodies are intelligences for Aristotle.

φύσιν τῆς ὕλης· πεφιληκέναι δὲ τὴν Ἠχώ· ποθεῖ γὰρ ἡ οὐρανία σφαῖρα τὴν ἐναρμόνιον κίνησιν καὶ τῇ ταύτης ἐφέσει κυκλοφορικῶς ἀεὶ φέρεται.

ἀναλογεῖ δὲ τῷ οἴνῳ τὸ αἷμα κατὰ κρᾶσιν καὶ οὐσίαν· ὥσπερ γὰρ ἐξεθέμεθα τὸν οἶνον τετραμερῆ, τὸν αὐτὸν τρόπον καὶ τὸ αἷμα· ἔχει γὰρ τὸ ὑδατωδέστατον μέρος ὅπερ ἕλκουσιν οἱ νεφροί· ὁμοίως δὲ καὶ ἀναλογοῦν ὡς ἄνθος, ὅπερ ἕλκει χοληδόχος κύστις, καὶ τὸ παχὺ καὶ τρυγῶδες ὅπερ ὁ σπλὴν ἕλκει, καὶ αὐτὸ τὸ οἰνῶδες καὶ καθαρὸν τοῦ αἵματος ὅπερ ἀναδιδόμενον· καθάπερ καὶ οἶνος ἐν τῷ σώματι τρέφει τὰ μόρια.

1.18. Διὰ τί τρωθεῖσα βάλανος καὶ γνάθου τὰ λεπτὰ οὐ κολλᾶται; καθόλου ὅσα ἀπὸ σπέρματος γέγονεν, ἀπὸ φύσεως ταῦτα παλιγγενεσίαν οὐχ ὑπομένει τὴν αὐτήν, ὡς νεῦρον, φλέψ, ὑμένες καὶ τὰ παραπλήσια· εἰκότως οὖν καὶ ταῦτα νευροειδῆ ὄντα τελείαν οὐχ ὑπομένει τὴν γένεσιν· τὰ μέντοι ἐξ ἀρχῆς ἐξ αἵματος διαπλασθέντα πάλιν γεννᾶται τῷ διὰ παντὸς τοῦ βίου αἷμα κεκτῆσθαι σπέρμα δὲ οὔ.

φοβερᾶς RSbem^{pc} 43 *πεφιληκέναι* ex *πεφικέναι* corr. O || *τὴν*^{1}] *τῇ* BL 44 *ἐφέσει*] *ἐν φύσει* BL || *κυκλοφορικῶς*] *κυλλοφορκῶς* L || *φέρεται* M Us.: *φέρεσθαι* cett. Buss. 45 *ἀναλογεῖ δὲ τῷ οἴνῳ τὸ αἷμα*] *τὸ αἷμα ἀναλογεῖ τῷ οἴνῳ* Go || *γὰρ* om. BL 45-46 *ὥσπερ – τετραμερῆ*] *ὥσπερ γὰρ ὁ οἶνος τετραμερὴς* Go 46 *τὸ* om. M 47 *οἱ* BLMo Buss. Us: om. cett. || *νεφροὶ* ex *μεφροὶ* corr. e || *καὶ* om. G || *τὸ ἀναλογοῦν* BL Buss. Us.: *τὸ* add. sup. lin. R 47-48 *ἀναλογοῦν ὡς ἄνθος*] *ὡς* add. supra lin. B: *ὡς ἄνθος ἀναλογοῦν* G: *ἄνθος ἀναλογοῦν* o 48 *ὅπερ*^{1}] *ὥσπερ* M || *χοληδόχος*] *ἡ χοληδόχος* Po*p* Buss. Us.: *χολιδόχος* mw: *ἡ χολιδόχος* G: *χολοδόχος* B || *ἕλκει* post *χολιδόχος* collocant Go || *τὸ* om. Oi || *τρυγῶδες*] *τριχῶδες* o || 48-49 *ὅπερ ἕλκει ὁ σπλὴν* BL 49 *σπλὴν*] *πλὴν* N^{ac}O^{ac}i || *ἕλκει*] *ἔχει* Se^{mg}o 50 *καὶ*] ὁ Go

1.18 (BLMNOPRSTbeim*p*w Buss. Us.; *g* 2.95)

1 *γνάθου* ex *γράθου* corr. T || *κολᾶται* BL 2 *καθόλου*] *καθὰ* M Buss. 4 *νευροειδεῖς* L || *νευροειδῆ ὄντα*] *cum semine orta sint g* || *τελείαν*] *ταχείαν* P: om. *g* 5 *μέντοι*] *μὲν* T || *διαπλασθέντα* ex *διπλασθέντα* corr. R: *διπλασιασθέντα* T: *διὰ πλασθέντα* e: *διαπλασθῆναι* BLNOi || *τῷ δὲ* BL Buss.: *καὶ τῷ* w 6 *τοῦ βίου* PRTemw Us.: *βίου* BLMNOSbi: *βίον* S Buss. || *αἷμα* om. BL Buss. || *σπέρμα δὲ οὔ* om. *g* || *οὔ* MRSTbemw Us.: *ὄν* BLNOPi: *σπέρματος δέονται* coni. Buss.

of matter. [They say that] he loved Echo: for the heavenly sphere desires harmonious movement, and by desire for this[201] it is continuously moved in a circle.

Blood is analogous to wine in blending and substance. For as we have explained[202] that wine has four parts, in the same way also does blood. For it has the watery part which the kidneys attract, and similarly what is analogous to a bloom, which the gall-bladder attracts, and the thick [part] like dregs which the spleen attracts, and the wine-like and pure part [itself] of the blood which is given off in the body like wine and nourishes the parts.

1.18. Why do a wounded *glans penis* and the rarer [parts] of the jaw not fuse?[203] [Because] in general what comes from seed does not admit of an identical natural regeneration, for example sinews, veins, membranes and the like. So it is reasonable that these too, being of the nature of sinew, do not admit of complete [re]generation. But the [parts] that were originally fashioned from blood are regenerated, because blood is possessed throughout life, seed not.[204]

---

201 See above, Introduction n. 109.

202 Usener saw this as a reference back to *Probl.* ii.70 Ideler, and used this as an argument for books 1-2 having originally formed part of the same collection as i-ii.

203 Cf. Aristotle, *HA* 3.11 518a1 (where however the reference is to the *foreskin* rather than the *glans*).

204 That is, seed is not present in the parts once formed; the creature possesses seed from which offspring may be produced, but that is irrelevant. Psellus' paraphrase has rather "for blood is everywhere [in the body], seed is not." – For Aristotle *all* parts including sinews are formed from blood: *GA* 2.6 743a18; his answer to our problem in *HA* 3.11 is simply that these parts lack flesh. Galen, *Ars medica* 26 (358.8-10 Boudon) asserts that what comes from blood is easily regenerated, while what comes from seed is not; similarly, specifying arteries, veins, *neura,* bones and membranes as from seed, Oribasius, *Coll. med.* libri incerti 15.17 (*CMG* 6.2.2 105.14-17 Raeder). *neura* is ambiguous between "sinews" and "nerves"; in Oribasius the reference seems to be to nerves, but here a reference to sinews seems more appropriate. For nerves coming from seed alone cf. Galen, *De semine* 1.8.5 (*CMG* 5.3.1 90.19-21 De Lacy); arteries, veins, nerves, bones and membranes, ibid. 1.10.1 (98.3). (See also Moraux 1984, 745-6.) Stephanus, *In Hippocr. prognost.* 1.14 (*CMG* 11.1.2 132.22-26 Duffy) says that cartilage, nerves, membranes, veins and arteries and come from seed, but explains this by saying that they are nourished by blood which is *turned into* seed – which undermines the explanation in our text of why parts from seed cannot be regenerated.

1.19. *Διὰ τί τὰ γίγαρτα καὶ τὰ σήσαμα καὶ κεγχραμίδες ὑγιῆ καταπινόμενα πάλιν ὠμὰ διὰ γαστρὸς ἐκκρίνεται; ἢ διὰ τὴν πολλὴν σμικρότητα καὶ ξηρότητα οὐ πέττεται, ἢ ὅτι τὸ τέλειον περιέχουσι τῆς γεννητικῆς σπορᾶς διὸ καὶ σφαιροειδῆ, ὡς ἐπὶ τὸ πλεῖστον σῶα παραμένουσιν· τοῦτο γὰρ τὸ σχῆμα τελειότερόν ἐστι παρὰ τῷ θεῷ παρὰ πάντα τὰ σχήματα, λέγω δὴ τετράγωνα, κυβοειδῆ, πυραμοειδῆ, κυλινδροειδῆ, ὠοειδῆ, φακοειδῆ καὶ τὰ ὅμοια.*

1.20. *Διὰ τί ἐπὶ τῆς ἐμφράξεως τῶν κυρτῶν τοῦ ἥπατος ἡ κατακλεὶς κατασπᾶται; ὅτι ἡ ἀπὸ τῶν κυρτῶν τοῦ ἥπατος στελεχιαία φλὲψ ἀνιοῦσα πρὸς τὴν κατακλεῖν ἀπὸ τῆς ἐμφράξεως καὶ πληρώσεως εὐρυνομένη τὸ μῆκος μειοῖ καὶ σπᾷ ἐπὶ τὸ κάτω, καθάπερ καὶ ἔντερον πληρούμενον σαρκῶν ἢ ἄλλων σωμάτων τινῶν· αὐτὴ δὲ καθελκομένη συγκατασπᾷ καὶ τὴν κατακλεῖν· καὶ οὕτω τὴν ὀδύνην ἐπιφέρει. ἔχουσι δὲ ξηροβήχιον· ἡ γὰρ δυσκρασία τοῦ ἥπατος ἀνιοῦσα πρὸς τὰ ἀναπνευστικὰ ἐπεγείρει ταῦτα πρὸς ἀπόκρισιν καὶ ποιεῖ ἐπ' αὐτὰ βῆχα.*

1.21. *Διὰ τί ἡ μήτρα παρὰ πάντα τὰ μόρια ἔνδοθεν τραχεῖα γέγονεν, ἐξοχὰς ἔχουσα ἃς καλοῦσι κοτυληδόνας; διὰ τὴν ἀντοχὴν τῶν σπερμάτων· πᾶν γὰρ ὑγρὸν εὐόλισθόν ἐστιν ἐν λείῳ σώματι· οὕτω γὰρ καὶ οἱ κονιῶντες τοὺς τοίχους προσκολάπτουσιν αὐτοὺς πρὸς τὸ τὰ κοιλώματα γίνεσθαι τῆς*

1.19 (MNOPRSTbeim*p*w Buss. Us.; *g* 2.96)
1 *γίγαρτα* ex *γίγαντα* corr. m || *σίσαμα* N^{ac}Oi || *αἱ κεγχραμίδες* M*p* Us. || *κεχραμίδες* R 2 *κατὰ πινόμενα* R 3 *ὅτι* add. sup. lin. T || *οὐ περιέχουσι* P 4 *γεννητικῆς* ex *σπερματικῆς* corr. m || *σφαιροειδεῖ* O 5 *τελεώτερον* Buss. 5-7 *παρὰ τῷ θεῷ – ὅμοια* om. *g* 6 *τὰ* scr. sup. lin. O || *δὴ*] *δὲ* Buss. || *τὰ τετράγωνα* Buss. || *καὶ κυβοειδῆ* m^{ac} 7 *ὠοειδῆ*] *ὠοδῆ* NOSbi || *φακοειδῆ*] *φακκοειδῆ* b: *φαλκοειδῆ* NOi: *κοκοειδῆ* P
1.20 (BLMNOPRSTbeim*p*w Buss. Us.; *g* 2.97)
1 *τοῦ* om. BL || *ἥπατος*] *ὕπατος* e 2 *ἡ* om. Oi, add. supra lin. R || *κυρτῶν* ex *κουρτῶν* corr. O || *ἥπατος*] *ὕπατος* e: *ἤπατος* O 3 *πρὸς*] *παρὰ* Tmw 3-4 *τὸ μῆκος μειοῖ*] *gravius ponderat g* 4 *ἐπὶ τὸ*] *ἐπί τι* BL || *καὶ*^{2} om. Buss. || *τὸ ἔντερον* Buss. Us. || *πλειρούμενον* Oi 5 *αὔτη* M Buss. Us: *αὐτῆ* e || *καθελκομένη* P*g* Us.: *καὶ ἑλκομένη* Tmw: *ἑλκομένη* cett. Buss. 6 *ἐπιφέρει*] *φέρει* NOi: *φέρειν* BL 7 *ἥπατος*] *ὕπατος* e
1.21 (BELMNOPRSTXbceim*p*w Buss. Us.; *g* 2.98)
2 *κοτυληδόνας*] *κοτυλυδόνας* BL: *κοτυλιδόνας* ETXc || *κοτυλιδόνας καλοῦσι* T || *ἀντυχὴν* B: *ἀντοχης* N 3 *ὑγρὸν* om. Ec || *εὐόλιστον* B || *κονιῶντες*] *χρείοντες* EXc 4 *προσκολάπτουσιν* MP*g*w Buss. Us.: *προσκωλάπτουσιν* B: *προκολάπτουσι(ν)* LNORSTXbc^{pc}eim: *πρὸκολάπτουσιν* sic E: *προκόπτουσι* c^{ac}: *κολάπτουσι* (ante *τοὺς τοίχους*) *p* || *τὸ*] *τῶ* N^{ac}Tw: om. BL || *τὰ* om. EXc 5 *κονιῶν*] *κοιλῶν* BL ||

1.19. Why are grape-pips and sesame seeds and fig-seeds, which are sound[205] when swallowed, excreted again through the gut without being digested?[206] Is it that they are not digested because of their considerable smallness and dryness? Or is it because they contain the perfection of generative sowing (which is why they are spherical), and for that reason remain intact? For this shape is more perfect in god's eyes than all [other] shapes, I mean square cubic pyramidal cylindrical egg-shaped lens-shaped and the like.

1.20. Why when the convex parts of the liver are obstructed is the collar-bone pulled downwards? Because the portal vein going up from the convex parts of the liver towards the collar-bone is widened by the blocking and [consequent] filling-up, and so reduces in length and pulls downwards, just as does the gut when filled with meat or other bodies. [The portal vein] being dragged downwards itself pulls the collar-bone down along with it and so produces the pain. They have a dry cough; for the bad tempering of the liver goes up to the respiratory [parts], arouses these to elimination [of the excess] and causes a cough in them.

1.21. Why is the womb rough inside more than all the [other] parts, having protuberances which they call cotyledons? For the seed to cling to; for everything fluid slides easily on a smooth body. For in this way those who plaster walls make peck-marks in them first so that the recesses can become

---

205 Psellus' summary has: "raw".

206 As is noted by Erasistratus fr.132 Garofalo, cited by Galen, *De locis affectis* 1.3 (8 34.15 Kühn).

διαμονῆς τῶν κονιῶν ἀσφαλίσματα· τούτου γὰρ χάριν καὶ αὐτὸ τὸ σπέρμα κολλῶδες κατεσκεύασεν ἡ φύσις πρὸς ἀντοχὴν τῆς διαμονῆς.

1.22. Διὰ τί ἐν ἀρίστῳ πολλοὶ ὑπερπλησθέντες οὔρου καὶ αἰδεσθέντες ἐξουρῆσαι παρελύθησαν καὶ ἰσχουρίαν ὑπέμειναν; ὅτι τῇ ὑπερπληρώσει σφόδρα διαταθεῖσα ἡ οὐρηδόχος κύστις καὶ ἄγαν ἐξογκωθεῖσα τὴν ἐν τῷ χιτῶνι κατὰ φύσιν ὑγρότητα ἐξέθλιψε καὶ διὰ τοῦτο κατεξήρανε ταύτην καὶ εἰς ξηρὰν δυσκρασίαν μετερρύθμισε καὶ οὐκέτι ἠδυνήθη συσταλῆναι, καθάπερ καὶ βύρσα ὑπερξηρανθεῖσα οὐκέτι ἐκταθῆναι ἠδυνήθη, εἰ μὴ ἐν ὑδρελαίῳ ἐνυγρανθείη. ἡ δὲ στραγγουρία τὸ ἐναντίον, ἥτις γίνεται διὰ παράλυσιν τοῦ μυὸς τοῦ κλείοντος τὸν τράχηλον τῆς οὐρηδόχου κύστεως ἢ δι' ἕλκωσιν αὐτῆς ἢ διὰ λίθου παρουσίαν, δακνομένης ἢ βαρουμένης καὶ ἐρεθιζομένης ἐκκρίνειν αὐτὸ τὸ πρὸς ὀλίγον συναγόμενον οὖρον ἐν αὐτῇ· δυσουρία δὲ κοινόν ἐστιν ὄνομα, τουτέστι τὸ δυσχερῶς οὐρεῖν, ὡς δυσπεψία καὶ δυσηκοΐα, καὶ δύσπνοια.

ἀσφαλίσματος E || *τὸ* om. w 6 *τῆς* add. sup. lin. O

1.22 (CGMNOPRSTbeimo*p*w Buss. Us.; *g* 2.99)

1 *ἐν ἀρίστῳ πολλοὶ*] *τινὲς* Go: *πολλοὶ* tantum *g* || *ὑπερπλησθέντες*] *ὑπερπλεισθέντες* o^{ac}: *ὑπερπληθέντες* e: *ὑπερεμπλησθέντες* M Buss. 2 *ἐξούρίσαι* G || *καὶ ἰσχουρίαν ὑπέμειναν* om. Go: *destillatione urinae laborarint g* || *τῇ* om. Go 3 *οὐρηδόχος*] *οὐριδόχος* GT: *οὐροδόχος* CMSo, Buss. || *κύστις* ex *κύστης* corr. N || *ἐξογκοθεῖσα* C 4 *χειτῶνι* C || *κατεξήρανε*] *κατεξήνε* C 5 *μετερρύθμισε*] *κατερρύθμισε* Oi || *ἐδυνήθη* Tmw 6-12 *καθάπερ – δύσπνοια* om. Go 6 *βύρσα* ex *κύρσα* corr. T || *ἠδυνήθει* C 7 *ἐνυγρανθείη* Us.: *ἐνυγρανθῇ* codd. Buss. || *ἡ δὲ στραγγουρία τὸ ἐναντίον*] *fieri autem urinae destillatio solet g* 8 *παράλησιν* C || *οὐροδόχου* CMT^{ac} Buss. 9 *διέλκωσιν* C || *δακνουμένης* CNOPSbiw 10 *ἐκρίειν* C || *ἐν αὐτῷ* P Buss.: *ἐν αὐτᾶ* M^{ac}: *ἐν αὐτῇ οὖρον* m 11 *κοινὸν ὄνομα ἐστὶν* C || *τὸ* om. C || *οὐρεῖν*] *οὐρῆν* O^{ac}T || *οὐρεῖν δυσχερῶς* P 11-12 *ὡς – δύσπνοια* om. C 12 *δυσηκοΐα*] *δυσκοΐα* NOSbim^{pc}: *καὶ δυσηκοΐα* om. *p*

securities for the remaining of the plaster. For this reason nature made the seed itself, too, sticky so that it would hold on and remain.[207]

1.22. Why do many people who are excessively full of urine at lunch-time and embarrassed to urinate suffer paralysis and retention of urine? Because the bladder that receives the urine, being greatly distended by being over-filled and excessively inflated, squeezes out the moisture that is naturally in its membrane and for this reason dries this and causes it to be excessively dry in its tempering, [so that] it can no longer contract, just as hide which has been excessively dried can no longer be stretched out, unless it is moistened in a mixture of water and oil. Strangury is the opposite,[208] coming about through paralysis of the muscle which closes the neck of the bladder which receives the urine or through scarring of this or through the presence of a stone, when [the bladder suffers] piercing [pain] or is weighed down and provoked to excrete the small amount of collected urine in it. The common name [of both] is *dusouria*, that is "difficulty in passing urine", like dyspepsia and difficulty in hearing (*dusêkoïa*) and difficulty in breathing (*dyspnoia*).

---

[207] Cf. Lucretius, 4.1242.

[208] That is: difficulty in passing urine can be caused either by a lack of elasticity in the bladder, so that it cannot contract and expel the urine, or by a blockage in the outlet, "strangury" being the proper name for the latter.

NATURAL PROBLEMS[209]

<BOOK II>

2.1.[210] Why do those with a flux in their lower belly become very thirsty? Because when what is moist is carried down the region of the stomach becomes dry; and thirst is according to dryness.

2.2.[211] Why do those who are grieved cry? Because grief is an affliction of the *pneuma*: so it rises up to the membrane, and squeezes out in quantity the moisture present there. The same thing happens also to those who rejoice.[212]

2.3.[213] What is the difference between a *spadôn* and a eunuch?[214] That the eunuch has been deprived of his testicles while an adolescent, but the *spadôn* when he has already advanced further in age.

---

209 The title appears as follows in the MSS: *Aristotle, natural problems book 4* (MPRT); *Aristotle, medical problems and natural puzzles, volume 3, or Aristotle, natural problems volume 4* (em: e omits "or"); *Aristotle, natural questions* (k). b starts a new numbering of problems but gives no title; NOSi treat this book as a continuation of the previous one.

210 = *Probl. Bamberg*. 1.

211 = *Probl. Bamberg*. 2.

212 Cf. [Alex.] *Probl.* i.31, which argues that weeping through grief is through compression of channels *(poroi)* in the eyes squeezing out moisture, weeping through laughter due to their relaxation; Adelard of Bath, *Quaest. Nat.* 32 p.156 Burnett, which gives the two explanations for weeping through laughter that the soul "leaps up" through joy and that heat is diffused through the body and expels moisture from the brain; and *Quaestiones Salernitanae* Ba63 (p.179 Lawn 1979), which says that in laughter there is a boiling up of the humours in the head. We are grateful to Charles Burnett for the latter two references.

213 = *Probl. Bamberg*. 3.

214 Photius, *Lexicon* sv *eunouchos* (1 36.26 Porson) says rather that *eunouchos* is the Attic form, *spadôn* the non-Attic. In the Byzantine period the terms could be used as equivalent, but *spadôn* was also used in contrast to *castrati*; we are grateful to Georges Sidéris for this information. See Psellus, *Poemata* 8.1301-3 (173 Westerink); Suda, s.v. *spadôn*, σ 897.54 (4 .413.28-32 Adler). For the contrast drawn in the present passage, but not the terminology, compare (with a fuller explanation) [Alex.], *Probl.* i.9 Ideler; we are grateful to Lou Filius for this reference.

φυσικὰ προβλήματα

2.1. *Διὰ τί οἱ τὴν κάτω κοιλίαν ῥευματιζόμενοι κατάδιψοι γίνονται; ὅτι τῶν ὑγρῶν κάτω φερομένων ἀναξηραίνεται τὰ πρὸς τῷ στομάχῳ· κατὰ ξηρότητα δὲ τὸ δίψος.*

2.2. *Διὰ τί δακρύουσιν οἱ λυπούμενοι; ὅτι πνεύματος πάθος ἐστὶν ἡ λύπη· ἀνατρέχει οὖν ἐπὶ τὴν μήνιγγα, καὶ ἀθρόως ἐκθλίβει τὸ παρακείμενον ὑγρόν· τὸ δ' αὐτὸ καὶ τοῖς χαίρουσι γίγνεται.*

2.3. *Τίνι διαφέρει σπάδων εὐνούχου; ὅτι ὁ μὲν εὐνοῦχος μειράκιον ὢν τῶν διδύμων ἐστέρηται, ὁ δὲ σπάδων προκόψας ἤδη τὴν ἡλικίαν.*

---

2 Tit. *Ἀριστοτέλους φυσικὰ προβλήματα τοῦ* (*τὸ* R, *τόμος* T) *δʹ* MRT: *Ἀριστοτέλους φυσικῶν προβλημάτων τὸ δʹ* P: *ἀριστοτέλους ἰατρικῶν προβλημάτων καὶ φυσικῶν ἀπορημάτων ἐκλογαὶ τόμος τρίτος* (*τὸ γ* e)· *ἢ* (om. e) *ἀριστοτέλους* (*ἀριστόλους* m) *φυσικὰ προβλήματα τόμος τέταρτος* (*τοῦ δ* e) em: *ἀριστοτέλους φυσικὰ ἐρωτήματα* k: novam seriem problematorum sine titulo exhibet b: hunc librum et praecedentem in unum coniunguunt NOSi

2.1 (BLMNOPRSTbeikm*p* Buss. Us.; *g* 2.100; *r* 1)

1 *οἱ* om. NOi 2 *κάτω φερομένων*] *καταφερομένων* BLk || *ἀναξηραίνονται* P || *τὰ* om. BL || *τὸν στόμαχον* k 2-3 *κατὰ ξηρότητα*] *καταξηρότητα* BLe[pc]: *καταξηρότητος* e[ac]

2.2 (CGMNOPRSTbeikmo*p* Buss. Us.; *g* 2.101; *r* 2)

1 *ὅτι* P*r* Us: *διότι* Go: om. cett., Buss. 2 *μίνιγγα* GT[pc]: *μηνίγκαν* k: *μήνυγγα* m: *μήνιγκα* o: *μίνιγκα* T[ac] || *ἀθρόως*] *ἀθρῶος* k || *ἐκθλίβει*] *ἐκκρίνει* k 2-3 *τὸ παρακείμενον ὑγρόν*] *τὸ πάθος τὸ πᾶσι κείμενον ὑγρὸν* k

2.3 (BCLMNOPRSTbeim Buss. Us.; om. *g*, sed cf. ad 2.5; *r* 3; memorare videtur Petrus ap. Olivieri 1988, 196)

1 *μυράκιον* C: *μειράκειον* BT || *ὢν*] *ὂν* BLNOTbi 2 *δυδοίμων* C || *ὑστέρηται* BL

2.4. Διὰ τί οἱ σπάδοντες οὐ φέρουσι γένεια; ὅτι ὑστερεῖ ἐπ᾿ αὐτῶν τὸ θερμόν, ὑπὸ δὲ θερμασίας αὔξονται αἱ τρίχες.

2.5. Διὰ τί οἱ μὲν εὐνοῦχοι ὀξύφωνοι, οἱ δὲ γάλλοι οὔ; Ὅτι οἱ μὲν στενοὺς ἔχοντες τοὺς πόρους ἐκτέμνονται, οἱ δὲ ἤδη εὐρυτέρους ἀποκόπτονται· διὰ στενοῦ δὲ πνεῦμα φερόμενον λεπτὸν ἠχεῖ, δι᾿ εὐρυτέρου δὲ βαρύτερον.

2.6. Διὰ τί καὶ μετὰ τὴν ξηρὰν καὶ ὑγρὰν τροφὴν ἐρευγόμεθα; ὅτι κτᾶται τούτοις πνεῦμα· τῆς οὖν ἀσθενεστέρας τροφῆς κάτω χωριζομένης λεπτομερὲς ὂν τὸ πνεῦμα ἐπὶ τοὺς ἄνω τόπους ἑαυτὸ κουφίζει.

2.7. Διὰ τί πλέον τινές τινων συνουσιάζουσιν; ὅτι μᾶλλόν τινές τινων θερμότεροι.

2.8. Διὰ τί οἱ ἀπὸ φωτὸς εἰς συνηρεφεῖς οἴκους εἰσελθόντες οὐχ ὁρῶσιν εὐθέως; ὅτι τὸ μέλαν παχὺ ὂν ἀποκρύπτει τὴν βλεπτικὴν οὐσίαν ταύτην· ἐπιμείνασα γοῦν ἀνύει.

2.4 (BCLMNOPRSTbeikm Buss. Us.; *g* 2.102; *r* 4; paraphrasin tantum habet *p*)
1 *σπάδοντες*] *εὐνοῦχοι* k || *ὑστερεῖ*] *ὑστεροῖ* CTe[pc]: *οἱ* | *στεροὶ* k || *ἐπ᾿ αὐτῶν*] *ἐπ᾿ αὐτὸν* CT: *ἐπ᾿ αὐτοὺς* BL: *ἐπ᾿ αὐτῶ* P 1-2 *τὸ θερμόν*] *τῶν θερμῶν* k 2 *ὑπὸ δὲ θερμασίας*] *ὑπὸ δὲ θερμανσίας* RSb: *ὑπὸ θερμασίας γὰρ* BCL*r* Buss. || *τρίχες*] *αἱ τρίχες* Us.: *τρίχαι* k
2.5 (MNOPRSTbeikm*p* Buss. Us.; *g* 2.103)
1 *μὲν* om. k || *εὐνοῦχοι*] *qui pueri execti sunt g*, ex 2.3 || *οἱ δὲ – ὅτι οἱ μὲν*] *ὅτι* tantum k || *γάλλοι*] *γάλοι* N[ac]S: *σπάδοντες* T: *qui adulti iam* (sc. *execti sunt*) *g*, ex 2.3 2-3 *οἱ δὲ ... ἀποκόπτονται* om. gk, add. in marg. T 3 *φερόμενον* post *ἠχεῖ* exhibet k 3-4 *δι᾿ εὐρυτέρου δὲ βαρύτερον* om. *g* 4 *βαρύτερον*] *ταχύτερον* k
2.6 (BLMNOPRSTbeim*p* Buss. Us.; *g* 2.104; *r* 7)
1 *καὶ*[1] om. BM, Buss. 1-2 *ὅτι χωριζομένης*] *quia spiritus vel cum cibo admittitur, vel post ex concoctione movetur g*: *quia permiscitur* tantum *r* 2 *κτᾶται τούτοις*] *τούτοις κτᾶται* B || *τῆς*] *τοῖς* BLT || *ἀσθενεστέροις* BL || *κάτω χωριζομένοις* BL: *καταχωριζομένης* P 3 *τὸ πνεῦμα* om. *r* cod. antiquissimus || *καὶ ἑαυτὸ* O[ac]: *ἑαυτῶ* BL
2.7 (MNOPRSTbeikm Buss. Us.; *g* 2.105; *r* 9: paraphrasin tantum habet *p*)
1 *τινῶν τινὲς* k || *τινές*[2]] *τιναῖς* k 1-2 *θερμότεροι τινῶν θερμότεροι* Oi
2.8 (MNOPRSTbeim*p* Buss. Us.; *g* 2.106; *r* 10)
1 *ἀποφωτὸς* m || *συνερεφεῖς* m || *οἴκους*] *τόπους* m 2 post *εὐθέως* add. *quoniam lux mediocris prop[r]ius ad tenebras, quam amplissima illa* (sc. *lux*) *accedit* g || *ἀποκρύπτον* PRTe[ac]m || *τὴν βλεπτικὴν οὐσίαν ταύτην*] *vim conspiciendi g*: *vivificum spiritum r* 2-3 *ταύτην· ἐπιμείνασα* om. NOSbig: *ἐπιμείνασα* om. M: *ταύτῃ ἐπιμεῖναν* coni. Buss. 3 *γοῦν ἀνύει* PTm*r* Us.: om. cett. Buss. || totam solutionem ita exhibet *r*: *quia niger in oculos constitus humor inpingatur si vivificum spiritum et vivificum spiritus tenuens manens aperiens porus*

2.4.[215] Why do *spadônes* not have beards? Because the hot in them fails, and hairs grow because of heat.

2.5.[216] Why do eunuchs have high voices, but *galli*[217] not? Because the former are castrated when the passages are narrow, the latter when they are already wider. The breath when it passes through a narrow passage has a high sound, through a wider one a deeper.

2.6.[218] Why do we belch after both dry and liquid nourishment? Because *pneuma* is acquired through these?[219] When the weaker nourishment is separated downwards the *pneuma* which is rare lifts itself up to the upper regions.

2.7.[220] Why do some people have intercourse more than others? Because some are hotter than others.

2.8.[221] Why do those who go from the light into shaded houses not see at once? Because the dark, which is dense, conceals this substance which enables vision. At any rate, after waiting it becomes effective.

---

215 = *Probl. Bamberg*. 4. Cf. Macrobius, *Sat.* 7.7.8 (Linke 1880, 54 n.2; also cited by Lawn 1979 222, on the next text) and, in much more detail, *Quaest. Salern.* P43 (pp.222-223 Lawn 1979: lack of heat narrows the pores through which the hairs emerge).

216 Cf. [Alex.] *Probl.* i.8 and i.97, and Macrobius, *Sat.* 7.10.13 (cited by Linke 1880, 54 n.2) which argue that moisture narrows the windpipe of eunuchs and makes their voice higher, but do not have the contrast with *galli*; *Probl. Bamberg.* 5, which similarly refers only to eunuchs, not *galli,* and where a reference to narrowing of channels is less explicit in the oldest MS, A; *Quaest. Salern.* P40 (p.221 Lawn 1979) which cites Galen for lack of heat as the cause, with no reference to the windpipe. Aristotle himself explains the phenomenon rather by castration weakening "the part which moves the air", because the testicles act like loom-weights and their removal reduces the tension of the blood-vessels (*GA* 5.7 787b20-788a15). Aristotle's theory does not, straightforwardly at least, explain the specific point in our text, which is why the effect depends on the age at which castration occurs.

217 Castrated priests of Cybele. Georges Sidéris points out to us that this reference would be anachronistic after the fifth century AD.

218 = *Probl. Bamberg.* 7.

219 "along with these", Buss. The passive use of *ktasthai* in the present seems late.

220 = *Probl. Bamberg.* 9.

221 = *Probl. Bamberg.* 10. Cf. [Alex.] *Probl.* ii.37, which appeals to the darkness making the visual *pneuma* dense and turbid, or else to the suddenness of the change; *Quaest. Salern.* W3 (p.267.6-8 Lawn 1979), saying that we cannot see again until the scattered rays of visual *spiritus* have gathered together once more. See above, Introduction nn.81, 116.

2.9. Διὰ τί γυναῖκες τὴν ὑστέραν παθοῦσαι παρακόπτουσιν; ὅτι νευρώδης οὐσία, ἔχει δὲ τὰς ἀρχὰς ἀπὸ στομάχου καὶ μηνίγγων, ὅθεν καταπέπλεκται.

2.10. Διὰ τί ἔμμηνα γίνεται; ὅτι ταῖς θηλείαις ἀργότερον βίον ζώσαις τὸ περιττὸν σῶμα πολὺ πλέον, πολὺ δὲ τὸ αἷμα πανταχοῦ· τὸ περισσὸν οὖν ἐκκρίνεται κουφισμοῦ χάριν.

2.11. Διὰ τί ἐπὶ τοῖς λοιμοῖς ὡς ἐπὶ τὸ πλεῖστον ὀφθαλμίαι προκατάρχονται; ὅτι τοῦ πνεύματος πεῖσίς ἐστιν ὁ λοιμός· τροπῆς οὖν περὶ αὐτὸ γινομένης ἄνω φερόμενον τὸ πνεῦμα, τὴν ὅρασιν ταράσσει· ἐκεῖ γὰρ μᾶλλόν ἐστι λεπτὸν καὶ εὐρούστερον.

2.12. Διὰ τί οἱ τρομικοὶ πίνοντες περιφέρονται; ὅτι πολύθερμος ὁ οἶνος· πολὺς οὖν γενόμενος τοῖς νευρώδεσιν ὑπεισέρχεται καὶ ταράσσει τὸ πνεῦμα τὸ ἐν αὐτοῖς, τοῦτο δὲ ἀπὸ μηνίγγων φέρεται.

2.9 (BLMNOPRSTbeim*p* Buss. Us.; *g* 2.107; *r* 8)
1 *αἱ γυναῖκες* BLm Buss. Us.: *γυναῖκα* P || *παθοῦσαι*] *ποθοῦσαι* BL: *παθοῦσαν* P || *παρακόπτουσιν*] lacunam habet *r* cod. antiquissimus: *παρασκόπτουσιν* O[ac] || *νευρώδεις* B 2 *οὐσία*] *οὐσίας* BL: *οὐσία ἐστὶν* R Buss: *ἐστὶ* M || *ἀπὸ τοῦ στομάχου* NOi || *μήνιγγαν* B: *μινίγκων* T || *ὅθεν* om. BL 2-3 *ὅθεν καταπέπλεκται* om. *r*
2.10 (BLMNOPRSTbeikm*p* Buss. Us.; *g* 2.108; *r* 11)
1 *καὶ διατί* k || *ἔμμηνα*] *ἔμηνα* BL: *ταῖς θηλίαις ἔμμινα* k || *θηλείαις*[2]] *θηλύαις* e: *θηλίαις* k || *ἀργότερον*] *ἀργώτερον* k || *ἀργότερον βίον ζώσαις*] *naturae plus pinguioris* (sic) *sunt r* || *ζώσας* B: *ζώσες* k 2 *περιττὸν σῶμα*] *περίττωμα* Buss. Us.: *περὶ πάντων σῶμα* k || *τὸ περισμὸν οὖν* O[ac] || *οὖν*] *ὃν* NOgi: *ὃν* BL
2.11 (MNOPRSTbeim*p* Buss. Us.; *g* 2.109)
2 *πεῖσίς ἐστιν*] *πίσης ἐστὶν* m: *πῆσις ἐστὶν* T 2-3 *περὶ αὐτὸ* add. in marg. T 3 *γενομένης* m || *ὅρασιν* NOi 4 *ἐστι* om. NOi
2.12 (BLMNOPRSTbeim*p* Buss. Us.; *g* 2.110; *r* 17)
1 *περιφέρονται*] *παραφέρονται* B Buss.: om. *r* qui sic vertit: *quare multum bibentes vinum tremuli facti sunt* || *πολὺ θερμὸς* MPRTem 2 *γενόμενος*] *γινόμενος* MNORSTbei || *ταράσση* BL 3 *τὸ* om. T || *ἀπὸ μηνίγγων*] *ἀπὸ μινίγγων* Tg: *ἐπὶ μηνίγγων* B: *ἐπὶ μηνίγκων* L: *ἐπὶ μήνιγγας* Buss.: *hunc* ... in *mininga* ... *contingit r*

2.9.[222] Why do women who are afflicted in their womb become deranged?[223] Because it is a sinewy substance, having its origins in the stomach and the membranes from which it is interwoven.

2.10.[224] Why do menses occur? Because females live a life with less exercise and have much more residual matter,[225] and much blood everywhere.[226] So the excess is excreted for the sake of relief.

2.11. Why do eye diseases for the most part begin in times of plague? Because plague is an affliction of the *pneuma*. So when a change occurs concerning it the *pneuma* is carried upwards and disturbs the vision; for there it is wider and flows more easily.

2.12.[227] Why are people who are inclined to tremble unsteady on their feet when drunk?[228] Because wine is very hot; so when there is a lot of it enters into the nerves and disturbs the *pneuma* in them, and this is carried away from[229] the membranes [of the brain].

---

222 = *Probl. Bamberg*. 8.

223 We are grateful to Lou Filius, whose translation of the start of this problem in the Arabic version has helped to clarify its meaning.

224 = *Probl. Bamberg*. 11.

225 Literally "residual body".

226 Aristotle attributes menstrual blood to the natural coldness and weakness of women, but *not* to their less active life: *GA* 1.19 726b31.

227 = *Probl. Bamberg*. 17. Cf. Theophrastus, *Vert.*1 and 11 (part of the latter = 2.43 below), though not mentioning *heat* in connection with wine. Conversely *Quaest. Salern.* P21 (p.216 Lawn 1979), compared by Lawn with *Probl. Bamberg*. 17, asks why wine which is held to be hot produces effects linked with cold, including trembling, and replies that it is not the heat but the quantity of wine that is responsible, because the moisture in it overcomes the heat that is needed for digestion, and rises to the brain. Lawn also cites [Ar.] *Probl.* 3.1, 3.6, which similarly appeals to the effect of moisture overcoming heat, but without specifically mentioning trembling, and Macrobius, *Sat.* 7.6.9, which mentions trembling but in the context of an argument that wine should in fact be regarded as *cold*.

228 *Probl. Bamberg*. just has "why do those who drink much become unsteady?". Lou Filius informs us that the Arabic version of this problem (for which see his forthcoming edition) makes the question why those who tremble become *steady* on their feet when drunk, which has more point to it as a problem.

229 On the face of it Bussemaker's emendation "to the membranes"gives a better sense, but the evidence of the MSS is against it, since the two (late) MSS that do have *epi* agree with the other MSS in having the genitive. Theophrastus locc.

2.13. Διὰ τί τὰ μὲν ἄλογα ζῷα τεχθέντα περιπατεῖ εὐθέως, οἱ δὲ ἄνθρωποι οὔ; ὅτι ἐπὶ μὲν τῶν ἀλόγων ἴσον ἐν παντὶ μέρει τὸ θερμόν, ἐπὶ δὲ τῶν βρεφῶν πλεῖον περὶ τὴν κεφαλήν· ὅταν οὖν ὁμαλισθῇ πανταχοῦ τὸ θερμόν, ἀνίσταται.

2.14. Διὰ τί οἱ σπώμενοι πυρέξαντες ἀπαλλάσσονται τοῦ σπᾶσθαι; ὅτι θλιβομένου τοῦ ἐν τοῖς νεύροις πνεύματος καὶ ψυχομένου ὁ σπασμὸς γίνεται, ὑπὸ θερμασίας δὲ λύεται, θερμὸς δὲ ὁ πυρετός.

2.15. Διὰ τί ἐπὶ ταῖς χολαῖς οἱ μέλαιναν χολὴν ἐμέσαντες κινδυνώδεις; ὅτι μηκέτι τῆς χολώδους ὑποστάσεως οὔσης τὸ ἀπὸ τοῦ ἥπατος αἷμα ἕλκεται.

2.16. Διὰ τί ἀπουρήσαντες ἀποπνέομεν; ὅτι πληρωθεῖσα ἡ κύστις,

---

2.13 (BCGLMNOPRSTbeikmo*p* Buss. Us.; *g* 2.111; *r* 18)
1 *ἄλογα* om. k || *τεχθέντα*] *τεχθέντα ἀθρόως* o: om. k || *περιπατεῖ εὐθέως*] *εὐθέως περιπατεῖ* G: *ἀθρόως περιπατεῖ* o 2 *οὔ*] *οὐχὶ* k || *μὲν* om. Oi || *ἀλόγων ζώων* Co || *ἴσον* om. *r* || *ἐν*] *ἐπὶ* Go || *μέρει*] *μέλει* BL 3 *τῷ βρεφῶν* m || *πλείω* G: *πλέον* C: *τὸ πλεῖον* k: *solum r* || *περὶ*] *ἐπὶ* Ck || *κεφαλὴν ἐστὶν* k 4 *ἀνίσταται*] *ἀνίστανται p* Us. (*τὰ βρέφη* tamen neutri generis est): *συνίσταται* C
2.14 (BLMNOPRSTbeim*p* Buss. Us.; *g* 2.112; *r* 28)
1 *ἀπαλλάσονται*] *ἀπαλλάττονται* T: om. BL: *bonum est r* || *τοῦ σπᾶσθαι*] *τοῦ πάθους* T: om. *r* 2 *νεύροις*] *pori r* 3 *θερμανσίας* Ni: *ὑποθερμανσίας* O || *λύεται*] *γίνεται* BL
2.15 (BCLMNOPRSTbeikm*p* Buss. Us.; *g* 2.113: *r* 21)
1 *ἐπὶ ταῖς χολαῖς* om. k*r* || *οἳ* C || *μέλαινα* BLT: *τὴν μέλαιναν* k || *ἐμέσαντες*] *αἱμέσαντες* C || *κινδυνώδης* BL: *κινδυνεύουσι* k 2 *οὐκέτι* k || *ἀποστάσεως* T: *non sedit ad epate r* || *ἥπατος*] *ὕπατος* e
2.16 (MNOPRSTbeikm*p* Buss. Us.; *g* 2.114)
1 *ἀποπνέομεν* Buss. (in var. lect.) Us.: *spiritus alvo excedere soleat g*: *ἀναπέρομεν* sic *p*: *ἀναπνέομεν* cett. || *πληρωθείσης τῆς κύστης* k 2 *ἀπηυθυσμένον*

2.13.[230] Why do irrational animals walk around as soon as they are born, but humans do not? Because in the case of irrational animals the heat is equal in every part, but in the case of [human] infants it is more around the head. So when the heat is made even everywhere, they stand up.

2.14.[231] Why are those who suffer spasm relieved from it when they suffer fever? Because spasm occurs when the *pneuma* in the sinews is crushed and chilled, but it is released by heating, and fever is hot.

2.15.[232] Why in the case of bile are those who vomit black bile in danger?[233] Because the bile-like substance is exhausted and the blood from the liver is drawn out.

2.16. Why do we pass wind[234] after urinating?[235] Because when the bladder

---

citt. speaks of alien *pneuma* disturbing and separating the movements of the *pneuma* in the brain; he does not specifically mention movement of it towards or away from the membranes. The explanation in the Arabic version seems to be that the vapour from the wine blocks the passage of the psychic *pneuma* to the rest of the body and prevents it from causing the trembling. For *pneuma* causing trembling cf., perhaps, [Ar.] *Probl.* 5.15.

230 = *Probl. Bamberg*. 18. Cf. Adelard, *Quaest. Nat.* 37, giving an explanation in terms of the superiority of human beings in being rational; *Quaest. Salern.* B228 (p.115-116 Lawn 1979), which gives an explanation in terms of the influence of menstrual blood on the embryo. (We are grateful to Charles Burnett for the latter two references). Cf. Lawn 1963, 39, 153-4.

231 cf. *Probl. Bamberg*. 28. Hippocrates, *Aph.* 2.26 (4 478.4-5 Littré), says that it is better for fever to follow spasm than vice versa; cf. also id. 4.57 (4 522.12-13 L.), 5.5 (4 534.1-3 L.), *Coac.* 2.348, 352 (5 658.10,14-18 L.), *Loc. hom.* 39 (75.23-4 Craik), *Iudic.* 61-2 (9 294.7-10 L.), *Ep.* 24 (9 398.20-21 L.). Galen, *In Hippocr. Aph.* 2.26 (17.2 514.11-515.4 Kühn), explains *Aph.* 2.26 by arguing as here that fever dissipates the sticky and cold humours that cause the spasm. Cf. also id. 4.57 (17.2 636.2-6 K.), 5.5 (17.2 788.6-7 K.), 6.54 (18.1 93.1-5 K.) ; *In Hippocr. Prorrhet.* 2.49 (*CMG* 5.9.2 90.15-20 Diels), *In Hippocr. Epidem. 6* 6.8-9 (*CMG* 5.10.2.2 350.4-6, 351.5 Wenkebach and Pfaff); [Galen], *De optima secta ad Thrasybulum* 13 (1 138.1-3 Kühn); Palladius *In Hippocr. Epidem. 6* 3.6 (2 79.5-13 Dietz); Theophilus Protospatharius *In Hippocr. Aph.* 2.26, 4.57, 5.5, 5.6 (2 321.1-3, 420.2-6, 441.4-6, 442.9-10 Dietz).

232 = *Probl. Bamberg*. 21.

233 This seems to be the required sense, though *kindunôdês* = "dangerous".

234 Usener's emendation; so Gaza and Bussemaker in his translation, though not in his text. For *apopnein* in this sense cf. *Lexica Segueriana* 6 (*Collectio vocum utilium*) s.v. *apopsophein* (439.29-30 Bekker), Photius, *Lexicon* s.v. *apopsophein*

ὑποκείμενον τὸ ἀπηυθυσμένον ἔντερον θλίβει· ἀνεθείσης οὖν αὐτῆς τὸ ἐπισχεθὲν ἐν τῷ ἐντέρῳ πνεῦμα ὀλισθαίνει.

2.17. Διὰ τί χασμησάμενοι ἢ πολὺ οὐρήσαντες ὑποφρίττομεν; ὅτι τοῦ θερμοῦ πνεύματος, μέρος τι συνεκτρέχει· ἀκολούθως οὖν ἐκ παραδρομῆς ὑποψυχόμεθα.

2.18. Διὰ τί ἐπὶ τῶν λευκῶν ἢ ἀλφῶν αἷμα οὐχ ὑπακούει κεντουμένων; ὅτι ἐκ τῶν κατὰ τὴν ἐπιφάνειαν πόρων τὸ θερμὸν ὑποκεχώρηκεν ἢ διέπνευσεν.

2.19. Διὰ τί αἱ συλλαβοῦσαι κισσῶσιν; ὅτι τῶν παρὰ φύσιν ἐστὶν ἡ σύλληψις, τὸ δὲ πάθος περὶ τὴν μήτραν· νευρῶδες δὲ τὸ σπλάγχνον· στομάχου δὲ πάθος ἡ κίσσα, ὥσπερ ὄρεξις καὶ ἀνορεξία.

T^{pc}*p* Us.: *ἀπηυθισμένον* NO^{pc}PST^{ac}bi: *πηυθισμένον* O^{ac}: *ἀπευθυσμένον* k Buss.: *ἀπηνθισμένο* Me: *ἀπηνθισμένον* m: *ἀπηνθυσμένον* R || *ἀνεχθείσης p* || *τὸ*[2]] *τῶ* k 3 *ἐν* om. POi^{pc}k || *τῶ* om. Oi^{pc} || *ὀλισθάνει* T: *effluit cum sono g*

2.17 (MNOPRSTbeikm*p* Buss. Us.; *g* 2.115)

1 *ἢ*] *οἱ* k || *ὑποφρίττομεν*] *φρύττομεν* T 1-3 *τοῦ θερμοῦ – ὑποψυχόμεθα* omissis ad *ἐκ τῶν* (2.18.2) transilit k 2 *τι* Buss. Us.: *τί* codd. || *συνεκτρέχει*] *συντρέχει* M Buss. 3 *ὑποψυγόμεθα* RTe^{ac}m: *ὑποψύγομεν* M

2.18 (MNOPRSTbeikm*p* Buss. Us.; *g* 2.116)

1 *διὰ τί – ὅτι* om. k, cf. ad 2.17.1-3 || *αλφῶν* O: *ἀλοῶν* P || *οὐχ* om. m || *οὐχ ὑπακούει*] *οὐχ ὑπεκρέει p* || *κεντουμένων p* Us.: *κεντούμενον* NOim Buss.: *κεντώμενον* MPRSTbe

2.19 (BLMNOPRSTbeim Buss. Us.; *g* 2.117; *r* 23)

1 in margine *hunc appetitum sive morbum malaciam vocat Plinius* [*Nat. hist.* 23.105, 107] *g* || *παραφύσιν* T 2 *σύληψις* BL || *τὸ δὲ πάθος περὶ* (*παρὰ* T^{ac}) *τὴν μήτραν* PTm: add. N in marg.: *quia contra naturam concipiunt p a s s i o n e m i n m a t r i c e m et quia nervosa sunt eius viscera r*: om. BLMORSbeig Buss. || *νευρῶδες δὲ τὸ σπλάγχνον* om. *g* 3 *ὥσπερ – ἀνορεξία* om. *r*

is full it compresses the rectum which lies beneath it. So when [the bladder] is released it the *pneuma* which was trapped in the gut slips out.

2.17. Why do we shiver slightly after yawning or passing much urine?[236] Because part of the hot *pneuma* passes out with it. Consequently we are slightly chilled as a result of its passage.

2.18. Why in the case of white leprosy or dull-white leprosy does blood not come[237] when they are pricked? Because the heat has withdrawn from or evaporated through the pores on the surface.

2.19.[238] Why do women who have conceived have cravings for strange food? Because the conception is unnatural, and the effect is in the region of the womb. The inward part[239] is sinewy. Craving is an affection of the

---

(α 2732, 1 251.10-11 Theodoridis).

235 A shorter version of [Alex.], *Probl.* i.33 Ideler. We are grateful to Lou Filius for pointing this out.

236 Cf. [Ar.] *Probl.* 8.8, 8.13, 33.16 (cold air enters the "veins" (so 8.8, 33.16) or the passages.).

237 Literally "respond", *hupakouei* (a term used *inter alia* of servants). p's *hupekrheei* ("flow out from under") is more likely to be a misreading of *hupakouei* than the reverse; for a reference to "flowing" is more obvious in the context.

238 = *Probl. Bamberg.* 23; cf. [Alex.], *Probl.* ii.73 (explanation in terms of the effect of the retained menstrual blood on the mouth of the stomach – cf. Galen, *De causis sympt.* 1.7, 7 134.9 Kühn, *De locis affectis* 5.6, 8 343.2 Kühn – during the second and third months, after which it is no longer a problem as it is used up by the embryo); *Quaest. Salern.* B294-295 (p.140 Lawn 1979), which argues that after the second month either the distemper produced by the retained blood is now so great that it does not produce an appetite at all, or else the body overcomes the retained blood. Galen, *De causis sympt.* 1.7 similarly argues that the cravings are caused by a bad residue (7 132.19 Kühn) and that by the fourth month this has either been removed by vomiting or concocted by the body (133.7), but also that after the second month more of the blood is used by the embryo itself (133.10).

239 I.e. the womb, for the sinewy nature of which cf. 2.9 above and [Alex.] ii.64 Ideler. The relevance of this sentence is at first sight unclear; but *Quaest. Salern.* B295 (see the preceding note) explains the fact that the effect of retained menstrual blood is felt *in the stomach* by arguing that the womb and the stomach are close to each other, cold, and *sinewy* (*nervosi*); and no.23 of the *Problemata Bambergensia*, though not specifically mentioning the retained blood, does say slightly more than our present text, in observing that the women experience an effect in the womb and "because its tissues are sinewy, therefore the stomach is

2.20. Διὰ τί οἱ τεταρταῖοι χρονιώτεροι τῶν ἄλλων τύπων; ὅτι περὶ τὸ γεῶδες τοῦ σώματος σύγκριμα γίνεται, βαρὺ δὲ τὸ γεῶδες, ὅθεν βραδεῖς οἱ τεταρταῖοι.

2.21. Διὰ τί τῶν περὶ τὰ ἄρθρα ἀποπρησθέντων οὐ κατουλοῦται τὰ ἕλκη; ἐπί τινων τετήρηται τοῦτο, ὅτι πλείων ἐν τοῖς ἄρθροις ἡ ὕλη· φερομένη οὖν ἡ ὕλη ἐκεῖσε δριμυτέρα γίγνεται καὶ ἀναξηραίνει. ἢ ὅτι τῶν ἄρθρων τὸ πέρας χόνδρος ἐστίν; ἀπολισθαίνουσα οὖν ἡ κατούλωσις οὐ κρατεῖ.

2.22. Διὰ τί τὸ ὄμβριον ὕδωρ ἐλαφρόν ἐστι, καίπερ ἀπὸ θαλάσσης ἁρπαζόμενον, ἔστι δὲ τοῦτο βαρύ; ῥητέον ὅτι ἀναληφθὲν τοῦτο καὶ γυμνασθὲν ὑπὸ τοῦ πνεύματος λεπτύνεται καὶ ὑπὸ τοῦ περιέχοντος αὐτὸ ἀέρος γίνεται κοῦφον· τὸ δὲ κοῦφον ἄνω φερόμενον ἐκχεῖται κατὰ πλείονα συναθροισμόν.

2.23. Διὰ τί τὸ ψυχρὸν πολέμιον ὀστέοις ὀδοῦσι νεύροις; ὅτι τῆς γεώδους συγκρίσεώς ἐστι ταῦτα πάντα, ψυχρὸν δὲ τὸ γεῶδες· ἐπιψυχόμενα οὖν βλάπτεται.

2.20 (BCLMNOPRSTbeikm*p* Buss. Us.; *g* 2.118; *r* 25)
1 *καὶ διατί* k || *χρονιώτεροι* om. *r* || *τύπων*] *τούτων* C: *πυρετῶν* Mk 1-3 *ὅτι – τεταρταῖοι*] *quia terrestris est humor melancolicus putrefactus accendit et diu permanet r* 2 *σύγκριμα*] *σύγκρισις* M 2-3 *ὅθεν – τεταρταῖοι* om. k
2.21 (MNOPRSTbeim*p* Buss. Us.; *g* 2.119; *r* 30)
1 *ἀποπρησθέντων* om. *gr* 2 *ἐπὶ – τοῦτο* om. *r* || *τινων*] *τι* T^{ac}: *plerisque* g || *τετήρηται* Us.: *τετηρεῖται* Tem Buss.: *τὲ τηρεῖται* cett. || *πλεῖον* NOi 3 *ἐκεῖσε*] *ἐκεῖ* P || *δριμύτερον* Me^{ac} Buss. || *ἀναξηραίνει*] *ἀναξαίνει* $MR^{pc}e^{ac}$: *ἀναξαιξηραίνει* R^{ac}: om. *r* || *ἢ* om. *r* qui pro *καὶ ἀναξηραίνει ἢ ideo non sanatur* habet 4 *χόνδρον* O || *ἀπολισθάνουσα* Tm || *ἀπολισθαίνουσα – κρατεῖ* om. *r* || *κατούλωσιν* O: om. S
2.22 (CMNOPRSTbeim*p* Buss. Us.; *g* 2.120; *r* 34)
1-2 *καίπερ – βαρύ* om. *r* 2-3 *ἀναληφθὲν – λεπτύνεται* om. r 3 *λεπτύνεται*] *λεπτὸν γίνεται* $CNOSbe^{mg}i$ 4 *καὶ γί(γ)νεται* MPRTem || *καταπλείονα* Cem 4-5 *τὸ δὲ κοῦφον – συναθροισμόν* om. *gr* 5 *συναθροισμοῦ* O^{ac}
2.23 (DMOPRSTYbeiklmo*p* Buss. Us.; *g* 2.121; *v* 5.65)
1 *νεῦρεις* m: *νεύρεις* e^{ac}: *καὶ νεύροις* DGYko*p* 2 *συγκράσεως* D || *συγκρίσεώς ἐστι*] *natura constant g* || *τὸ*] *τῷ* k 2-3 *ἐπιψυχόμενα οὖν βλάπτεται*] *ἐπιψυχόμενα οὐ βλάπτεται* o: *ἔτι οὖν ἐπιψυχόμενα βλάπτεται* D: *frigefacta igitur distemperamentum subeunt atque laeduntur v*

stomach, like appetite and loss of appetite.

2.20.[240] Why are quartan fevers more long-lasting than the other types? Because the coalescence occurs in the earthy part of the body,[241] and the earthy is heavy, and as a result of this quartan fevers are slow.

2.21.[242] Why when people are burned around the joints do the wounds not form scars? In some cases this has been observed, that there is more matter in the joints; so being conveyed there the matter becomes more acrid and dries out. Or is it because the boundary of the joints is cartilage? So the scarring slips off and does not take hold.

2.22.[243] Why is rain water light, although it has been taken up from the sea, and this is heavy? We should say that when this has been taken up and exercised by the wind and by the air that surrounds it it becomes light; and the light [part], being carried up, pours [down] when there is a greater collection [of it].

2.23. Why is cold an enemy to bones, teeth and sinews?[244] Because all these are of an earthy composition, and what is earthy is cold; so they are harmed when they are chilled.

---

affected in sympathy through the sinews that extend to it". It looks very much as if our text is a compressed version of a fuller original better preserved in the *Problemata Bambergensia,* and as if the *Salernitan Question* has either preserved even more of the original or (more probably) has coupled it with the point about retained blood from another source.

240 = *Probl. Bamberg*. 25.

241 For the connection between quartan fevers and earth cf. Plato, *Timaeus* 86a, cited by Galen, *In Hippocr. De nat. hom.* 2.22 (*CMG* 5.9.1 86.8-24 Mewaldt): also Alexander of Tralles, *De febribus* 419.7,21 Puschmann.

242 = *Probl. Bamberg*. 30.

243 Cf. *Probl. Bamberg*. 34 (which simply says the heavy part is lost in the air); [Alex.], *Probl.* i.55 (see below on 2.34), which says that only the fresh part is drawn up; Cassius, *Probl.* 63 (which explains why rainwater is not *salt*; either the salt part is not drawn up from the sea, or it is but is removed by the force of the air, as here). Cf. also Sharples 1998b, 206.

244 [Alex.], *Probl.* i.50 Ideler discusses why hot water is harmful to sinews, given that cold water also is. We are grateful to Lou Filius for drawing our attention to this parallel. For *sunkrisis* in the next sentence in the sense of "composition" cf. e.g Soranus, *Gyn.* 1.22.1.2, 2.42.1.8, 5.2, of bodily consititution. MS D's *sunkrasis* is either a slip of the pen or a misguided emendation.

2.24. Διὰ τί ἡ χιὼν ψυχρὰ οὖσα καίει; καθ' ὑπερβολὴν πυκνώσεως· ἔμπυκνος γὰρ ἡ χιών, αἱ δ' ἀκρότητες ἰσότητες θερμή τε καὶ ψυχρά.

2.25. Διὰ τί οἱ φρενιτικοὶ μὲν μικρόσφυκτοι, μεγαλόσφυκτοι δὲ οἱ ληθαργικοί; ὅτι οἱ μὲν ἀραιουμένου τοῦ ἐν ταῖς μήνιγξι πνεύματος παραφέρονται, οἱ δὲ παχυνομένου βαρύνονται.

2.26. Διὰ τί τινων πυρετῶν λυομένων ὑπὸ ἱδρῶτος οἱ καρδιακοὶ διαφοροῦνται; ὅτι ἐφ' ὧν μὲν διαπνεῖ τὰ νοσοποιοῦντα, ἐφ' ὧν δὲ διαλύεται τὸ σύγκριμα.

2.27. Διὰ τί χειμῶνος πειναλέοι γινόμεθα μᾶλλον καὶ καταπέσσομεν τὴν τροφήν; ὅτι πυκνουμένου τοῦ σώματος ἔξωθεν τὸ θερμὸν εἰς τὸ βάθος ἀποκλείεται· ὑπὸ θερμασίας δὲ ἡ τροφὴ πέσσεται.

2.28. Διὰ τί οἱ πλείονα συνουσιάσαντες αἷμα ἔσθ' ὅτε ἀποκρίνουσιν; ὅτι τῆς σπερματικῆς οὐσίας ἐκκενωθείσης προχειρότερον αἵματος ὑγρὸν οὐκ

2.24 (DGMOPRSTYbeiklmo*p* Buss. Us.)
1 *ψυχρὰ οὖσα* om. T || *καίει τοῦτο ἑρμηνεία ἐστὶ* Re (sed linea subduxit e) || *καθ' ὑπερβολὴν*] *καθάπερ βολὴν* T: *διὰ τὴν ὑπερβάλλουσαν* D: *καθ' ὑπεροχὴν* o || *πυκνώσεως*] *πύκνωσιν* DGYo 1-2 *καθ' ὑπερβολὴν – ψυχρά*, i.e. totam solutionem, om. k 2 *ἔμπυκνος*] *ἔκπυκνος* SY: *πυκνοτάτη* D || *γὰρ* D Buss. Us.: *δὲ* cett. || *ἰσότητες*] *ἴσαι* D Buss. || *θερμή τε καὶ ψυχρὰ* OSbil*p* Us.: *θερμοί τε καὶ ψυχροὶ* R^{ac}Tm: *θερμαί τε καὶ ψυχραὶ* DGMPR^{pc}Yeo Buss.
2.25 (MOPRSTbeiklm*p* Buss. Us.; *g* 2.122)
1 *φρενητικοὶ* MST*p* || *μικρόσφικτοι* Oikl: *σμικρόσφυκτοι* M*p* || *μεγαλόσφηκτοι* S: *μεγαλόσφυμοι* T^{ac} 2 *ληφαργικοὶ* M: *λιθαργικοὶ* k || *ἀραιούμενοι* M || *ταῖς*] *τοῖς p* Buss. || *μήνιξι* Re*p* 3 *παραφέρωνται* k || *παχυνομένου βαρύνονται*] *βαρυνομένου παχύνονται* k
2.26 (BLMOPRSTbeilm Buss. Us.; *g* 2.123)
1 *τινων*] *τινὰς* M: *τῶν* Buss. 2 *διαφέρονται* BOi || *μὲν* om. BL Buss. || *διαπνεῖται* BL || *τὰ* om. L 3 *σύγκριμα* Buss. Us.: *σύγκριμμα* BL: *σύγκραμα* MOPRSbeilm: *σύγραμμα* T: om. *g* (*quoniam materia vitiosa partim difflatur, partim resolvitur*)
2.27 (CDGMOPRSTYbeiklm*p* Buss. Us.: *g* 2.124)
1 *πειναλέος* e: om. DGY || *γινόμεθα*] *γενώμεθα* k: om. DGY 1-2 *μᾶλλον καὶ καταπέσσομεν τὴν τροφὴν*] *μᾶλλον τὰς τροφὰς πέττομεν* DGY: *μᾶλλον τοῦ θέρους* tantum k 2 *πυκνουμένου*] *πυκνωμένου* T: *πυκνοῦ* Oi || *τῶν ἔξωθεν* k || *τὸ θερμὸν*] *θειμῶν* ut vid. k || *τὸ*^{2} om. DY 3 *ἀποκλείεται*] *ἐκλείεται* G: *κατακλείεται* DY || *ὑπὸ θερμασίας δὲ ἡ τροφὴ*] *καὶ ὑπὸ θερμασίας ἡ τροφὴ* COSbil: *ἡ τροφὴ δὲ ὑπὸ θέρμης* D: *ὑπὸ δὲ θερμασίας* k || *πέσσεται*] *ὀρεγόμεθα καὶ πέττωμεν τὴν τροφήν* k
2.28 (BCLMOPRSTbeiklm Buss. Us.; *g* 2.125; *r* 37)
1 *πλείονα*] *πλείωνα* B: *πλεῖον* Mk Buss. || *συνουσιάζοντες* km || *ἔσθ' ὅτε*] *ἔσθονται* k || *ἀποκρίνεται* m 1-3 *ὅτι – ἔστιν*] *quia semine substantiae effusum excernitur r* 2 *τῆς* om. k || *προχειρότερον* BCLOPSbil: *πρόχειρον* MRTekm ||

2.24. Why does snow, though cold, burn? By excessive compression; for snow is compressed, and the extremes are equalities, hot and cold.[245]

2.25. Why do those with inflammation of the brain have a weak pulse, those who are lethargic a strong one?[246] Because the former are deranged when the *pneuma* in the membranes becomes rarer, the latter are weighed down when it becomes dense.

2.26. Why, when certain fevers are resolved by sweating, are those who suffer from disease of the heart weakened?[247] Because in the former cases what causes disease is evaporated, in the latter what coalesces is dissolved.

2.27. Why are we more hungry in winter[248] and digest our nourishment thoroughly [then]? Because when the body becomes dense on the outside it shuts up the heat in the depths, and nourishment is digested by heat.

2.28.[249] Why do those who engage in intercourse more often sometimes emit blood? Because when the substance of seed has been exhausted there is no

---

245 Aristotle, *Meteor.* 4.4 382b8, says that cold is said to burn because it concentrates *heat*; similarly Alexander ad loc. (*In Meteor.* 202.28ff.)

246 The same question in *Problemata Hippocratica* 41, as noted by Jouanna 1997, 531.

247 Buss., reading *diapherontai* and taking *kardiakoi* to refer to "fevers" rather than to people suffering from illness, has "[fevers] of the heart are dissipated *by breathing*" (our emphasis). For "fevers of the heart" cf. Oribasius, *Ecl med.* 37.1.1; Aëtius Amidenus 9.1.7; Aretaeus, *De cur. acut. morb.* 2.3. [Galen], *Introd. seu medicus* 13, 14 735.2-6 Kühn says that 'The "cardiac" condition was not so called because the affliction is about the heart, but because the ancients called the stomach "heart". When this [presumably the stomach] is resolved in acute fever the result is that those who are sick διαφορεῖσθαι' – which LSJ index here under "are exhausted by dissipating, weakened". Cf. also Diogenes of Oenoanda, fr. 117.1.7 Smith, "an affliction of the heart weakens me" (καρδιακόν με διαφορεῖ πάθος), in the context of life-threatening illness; and below, n.254).

248 Cf. [Ar.] *Probl.* 8.9, which gives a similar explanation but refers to *boulimia* rather than just to hunger, and discusses the condition at much greater length.

249 Cf. *Probl. Bamberg.* 37, which simply says "the substance of semen (i.e., blood) is discharged"; Aristotle, *GA* 1.19 726b7-11, citing the phenomenon as an argument for semen being *derived from* blood – an incorrect explanation which is absent from our present text, though the wording of the *Bamberg Problem* suggests it may be implied; Nemesius, *De nat. hom.* 25 86.15-18 Morani, saying that the phenomenon is evidence that semen comes from a vein. – For the phenomenon cf. Theophrastus, *HP* 9.18.9.

ἔστιν.

2.29. Διὰ τί τὸ μέλι τοῖς μὲν ἕλκεσι δριμὺ τυγχάνει, τῇ δὲ γεύσει γλυκύ; ὅτι τοῖς ἕλκεσιν ἀνεξαμμένοις ἡ ἀπὸ τοῦ μέλιτος προσπίπτει λεπτότης, δριμὺ δὲ τὸ λεπτόν· ἡ δὲ γεῦσις ὑγρὰ οὖσα τῆς ἐν τῷ μέλιτι λειότητος ἀντιλαμβάνεται.

2.30. Διὰ τί τὰ νεῦρα νυσσόμενα μὲν φλεγμονὰς ἐπιφέρει, κοπτόμενα δὲ οὐκέτι; ὅτι τὸ μὲν πνεῦμα, διὰ στενῶν τῶν πόρων ἀνωμάλῳ τῇ διαιρέσει προσπῖπτον ἀποστρέφεται· καὶ εἰς αὐτὸ πιλοῦται, ἐπὶ δὲ τῆς τελείας διακοπῆς διαπνεῖ.

2.31. Διὰ τί οἱ πρὸ τῆς σημασίας ἱδρῶτες χαλεποί; ὅτι οἱ μὲν μετὰ τὴν ἐπισημασίαν γίνονται κατὰ ἄνεσιν, οἱ δὲ πρὸ τῶν ἐπισημασιῶν κατ' ἔκθλιψιν.

2.32. Διὰ τί οἱ νήστεις δυσώδες ἔχουσι τὸ στόμα; ὅτι τῆς ἐν τῷ στομάχῳ δριμύτητος μὴ κατακιρναμένης νεαρᾷ τροφῇ ἡ ἀναθυμίασις ἀηδής.

ὑγρὸν] ὑγρὰν P: εἰσφορᾶ k 2-3 οὐκ ἔστιν om. k

2.29 (BDGLMOPRSTYbeiklm*p* Buss. Us.; *g* 2.135; *r* 38)

1 καὶ διατί k || ἕλκεσι] ἕλκουσι k 2-4 ὅτι – ἀντιλαμβάνεται] *quia quod ipsum tenuem est acer est et quod ipsum humidum est dulce est* tantum *r* 2 ἕλκεσιν] ἕλκεσι elm: ἕλκυσιν L: ἕλκουσι k qui ante hoc verbum ex 2.32.1-2 ἐν τῷ στομάχῳ θερμότητος per errorem transtulit, deinde ἐν τῷ στομάχῳ linea subduxit || ἀνεξαμμένοις] ἀπεξηραμμένοις Buss.: ἀναξηραμένοις BL: ἀναξαμμένοις οὖσιν GY: ἀναξανθεῖσι D: τοῖς ἀνεξαμμένοις l: ἀναδεξαμμένοις Oi: *horridis resiccatisque g* || ἡ om. Yk || ἀπὸ M Buss. (in var lect.) Us.: om. D: ὑπὸ cett. || προσπίπτει] ἐμπίπτει Dp || λεπτότης] λεπτότατα Y 3 δὲ om. k || τῆς] τοῖς Tk

2.30 (BLMOPRSTbeilm*p* Buss. Us.; *g* 2.126; *r* 39)

1 μὲν om. BL*p* 2 στενῶν τῶν πόρων] τῶν στενῶν πόρων MRe*p* Buss. || ἀνωμάλῳ P*g* Us.: ἀνωμάλως MORSTbeilm Buss.: ἀνομάλως BL || post τῇ διαιρέσει add. τουτέστι ORSbeil: τουτέστι τὸν χωρισμὸν BL: τουτέστι τῇ νύξει PTm: om. M*g* Buss. Us. 2-4 ὅτι – διαπνεῖ] *quia per incisus solvitur spiritus nam punctus subcluditur et indignatur r* 3 αὐτὸ] αὑτὸ BLMOPSTeil: αυτὸ R || post πιλοῦται add. *et recalescit g* || δὲ om. BL 4 διαπνοῖ T

2.31 (BLMOPRSTbeilm*p* Buss. Us.; *g* 2.127; *r* 40)

1-2 ἱδρῶτες – ἐπισημασίαν] in marg. l: γίνονται κατὰ ἄνετι· οἱ μὲν μετὰ τῆς ἐπισημασίας Oi || τὴν ἐπισημασίαν P Us.: τῆς ἐπισημασίας OSbeilm*p*: τῆς σημασίας BLMRe Buss.: τὰς ἐπισημασίας T[pc]: τὰς σημασίας T[ac] || 2 ἐπισημασιῶν] ἐκσημασιῶν BL: σημασιῶν Buss. 2-3 οἱ δὲ – ἔκθλιψιν om. *r* || κατ' ἔθλιψιν T

2.32 (BCDGLMOPRSTYbeiklmo*p* Buss. Us.; *g* 2.128; *r* 41)

1 νήστεις] νίστεις G: νηστεύοντες D: νιστηκοῖ k || τὸ] εἰς τῶ k: om. C || τῆς] ταῖς BL 2 δριμύτητος] δριμήτηταις L: θερμότητος k || μὴ κατακιρναμένην BL: μὴ κατακιραμένης O: μὴ καταρκιναμένης Y || δριμύτητος – τροφῇ] *non est degestio*

moisture more ready than blood.

2.29.[250] Why is honey piercing to wounds, but sweet to the taste? Because the thinness from the honey touches the wounds when they are inflamed, and what is thin is piercing; but taste is moist and grasps the smoothness in the honey.

2.30.[251] Why do sinews produce inflammation when pierced, but not when cut? Because the *pneuma* encountering the division which is uneven through narrow passages is turned back and compressed upon itself, while in the case of a complete cut it passes through.

2.31.[252] Why are sweats before the appearance of a disease dangerous? Because those after the appearance of the disease result from relaxation, those before from squeezing out by compression.

2.32.[253] Why do those who are fasting have foul-smelling mouths? Because when the pungency in the stomach is not mixed with fresh nourishment the exhalation is unpleasant.

---

250 Cf. *Probl. Bamberg.* 38 (same solution, abbreviated).

251 = *Probl. Bamberg.* 39.

252 = *Probl. Bamberg.* 40.

253 = *Probl. Bamberg.* 41. Cf. [Ar.], *Probl.* 13.7, where the explanation is that the heat of an empty stomach causes putrefaction, while when we eat the heat of the food overcomes the internal heat.

2.33. Διὰ τί ἐν ταῖς καρδίαις οὐδὲν γίνεται πάθος; ὅτι διὰ τὴν κυριότητα αὐτῆς φθάνει τοῦ πάθους ὁ κίνδυνος.

2.34. Διὰ τί ἐν τοῖς αἰγιαλοῖς σκάψαντες γλυκὺ εὑρίσκομεν ὕδωρ; ὅτι παχὺ τὸ τῆς θαλάσσης ὕδωρ, ἐκ πολλῶν συνεστός· τὸ λεπτομερὲς οὖν αὐτῆς διὰ τῆς γῆς παρεισδύνει· γλυκὺ δὲ τὸ λεπτόν, ὥσπερ τὸ παχὺ ἁλμυρόν.

2.35. Διὰ τί τὰ πυρίκαυστα δυσκατούλωτα; ὅτι πάντες ἀθρόως ἀφανίζονται οἱ πόροι καὶ πάσχουσιν ὅμοιον τῇ νεκρώσει, δυσέκπλυντος δὲ ἡ ἀποκεκλεισμένη κατ᾽ αὐτὰ ὕλη.

2.36. Διὰ τί οἱ πρεσβῦται φίλοινοι; ὅτι χρόνῳ κατεψυγμένοι χρείαν ἔχουσι τοῦ θερμοῦ, θερμαντικὸς δὲ ὁ οἶνος.

---

bona r || ἡ ἀναθυμίασις] *agrito* (leg. *aegritudo,* Stoll) *ex hoc fumum sursum ascendit r* || ἀηδὴς] ἀειδὴς k: ἀηδὴς γίνεται DGYo Buss.

2.33 (BLMOPRSTbeiklm*p* Buss. Us.; *g* 2.129)

1 οὐδὲν] ἡμῶν k || πάθος add. in marg. O || κυριότητα] κιριότιτα k: καιριότητα OSbil: *opportunitatem g* 2 ὁ κίνδυνος Me^{ac} Buss. Us.: τὸν κίνδυνον cett.

2.34 (MOPRSTbeiklm*p* Buss. Us.; *g* 2.130)

1 σκάψαντες] σκάπτοντες k || γλυκὺ εὑρίσκομεν ὕδωρ] εὑρίσκηται(?) ὕδωρ γλυκύ k: εὑρίσκεται ὕδωρ γλυκύ *p* 2 παχὺ om. k || ἐκ πολλοῦ P || ἐκ πολλῶν συνεστὸς om. *g* || συνεστὸς] συνεστῶς PRS^{ac}bekm 3 παρεισδύει PT 3-4 γλυκὺ – ἁλμυρόν om. *g*: ὥσπερ – ἁλμυρόν om. k || τό παχὺ ἁλμυρὸν OSb^{pc}il*p* Us.: παχὺ τὸ ἁλμυρὸν MPRTb^{ac}em Buss.

2.35 (BLMOPRSTbeilm*p* Buss. Us.; *g* 2.131)

1 πυρίκαυτα RTe^{ac}m Buss. || δυσκατούλωτα ἐστιν BL Buss. 2 δυσέκπλυντος Us.: δυσέκληπτος P Buss.: δυσέκλιπτος m: δυσέκλειπτος BLMORSTbeil: δυσέκλυτος *p*: *difficulter educatur* g || δὲ] δὲ αὐτῶν PTm 3 κατ᾽ αὐτὰ] κατὰ τὰ αὐτὰ BL

2.36 (DGMOPRSTYbeiklmo*p* Buss. Us.; *g* 2.132)

1 πρεσβῦται] πρεσβύτεροι k: γέροντες D || τῶ χρόνω DGYk || κατεψυγμένοι ὄντες D 2 θερμαντικὸς] θερμαντικὰ Y θερμαντικὸν o: τοιοῦτον D || δὲ] γὰρ Yko

2.33. Why is no [pain] felt in the heart? Because of its importance danger comes before the feeling.[254]

2.34. Why do we find fresh water when we dig on the sea-shore?[255] Because sea-water is dense, composed of many ingredients. So the rare part of it enters in through the earth; and the rare part is fresh, as the dense is salty.

2.35. Why do [places] that are burned by fire form scars with difficulty? Because all the pores are destroyed together and undergo something like mortification, and the matter shut up in them is removed[256] with difficulty.

2.36. Why are old men fond of wine? Because they have grown cold with time and need heat, and wine heats.[257]

---

[254] I.e., the patient dies before he has time to feel the pain. Alexander, *DA* 98.22-24 argues that the reason some people recover from so-called heart-disease is that it is in fact a disease of the stomach rather than the heart; cf. Aëtius Amidenus 9.1, p.273.18-21 Zervos 1911, arguing that the disease starts in the mouth of the stomach and then affects the heart, but also that "the ancients" called the mouth of the stomach "heart" because of the symptoms it produced; for the latter point cf. also Nemesius, *De nat. hom.* 21, 82.13-18 Morani, citing Thuucydides 2.49.3 (but Morani brackets the reference to the latter).

[255] The same explanation at [Alex.], *Probl.* i.55 (see above, on 2.22); cf. Flashar 1962b, 655, and for the general principle Pliny, *NH* 31.70; Lucretius, 2.476, 5.269, 6.635; Seneca, *NQ* 3.5, with Sharples 1998b, 220 n.628. [Ar.], *Probl.* 23.21 and 37 both say that the water is fresh at first, salty later; the former because the water found at first is that which is already there and *concocted* by the earth, the latter that the water found at first is the fresher and hence lighter part, but subsequently the passages become wider and the salt water overwhelms the fresh.

[256] Following Gaza's *difficulter educatur*, which gives the required sense. The majority of MSS have *dusekleiptos,* but this means "hard to escape from"; P and Bussemaker have *duseklêptos*, which could have the sense of "removed with difficulty" but is an accepted reading nowhere else, according to the TLG (on-line version, March 2006). Usener conjectured *dusekpluntos*, "hard to wash out", but this (or rather the form found elsewhere, *dusekplutos*) does not seem to be a medical term. Neither does *dusexaleiptos*, with a similar sense, which is an emendation for *duseklêptôs* in Galen, *Anim. Pass.* 1.5 19.11 Marquardt, nor *duseklutos* "hard to undo", the (conjectural?) reading of Psellus' summary, here even though the latter is used once by Galen (*Comp. med. secundum locos* 6.2, 12 906.11 Kühn).

[257] For the coldness of the elderly scc Aristotle, *Rhet.* 2.13 1389b31; Plato prescribes wine as a remedy for the *dryness* of old age at *Laws* 2 666b. Plutarch *Qu. conv.* 3.3 650a reports that Aristotle (fr.108 Rose^3) said in his *On Drunkenness* that old men are most affected by drink and women least, but gives no reason.

2.37. Διὰ τί μᾶλλον τὰ δεξιὰ τῶν ἀριστερῶν εὐσαρκότερα; ὅτι μᾶλλον αὐτῶν κινουμένων αἱ ὗλαι συνεπιδίδονται.

2.38. Διὰ τί ἐπὶ τοῖς ἀριστεροῖς ὤμοις τὰ φορτία φέρομεν; ὅτι τῶν ἀριστερῶν δυσκινήτων ὄντων πλείων ἡ ἐπιμονὴ τῆς ὕλης, ἐπὶ δὲ τῶν εὐκινητοτέρων ἐλάττων.

Λοιπὸν τοίνυν περὶ τῶν κοινῶν συμπτωμάτων διελθεῖν καιρὸς (κοινὰ δὲ συμπτώματα λέγομεν ὅσα πάσαις ὡς ἐπὶ τὸ πολὺ συμβαίνει ταῖς ἡλικίαις), οἷον περὶ ἰλίγγων καὶ πταρμῶν καὶ χάσμης, ἔτι δὲ λυγμοῦ καὶ σκοτοδινήσεως ἐρευγμῶν τε καὶ τῶν τοιούτων ἁπάντων, αὐτά τε διὰ τί γίγνεται καὶ πῶς διατεθέντων τῶν σωμάτων καὶ διὰ τί τούτων ἔνια μέν ἐστιν ἀλλήλων λύσις, ἔνια δ' οὔ, καὶ ποίοις ἐργωδῶς γίγνεται· μετὰ δὲ ταῦτα περί τε φαλακρῶν καὶ μαδίσεως τριχῶν καὶ ὅλως διὰ τί αὔξονται καὶ φθείρονται τρίχες καὶ ὄνυχες καὶ κατὰ ποίους καιροὺς καὶ διὰ τὶ μέλαιναι καὶ πυρραὶ καὶ λευκαὶ γίνονται αἱ τρίχες· πρὸς δὲ τούτοις, περί τε φωνῆς καὶ τῶν περὶ φωνὴν συμπτωμάτων, οἷον διὰ τί ὀξεῖαι καὶ βαρεῖαι γίγνονται καὶ πῶς ἀπορρήγνυνται καὶ διὰ τί ἔνιοι διὰ τῶν ῥινῶν φθέγγονται· καὶ πάλιν περὶ ἀκοῆς καὶ ὀσμῆς καὶ γέλωτος· εἶτα περὶ ναυτίας καὶ διὰ τί οἱ μὲν τοῦτο πάσχουσιν οἵ δὲ οὔ, καὶ ποῖοι μάλιστα.

2.37 (DGLMOPRSTYbeilmo*p* Buss. Us.; *g* 2.134)
1 *εὐσαρκότερα*] *εὐσαρκώτερα* LOil: *sanabilior* g 2 *κινούμενα* P || *καὶ αἱ ὗλαι* D Buss. || *συνεπιδίδονται* GMPRTYemo*p*: *συνδίδονται* LOSbil: *συνεπιδιδόασιν* D Buss. Us.: *compescuntur, reddunturque sanitati opportuniores g*
2.38 (BCDGLMOPRSTYbeilmo*p* Buss. Us.; *g* 133)
1. *τοῖς ἀριστεροῖς ὤμοις*] *τῶν ἀριστερῶν ὤμων* BL Buss. Us.: *τοῖς ἀριστεροῖς p* || *φέρομεν*] *φέρωμεν* M: *οὐ φέρομεν* DY[pc]: *φέρομεν* trasp. post *τὰ φορτία* C || *ὅτι ἐπὶ* D 2 *πλεῖον* G*p* || *ὕλης*] *ὕλης διὸ καὶ πιεζομένη οὐ δύναται φέρειν* D 2-3 *ἐπὶ δὲ τῶν εὐκινητοτέρων*] *ἐπὶ δὲ τῶν κιμονητέρων* B: *ἐπὶ δὲ τῶν εὐκινήτων* C: *ἐπὶ δὲ τῶν κινητετέρων* L: *ἐπὶ δὲ τῶν δεξιῶν εὐκινητοτέρων* D, sic etiam g vel potius *dextra* suppl. in transl. 3 post *εὐκινητοτέρων* add. *ὄντων* D || *ἐλάττων*] *ἐλάττονα* o
praefatio ad 2.39 et seq. (BLMOPRSTbeilm Buss. Us.; paraphrasin brevem tantum habet *p*)
3 *ἐλίγγων* BL || *παρμῶν* BL: *πταρμοῦ* P || *λυγμοῦ*] *καὶ λυγμοῦ* P: *γλυκασμοῦ* B 4 *σκοτοδυνήσεως* BL || *ἐρευγμῶν*] *ἐρεγμῶν* OTilm: *ἐρετμῶν* BL: *καὶ ἐρυγμῶν* Buss. || *τε*[2] om. BL 4-5 *αὐτά τε – καὶ*[2] om. B 5-6 *καὶ πῶς - γίγνεται* om. Oil 5 *διατιθέντων* L || *καὶ* om. L 6 *λύσις ἀλλήλων* P || *ἐργοδῶς* BL 8 *αἱ τρίχες* M 9 *πυρραὶ*] *πυραὶ* BL: *πικραὶ* S || *καὶ λευκαὶ καὶ πυρραὶ* m || *αἱ τρίχες* BLM Us. || *τε*] *τῆς* BL 10 *τῶν* om. L || *φωνῆς* T || *βαρῦαι* T 12 *ὀσμῆς*] *ὁρμῆς* m 13 *πάσχουσι τοῦτο* Oil

2.37. Why are the right-hand [parts] more fleshy than the left? Because they are moved more[258] and the matter develops accordingly.

2.38. Why do we carry loads on our left shoulders?[259] Because the left-hand [parts] are moved less easily and so the matter stays there more,[260] less in the case of those that are moved more easily.

For the rest, then, it is time to go through the common symptoms (by common symptoms we mean those which for the most part occur at any age), for example about dizziness and sneezes and yawning, also hiccoughs and giddiness and belching and all such things, as to why they themselves come about and in what condition of body and why some of these resolve each other and some do not, and in what sort of people they cause difficulty;[261] and after this about bald people and loss of hair, and in general why hair and nails grow and perish, and on what occasions, and why hair is black or red or white;[262] and in addition to these about voice and the things that happen to voices, for example why they are high or deep and how they are fractured and why some people speak through their nostrils; and again about hearing[263] and smell[264] and laughter,[265] and then about nausea and why some people suffer from it and some do not, and what sort of people [do so] most.[266]

---

258 Cf. Aristotle, *PA* 3.9 672a25, [Ar.], *Probl.* 5.32 884a17, 6.5 886a8, 31.12 958b21, 31.18 959a21.

259 Cf. Aristotle, *IA* 4 705b30, where the reason given is that the right-hand side, which starts the movement, is thus left free.

260 This directly contradicts the previous problem.

261 These are the topics of 2.39-53.

262 2.54-82.

263 "voice ... and ... hearing" corresponds to 2.83-2.97. No discussion of speaking through nostrils is included, however.

264 2.98-104.

265 No discussion of laughter is included. Tickling is the subject of ii.45 Ideler and of [Ar.], *Probl.* 35.2 and 6-8, laughter of [Ar.], *Probl.* 11.13, 15 and 28.8.

266 "Nausea" corresponds to 2.105-106, but these are only a part of the group of problems on excretion and purgation which occupies 2.105-26. – On the relation between this preface and the following texts see also Usener 1859, x n.8.

2.39. Διὰ τί κάτω μὲν βλέποντες ἰλιγγιῶμεν ἀφ' ὑψηλοῦ, ἄνω δὲ οὔ; ἢ ὅτι ὁ μὲν ἴλιγγός ἐστι κινουμένου τοῦ ἄκρου τῆς ὄψεως οἷον τοῦ ἐν τῷ ἐγκεφάλῳ· ἰλιγγιῶμεν γὰρ ὅταν παρατρέχωσι πολλοί, ἂν τὴν ὄψιν ἐφ' ἕκαστον ἐπιβάλλωμεν, ὑπὸ τῶν παρατρεχόντων κινουμένης τῆς ὄψεως, ὄντος δὲ τοιούτου τοῦ ἰλιγγιᾶν, ὅταν κάτω βλέπωμεν, ἐὰν μὲν ἀπὸ βραχέος, οὐκ ἰλιγγιῶμεν διὰ τὸ ἀπηρεῖσθαι τὴν ὄψιν· ἐὰν δ' ὑψόθεν φερομένη κάτω ἡ ὄψις πρὶν ἀπερείσασθαι κινῆται, καὶ τὸ ἄκρον ἐκεῖνο κινεῖται· ἄνω δ' ὅταν βλέπωμεν, ποῦ οἰσθήσεται ἰοῦσά τε †ᾗ οἱ† φέρεται· εἰκότως οὖν κάτω βλέπουσι γίνεται ἴλιγγος.

2.40. Διὰ τί μόνον ἡμῖν ὁ πταρμὸς τῆς νυκτὸς οὐ γίνεται, τὰ δ' ἄλλα ὡς εἰπεῖν ἅπαντα ἃ καὶ τῆς ἡμέρας καὶ τῆς νυκτός; ἢ ὅτι ὁ μὲν πταρμὸς γίνεται ὑπὸ θερμοῦ τινος κινήσαντος τὸν τόπον τοῦτον ἀφ' οὗ γίνεται; διὸ καὶ ἀνακύπτομεν πρὸς τὸν ἥλιον ὅταν βουλώμεθα πταρῆναι· καθευδόντων δὲ ἡμῶν, ἀντιπερισπᾶται τὸ θερμόν· διὸ καὶ γίγνεται τὰ κάτω θερμὰ τῶν

2.39 (BLMOPRSTbeilm*p* Buss. Us.)
1-3 *ἀφ' ὑψηλοῦ – γὰρ* om. BL ‖ *ἄνω* om. M 2 *ὁ* om. P ‖ *ἴλιγγος μὲν* P 3 *παρατρέχουσι* BLMRSe 3-4 *ἐφ' ἑκάστου* P 4 *ἐπιβάλλωμεν* OPil Buss. Us.: *ἐπιβάλωμεν* MRSbem: *ἐπιβάλλομεν* T: *ἐπιβάλομεν* BL ‖ *ὑπὸ*] *ἐπὶ* BL 5 *τοῦ*] *τὸ* BL ‖ *ὀλιγγιᾶν* L 6 *βραχέως* Bm ‖ *οὐκ ἰλιγγιῶμεν*] *οὐ λιγγιῶμεν* BL ‖ *ἀπηρεῖσθαι* OSbil Us.: *ἀπορεῖσθαι* BL: *ἀπερείσασθαι* M: *ἐπερειδέσθαι* e: *ἐπερείδεσθαι* Buss.: *ἐπηρεῖσθαι* PRm: *ἐπηρίσθαι* T: *ἡδραιῶσθαι ἐπηρεῖσθαι* sic *p* 7 *κάτω* om. BL ‖ *ἡ*] *ἢ* O ‖ *ἀπηρείσασθαι* P: *ἐπηρείσασθαι p* ‖ *κινῆται*] *κινεῖται* BLPTe*p* ‖ *ἐκεῖνον* B 8 *δ'* om. BL ‖ *ᾗ οἱ* PT: *ἢ αἱ* m: lacunam relinquit R: om. OSbeil Buss. ‖ *ποῦ οἰσθήσεται – φέρεται*] *ποῦ οἰσθήσεται ἰοῦσάν τε φέρεται* M: *ποῦ οἰσῶσα φέρεται p*: *οἰσθήσεται* tantum BL: *ἀπερείσασθαι* †*ἰοῦσά τε*†(coni. *ἰδοῦσί γε*) *ᾗ ἥλιος φαίνεται* Us. 9 *βλέπουσι*] *βλεπτοῦσι* e: *βλέποντες* BLOil
2.40 (BCLMOPRSTbeilm*p* Buss. Us.; [Ar.] 33.15, Cassius 44)
1 *μόνον*] *μόνος* Cass. ‖ *ὁ πταρμὸς ἡμῖν* [Ar.] Us. ‖ *τῆς νυκτὸς*] *καθεύδουσιν* [Ar.] ‖ *νυκτὸς*] *νυκτὸς ὡς ἐπὶ τὸ πολὺ* Me (per emend.?) Buss. ‖ *οὐ γίνεται*] *γίνεται κατὰ φύσιν ἔχουσιν ἢ σπανίως* Cass. 2 *ἃ* om. M [Ar.] Cass. Us. ‖ *μὲν* add. supra lin. C 3 *ὑπὸ*] *ἀπὸ* Me: *καὶ ὑπὸ* [Ar.] ‖ *τοῦτον* om. m*p* 4 *ἀνακύπτομεν*] *ἀνακύπτει* BL: *ἀνακύπτει μὲν* C ‖ *βουλόμεθα* ACm*p*: *βουληθῶμεθα* M ‖ *πταρῆναι*] *πτάρνυσθαι* BL: *πτάρειν* [Ar.] 5 *ἀντιπερισπᾶται*] *ἀντιπερισπᾶσθαι* BL: *ἀντιπεριίσταται* [Ar.] Buss. (in var. lect.) Us. ‖ *τὸ θερμόν*] *τὸ θερμὸν ἔντος* [Ar.]

[Dizziness, sneezes, yawning, hiccoughs, giddiness and belching]

2.39.[267] Why do we become dizzy when we look downwards from a height, but not [when we look] upwards? Is it because dizziness occurs when the extremity of the sight is moved, that is, that in the brain? For we become dizzy when many people are running alongside each other, if we fix our sight on each one, for the sight is moved by those who run alongside us. Well, dizziness being like this, when we look downwards, if it is from a short distance, we do not become dizzy, because the sight [has somewhere to] rest; but if the sight travels downwards from a height and is moved before it [has somewhere to] rest, that extremity too [in the brain] is moved. But when we look up, where will our sight go and be carried?[268] So it is reasonable that dizziness occurs when [we] look downwards.

2.40.[269] Why does sneezing alone not happen to us at night, when everything else, virtually, which [happens] by day also [happens] by night? Is it because sneezing occurs when some heat moves the place from which it originates? This is why we lift our heads up towards the sun when we want to sneeze.[270] But when we are asleep the heat is drawn[271] inwards; and for this reason the

---

[267] A version of the argument of Theophrastus, *Vert.* 8, incorporating also material from id. 6. *Quaest. Salern.* Ba 39 (p.171 Lawn 1979) gives as alternative explanations (i) fumes rising from the stomach to the brain (a general cause of dizziness recognised by Theophrastus, though not in this specific context), (ii) imagination of what it would be like to fall, and (iii) the effect of being able to see more clearly in the upper air.

[268] This is the general sense given by the MSS, though the text is repetitive (literally "where will our sight be carried and where is it carried when it goes?") and in any case seems corrupt. Theophrastus himself says that when we look upward our sight is cut short by the light. Bussemaker here conjectured "where will sight be carried and where is it carried for those who are seeing?" Usener suggests "it will find a place to rest for those who look where the sun appears". In Bussemaker's version and in Theophrastus' the point is that our sight does not travel far enough to be disturbed; Usener's introduction of the idea that it *has* a place to rest seems unwarranted.

[269] = [Ar.], *Probl.* 33.15 (first part) = Cassius, *Probl.* 44.

[270] Cf. below, 2.47. Bright sunlight can indeed cause sneezing in some people; but the effect is due to light rather than heat (see e.g. http://en.wikipedia.org/wiki/Photic_sneeze_reflex).

[271] The MSS of our collection (and of Cassius) have *antiperispatai,* "is drawn

καθευδόντων καὶ τὸ πνεῦμα πολύ· εἰκότως οὖν οὐ πταρνύμεθα· ἀπαλλαγέντος γὰρ τοῦ θερμοῦ ἐκ τῆς κεφαλῆς, ὃ κινεῖν πέφυκε τὸ ἐνταῦθα ὑγρὸν (οὗ ἐξωθουμένου πέφυκεν ὁ πταρμὸς γίνεσθαι), καὶ τὸ συμβαῖνον πάθος εἰκὸς μὴ γίνεσθαι.

2.41. Διὰ τί τρίψαντες τὸν ὀφθαλμὸν παυόμεθα τῶν πταρμῶν; ἢ ὅτι ἡ διαπνοὴ ταύτῃ γίνεται τῷ ὑγρῷ; δακρύει γὰρ ὁ ὀφθαλμὸς μετὰ τὴν τρῖψιν, ὁ δὲ πταρμὸς διὰ πλῆθος ὑγρότητος. ἢ ὅτι τὸ ἔλαττον θερμὸν φθείρεται ὑπὸ τοῦ πλείονος, ὁ δὲ ὀφθαλμὸς τριφθεὶς πλείω λαμβάνει θερμότητα τῆς ἐν τῇ ῥινί; διὰ τοῦτο δὲ καὶ ἄν τις αὐτὴν τὴν ῥῖνα τρίψῃ, παύεται ὁ πταρμός.

2.42. Διὰ τί ἰλιγγιῶσιν εἰς ταὐτὸ βλέποντες κἂν κύκλον κινούμενον ἴδωσιν; ἄτοπον γάρ τι τὴν ἠρεμίαν τῇ κινήσει ταὐτὸ ποιεῖν ἐναντία ὄντα. ἢ ὅτι διίστησι καὶ ἡ στάσις τὰ τῷ κινεῖσθαι σῳζόμενα; τῆς ὄψεως γὰρ στάσης, ἑνὸς μορίου, καὶ τὰ ἄλλα συνεχῆ ἐν τῷ ἐγκεφάλῳ ἵσταται· ὁ δὲ κυκεών, ὥσπερ καὶ Ἡράκλειτός φησιν, ἐὰν μή τις ταράττῃ, διίσταται· τοῦτο δὴ δεῖ νοῆσαι πάσχοντα καὶ τὰ ἐν τῇ κεφαλῇ. χωριστὰ μὲν γὰρ καὶ

|| *κάτω*] *κατὰ* BL 6 *πολύ*] *τὸ* (om. Ap.X[a]) *πολὺ αἴτιόν ἐστι τοῦ ἐξονειρώττειν ἡμᾶς* [Ar.] || *οὐ*] *καὶ* BL || *πταρνύμεθα*] *πταρνόμεθα* m: *πτάρνυμεν* Cass. 7 *ἀπαλαγέντες* BL: *ἀπαλλαγέντες* P || *ἐκ*] *τοῦ ἐκ* [Ar.] || *ὃ*] *τὸ* C || *κινεῖν*] *κινεῖ* BL 8 *οὗ*] *οὗ ὑγροῦ* C || *ἐξωθουμένου*] *ἐξαιρουμένου* [Ar.] || *πέφυκεν ὁ πταρμὸς γίνεσθαι*] *γίνεται ὁ πταρμός* [Ar.] Cass. || *γίνεσθαι*] *γίνεται* BL 8-9 *καὶ – γίνεσθαι* om. BL 9 *μὴ*] *οὐ* Buss.

2.41 (BCLMOPRSTbeiklm*p* Buss. Us.; [Ar.] 31.1, 33.8; Cass. 45)

1 *διατρίψαντες* B || *τὸν πταρμὸν* BL: *τὸ πταίρειν* k || *ἢ*] *ἡ* M: om. k || *ὅτι*] *διότι* BLTm Cass. || *ἡ* om. [Ar.] Cass. Us. 2 *διαπνοὴ* BCLOSbil Cass.(NSb) Us.: *ἀναπνοὴ* MPRTekm [Ar.] 31.1 Cass. (cett.) Buss.: *ἀνάπνοια* [Ar.] 33.8 || *ταύτῃ* BCLMNORSbeikl: *ταυτὸν* P: *ταυτί* Tm: om. Cass. || *τῷ ὑγρῷ*] *τῶν ὑγρῶν* k 3 *ὁ δὲ πταρμὸς*] *ὁδὲ ὁ πταρμένος* k || *διὰ* [Ar.] Cass. Us.: *ἐκ* Me Buss.: *ἐὰν* BLOPRSTbilm: *ἐπὶ* k: om. C || *πλῆθος* BCLOPRSTbeilm [Ar.] Cass. Us.: *πλήθους* M Buss.: *πλήθει* k || *τὸ ἔλαττον θερμὸν*] *τῶν ἐλάτων θερμῶν* k || post *φθείρεται* deficit k 4 *πλείω*] *πλεῖον* BL 5 *καὶ ἂν*] *καὶ ἢν* Buss.: *κἄν* [Ar.]: *καὶ ἐάν* Cass. || *αὐτὴν τὴν ῥῖνα*] *τὴν αὐτοῦ ῥῖνα* BCLOSbil: *τὴν αὑτοῦ ῥῖνα* Buss.

2.42 (BLMOPRSTbeilm*p* Buss. Us.: cf. Thphr. *Vert.* 9-10)

1 *εἴς τε* Us. 2 *ταὐτὸν* Us.: *ταυτῶ* P || *ποιεῖ* BL 3 *ἢ διότι* PTm || *ἐν τῷ* M Us: *τῶ* P 3-4 *γὰρ στάσης*] *διάστασις* Thphr. cod. A, quod in *δὲ στάσης* corr. Us. 4 *στάσης*] *στάσις* BL || *τοῦ ἑνὸς* BL Buss. || *τ' ἄλλα* BL 4-5 *ὁ δὲ – διίσταται*] *εἰ δὲ μή, καθάπερ Ἡ. φησι, καὶ ὁ κυκεὼν διίσταται κινούμενος* Thphr. cod. A 5 *κυκεὼν* Thphr. cod. A Us. Wegehaupt: *κυκλεύων* codd. Buss. || *ἐὰν μή τις ταράττῃ*] *κἂν μεταταράττει* BL: *κἂν μεταταράττῃ* Buss. || *διίσταται* Thphr. cod. A Us. Wegehaupt: *ἵσταται* codd. Buss. 6 *δὴ δεῖ*] *δεῖ δὴ* BLOSbil || *πάσχοντα*]

lower parts of those who are asleep are hot and they breathe heavily.[272] So it is reasonable that we do not sneeze. For when the heat, which naturally moves the moisture in the head – and it is when this is forced out that a sneeze naturally occurs – departs from there, it is reasonable that the affection that results from it should not occur.

2.41.[273] Why does rubbing our eyes stop us from sneezing? Is it because moisture is able to evaporate[274] by this route? For the eye weeps after the rubbing, and sneezing is caused by a large quantity of moisture. Or is it that the lesser heat is destroyed by the greater, and the eye, when rubbed, acquires a heat which is greater than that in the nose? And for this reason it [is also true that] if one rubs the nose itself the sneeze stops.

2.42.[275] Why do people become dizzy [both][276] if they look at the same object and if they see a moving circle? For it is strange that rest should have the same effect as motion [though] they are opposite. Is it because rest too separates things that are preserved by[277] movement? For when vision, which is one part, comes to rest, the other [parts] continuous with it in the brain also come to rest. But the posset, as Heraclitus says, separates if one does not stir it.[278] One should recognise that this happens to what is in the head

inwards". Those of [Ar.], *Probl.* 33.15 have *antiperihistatai*, "is compressed inwards", which is the standard Peripatetic technical term for this process; but *antiperispatai* makes good sense and may be the word originally used in our collection.

272 See below, n.424.

273 = [Ar.], *Probl.* 31.1 = 33.8 = Cassius, *Probl.* 45: cf. [Ar.], *Probl.* 33.2, which has only the second of the two explanations here.

274 The MSS of our text are divided between *diapnoê* and *anapnoê*. Cassius has the former, the more natural term for the removal of *moisture* in the form of vapour, but the parallel texts in [Aristotle] have the latter, rendered "evaporation" by Forster. The argument is in any case that the emission of tears caused by rubbing enables the moisture already present to escape (rather than that the rubbing produces the moisture, which would encourage sneezing rather than stopping it).

275 = Theophrastus, *Vert.* 9-10. On this, and on the interpretation of the reference to Heraclitus in both texts, see Sharples 2002b, 222-228.

276 "both" makes the point of the question clearer; it is actually printed in the Greek text by Bussemaker and Usener, but is present in only one of the 14 MSS (including Photius' paraphrase) which include this problem.

277 One MS, followed by Usener, has "in".

278 The MSS of our text have "the person moving in a circle is at rest if no-one

ἠθροισμένα εἰς ταὐτὸ τὰ βαρέα βαρύνει καὶ ποιεῖ τὸν ἴλιγγον· ἡ μὲν οὖν ἠρεμία διὰ τοῦτο ποιεῖ ἴλιγγον· ἡ δὲ δίνησις τὰ βαρέα καὶ τὰ κοῦφα διίστησι, τὰ μὲν βαρέα εἰς τὸ μέσον, τὰ δὲ κοῦφα ἐπὶ τὸ ἔσχατον, δέον ἅμα τούτω εἶναι· διὰ τοῦτο καὶ τὸ κύπτειν ἰλίγγους ποιεῖ· χωρίζεται γὰρ τὰ μὲν βαρέα κάτω, τὰ δὲ ἐλαφρὰ ἄνω· φανερὸν δ' ἐκ τούτων ὅτι οὐδεμίαν κίνησιν τὰ ἐν τῇ κεφαλῇ κινεῖται.

2.43. Διὰ τί ἰλιγγιῶσι καὶ δι' ἔνδειαν καὶ διὰ πλησμονήν; ἢ ὅτι δι' ἔνδειαν μὲν σύντηξις γίνεται εἰς τὰ κενά, διὰ πλήρωσιν δὲ περιττώσεως ὑπερβολὴ εἰς τοὺς πλήρεις πόρους, ταῦτα δ' ἀμφότερα βαρύνει καὶ ἐφίστησι τὰς περιόδους, τοῦτο δ' ἐστὶν ἴλιγγος;

2.44. Διὰ τί τὸ σκορδινᾶσθαι ἡδύ; ἢ ὅτι ἐξ ὕπνου πνεῦμα τῇ διατάσει καὶ τῇ κινήσει ποιούμενον ἰὸν ὁμαλῶς διὰ τῶν ἄρθρων ἡδονὴν ποιεῖ; ὅτι δὲ πνεῦμα ποιεῖ, δῆλον· χάσμην γὰρ ἅμα ποιεῖ.

2.45. Διὰ τί ὁ πταρμὸς λυγμὸν μὲν παύει ἐρευγμὸν δὲ οὐ παύει; ἢ διότι οὐ τοῦ αὐτοῦ τόπου γέγονεν ἑκάτερον τὸ πάθος, ἀλλ' ὁ μὲν ἐρευγμὸς κοιλίας, ὁ δὲ λυγμὸς τοῦ περὶ τὸν πνεύμονα κατάψυξις καὶ ἀπεψία πνεύματος καὶ ὑγροῦ, κοινωνοῦσι δὲ οἱ περὶ τὸν ἐγκέφαλον τόποι τῷ πνεύμονι; οἷον τοῖς τε ὠσί· φανερὸν δ' ἐπεὶ ἅμα ἐνεοὶ καὶ κωφοὶ γίνονται, καὶ αἱ νόσοι ἀντιπεριίστανται αἱ τοῦ ὠτὸς εἰς τὰ τοῦ πνεύμονος πάθη, ἐνίοις δὲ

φάσκοντα PTm || *χωριστὰ* PTm Us.: *χωρίς· τὰ* cett. Buss. 7 *βαρύνει καὶ* om. Buss. 8 *διατούτων* B || *ἴλιγγα* MRTe || *δίνησις*] *κίνησις* Tm^{pc}: *δίησις p* 9 *διίστησι – κοῦφα* om. B 10 *τούτω* Wegehaupt: *τούτῳ* Us.: *τούτῳ τῷ* Oi: *τούτου τῷ* l: *τούτου* M: *τούτου τὸ* cett.: *ταῦτα* Buss. (in var. lect.) 11 *φανερὰν* Buss. || *ὅτι* om. BLOil Buss. || *οὐδὲ μίαν* L

2.43 (BLMOPRSTbeilm*p* Buss. Us.: cf. Thphr. *Vert.* 11)

1 *ἢ* om. P || *διότι* BL 2 *μὲν* om. BL || *σύντηξις*] *σύμπηξις p* || *δὲ* om. BL || *περιτώσεως* L || *ὑπερβολὴ* Buss. Us.: *ὑπερβολῆς* BLMORSTbeilm: om. P 3 *πόρους*] *τόπους* BL*p*: *πόρους* cett. Thphr. || *ἐφίστησι*] *ἀφίστησι* B: *διίστησι ἀμφίστησι* L: Thphr. cod. A

2.44 (MOPRSTbeilm*p* Buss. Us.)

1 *διατάσει*] *διατήξει* Buss.: *διατάξει* MRe 2 *ἰὸν* Buss. Us.: *οἷον* codd.

2.45 (MOPRSTbeilm*p* Buss. Us.; [Ar.] 33.1)

1 *ἐρευγμὸν*] *ἐρυγμὸν* Oil [Ar.] || *ἐρευγμὸν δὲ οὐ παύει* om. m 2 *γέγονεν* PSe^{mg} Us.: *γε* MORTbe^{ac}ilm Buss.: om. [Ar.] || *τὸ πάθος* coll. post *τόπου* PTm [Ar.] 3 *καὶ*^{2} om. OSbil 4 *τε* om. [Ar.] 5 *ἐπεὶ ἅμα ἐνεοὶ καὶ κωφοὶ* Us.: *ἐπεὶ ἅμα ⟨ἐνεοὶ⟩ εἶναι οἳ καὶ κωφοὶ* Buss.: *τῷ ἅμα εἶναι* (add. in marg. *ἐνεοὶ καὶ κωφοὶ*) *οἳ καὶ κωφοὶ γίνονται* e: *τῷ ἅμα εἶναι οἳ καὶ κωφοὶ* cett.: *ἅμα γὰρ ἐνεοὶ καὶ κωφοὶ* [Ar.] 6 *πνεύμονος*] *πνεύματος* MR || *ἐνίοις*] *ἐν τοῖς* coni. Forster in [Ar.] 7 *τὸ*^{2} M [Ar.]

too; for when the heavy [constituents] are separate and gathered into a single place, they weigh heavily and produce the dizziness. So this is why rest produces dizziness; but rotation separates the heavy and the light [constituents], the heavy to the centre and the light to the edge, when the two[279] should be together. This is why stooping down, too, produces dizzy spells; for the heavy [constituents] are separated downwards, the light ones upwards. It is clear from this that what is in the head does not move in any way.[280]

2.43.[281] Why do people become dizzy both on account of lack and on account of surfeit? Is it because lack brings about colliquescence into the empty [parts], surfeit an excess of residue entering the passages that are full, and both of these weigh down on and stop the rotations, and this is dizziness?

2.44. Why is yawning and stretching pleasant? Is it because, after sleep, *pneuma* produced by the stretching and movement, passing evenly through the joints, produces pleasure? And that it produces *pneuma* is clear, for it produces yawning at the same time.

2.45.[282] Why does sneezing stop hiccoughs but does not stop belching? Is it because each does not affect the same place, but belching is [a matter] of the gut, hiccoughs a cooling of the region around the lung and a lack of concoction of *pneuma* and moisture, and the places around the brain are connected to the lung, as also to the ears? This is clear because people become dumb and deaf at the same time, and diseases of the ear change into

---

disturbs him" ("even if you disturb him" BL). Our text is however clearly dependent on Theophrastus (as was shown by Mouraviev 1996), and Usener, whose text we follow here, was clearly right to emend it on that basis.

[279] Reading *toutô* (dual) with Wegehaupt. One MS, followed by Usener, has "they should be together with this", but this is obscure; the majority of MSS have "the being of this should be together", but this is awkwardly expressed and reads like an attempt to repair a corrupt text.

[280] This final sentence does not appear in Theophrastus.

[281] = Theophrastus, *Vert.* 11.

[282] = [Ar.], *Probl.* 33.1; cf. 33.5 (which explains the effect of sneezing and holding the breath on hiccoughs in terms of effects on the breath, without reference to heating, and the effect of vingar on hiccoughs as here) and 33.17 (which has the contrast between hiccoughs and belching, and the same explanations as here in different words).

σκαλεύουσι τὸ οὖς βῆχες γίνονται· τὸ δὲ περὶ τὸν πταρνύμενον τῆς ῥινὸς τόπον δηλοῖ ἡ ἀναπνοὴ κοινὴ οὖσα, ὥστε πτάρνυται μὲν θερμαινομένου, ταὐτὸ δὲ συμπάσχει ὁ κάτω τόπος. ἡ δὲ θερμασία πέττει· διὸ ὄξος μὲν παύει λυγμὸν καὶ ἀπνευστισμοὶ ἐὰν ἠρεμαία ἡ λὺγξ ᾖ· ἐκθερμαίνει γὰρ τὸ πνεῦμα τὸ κατεχόμενον, ὥστε καὶ ἐν τῷ πταρμῷ ἥ τε κατάσχεσις γενομένη τοῦ πνεύματος τοῦτο ποιεῖ, καὶ τότε ἀθρόως ἔκπνευσις γίνεται καὶ τοῦ κάτω τόπου· ἀδύνατον γὰρ πταρῆναι ⟨μὴ⟩ ἐκπνέοντα· ἡ οὖν ὁρμὴ ῥηγνύει τὸ ἐγκατειλημμένον πνεῦμα ὃ ποιεῖ τὸν λυγμόν.

2.46. *Διὰ τί χασμησαμένου τινὸς ἀντιχασμῶνται, καὶ ὅταν οὐροῦντα ἴδωσιν, οὐροῦσι, καὶ μάλιστα τὰ ὑποζύγια; ἢ διὰ τὴν μνήμην; ὅταν γὰρ μνησθῇ, κινεῖται τοῦτο τὸ μέρος· τοῖς μὲν οὖν ἀνθρώποις διὰ τὸ εὐαισθητοτέροις εἶναι ἰδοῦσιν εὐθὺς συμβαίνει κινεῖσθαι καὶ ἀναμιμνήσκεσθαι, τοῖς δὲ ὑποζυγίοις οὐκ αὔταρκες τὸ ἰδεῖν, ἀλλὰ προσδέονται καὶ ἄλλης αἰσθήσεως· διὸ καὶ ὀσφρηθέντα, ὅτι εὐκινητοτέρα αὕτη ἡ αἴσθησις τοῖς ἄνευ λόγου· καὶ διὰ τοῦτο εἰς τὸν τόπον ἅπαντα οὐρεῖ, ὅπῃ τὰ πρῶτα οὔρησεν· τότε γὰρ μάλιστα κινοῦνται ὅταν ὀσφρανθῶσιν· ὀσφραίνονται δ᾽ ὅταν πλησιάσωσιν.*

2.47. *Διὰ τί πτάρνυνται μᾶλλον πρὸς τὸν ἥλιον βλέποντες καὶ κινοῦντες κάρφει τοὺς μυκτῆρας; ἢ ὅτι ὅ τε ἥλιος θερμαίνει καὶ τὸ κάρφος τῇ*

Buss. Us.: *τῶ* cett. 7-8 *τῆς ῥινὸς τόπον δηλοῖ*] *τόπον εἶναι τῆς ῥινὸς κοινωνίαν τῷ πνεύμονι δηλοῖ* [Ar.] 8 *τόπον* [Ar.] Buss. Us.: *τόπω* MOPRSTbeilm || *θερμαινομένην* P 10 *ἀπνευστισμοὶ* M Buss. Us.: *ἀναπνευστισμοὶ* OPRSTbeilm: *ἡ ἀπνευστία* [Ar.] || *ἠρεμαία*] *ἠρεμία* Oi: *ἠρεμία* Ap.: *ἠρεμέα* T || *λὺξ* P Buss. 11 *κατεχόμενον*] *κατερχόμενον* m || *ἥ*] *ἢ* O || *τε κατάσχεσις*] *ἀντικατάσχεσις* [Ar.] 12 *γιγνομένη*] *γενομένη* [Ar.] Us.: *γιγνομένη* MORSTi: *γινομένη* P [Ar.] cod. Y[a] Buss. || *ποιεῖ τοῦ πνεύματος καὶ* PTm || *τότε ἀθρόως*] *οἰκείως ἡ* [Ar.] 13 *τοῦ κάτω*] *ἐκ τοῦ ἄνω* [Ar.] || *μὴ* add. Buss. Us. ex [Ar.] qui *πτάρειν μὴ ἐκπνέοντα* exhibet; ita quoque e[mg]: om. codd.

2.46 (MOPRSTbeilm*p* Buss. Us.; [Ar.] 7.6)

1 *χασμησαμένου τινὸς*] *χασμησαμένοις* [Ar.] || *οὐροῦνται* T 3 *μὲν οὖν*] *μὲν γὰρ οὖν* T: *μὲν* m 4 *εὐαισθητοτέροις* OPRSbeil*p* [Ar.] codd. quidam, Buss. Wegehaupt: *εὐαισθητότερον* M: *εὐαισθητοτέρους* Tm [Ar.] Us. || *κινεῖσθαι καὶ* MPTelm [Ar.] Us.: om. ORSbi*p* Buss. 6 *ὀσφρηθέντα* MPRTem Buss. Us.: *ὀσφρηθέντες* OSbil: *ὀσφρανθέντα* [Ar.] 7 *τοῖς* om. Buss. || *τὸν αὐτὸν τόπον p* [Ar.] Us. 8 *ὅπῃ*] *ὅπου* T: *οὗ ἂν* [Ar.] || *τὰ πρῶτα* MOPRSbeilm Buss. Us.: *τὰ πρῶτον* T [Ar.] || *οὔρησεν* Us.: *οὔρησεν* vel *οὐρήσειε* Buss. (var. lect.): *οὐρήσει* codd. et [Ar.] cod. Y[a]: *οὐρήσῃ* [Ar.]

2.47 (MOPRSTbeilm Buss. Us.)

1 *κινοῦνται* Oil: *κινοῦντας* M 2 *κάρφει*] *κάρφη* OSil: *κάρφει* post *μυκτῆρας* P ||

afflictions of the lung, and some people cough when they scratch their ear. Concerning the part of the nose where sneezing occurs [the connection] is shown by respiration, which is common, so that one sneezes when it is heated, and the region below undergoes the same experience. Heat causes concoction; this is why vinegar stops hiccoughs, and holding one's breath, if the hiccoughs are persistent.[283] For the breath that is restrained causes heating, so that in the case of sneezing too holding the breath has this effect, and then there is an exhalation altogether from the region below as well; for it is impossible to sneeze <without>[284] breathing out. So the impulse breaks up the trapped *pneuma* which causes the hiccoughs.

2.46.[285] Why when someone yawns do people yawn back, and when they see someone urinating, they urinate, and especially yoke-animals [do this]? Is it because of memory? For when one remembers, that part is moved. Human beings, because they have keener sensation, are moved and remember as soon as they see, but for yoke-animals seeing is not sufficient, but they need another sense in addition too; so [they react in this way] also having smelled, because this sense is more easily moved in irrational [animals]. And for this reason they all urinate in the place where they did at first; for they are most moved when they smell [it], and they smell when they get close.[286]

2.47.[287] Why do people sneeze more when they look at the sun and when they stir their nostrils with a straw? Is it that both the sun heats and also the

[283] See below, 2.53.

[284] Added by Bussemaker and Usener from [Aristotle].

[285] = [Ar.], *Probl.* 7.6: cf. ibid. 7.1 (a shorter version of the same explanation, without the reference to animals), [Alex]. *Probl.* i.34 Ideler (a much shorter version of the answer given here).

[286] To the place (so Forster and Flashar in their translations of [Aristotle]), or to one another?

[287] Cf. [Ar.], *Probl.* 33.4 (the same explanation in different words), 35.8 (the same explanation for tickling with a feather, not mentioning the sun). See above, 2.40.

κινήσει, θερμαινομένου δὲ τοῦ ὑγροῦ πνεῦμα γιγνόμενον ἐκπίπτει θᾶττον (οὐ γὰρ δύναται ὁ ἴσος τόπος πλείω γιγνόμενον τὸν ὄγκον στέγειν), ἡ δὲ ἔκπτωσις πταρμὸς καλεῖται;

2.48. *Διὰ τί ἐν τῷ χειμῶνι ἀναπνέουσιν ἐλαττονάκις; ἢ ὅτι ἐν μὲν τῷ θέρει ὁ ἀναπνεόμενος ἀὴρ διὰ τὴν ἀλέαν οὐχ ἱκανῶς καταψύχει, εἶτα πάλιν δεῖται τοῦ καταψύχοντος, διὸ ἀναπνεῖ ταχύ (τοῦτο δὲ αἴτιον καὶ τοῦ περὶ τὰς τελευτὰς ἀναπνεῖν ταχύ· διὰ γὰρ τὴν ἀδυναμίαν τῆς ὁλκῆς τοῦτο γίνεται· ὀλίγος γὰρ ἑλκυσθεὶς οὐ διαψύχει, ὀλίγος δ' εἰσέρχεται δι' ἀσθένειαν τῆς σπάσεως)· ἐν δὲ τῷ χειμῶνι τοὐναντίον; ἡ γὰρ ὥρα συγκαταψύχει.*

2.49. *Διὰ τί ἀνισταμένοις ἴλιγγος μᾶλλον γίνεται ἢ καθίζουσιν; ἢ διότι ἠρεμοῦσιν αὐτοῖς τὸ ὑγρὸν συνίσταται ἐν τῇ κεφαλῇ πλέον, κινουμένοις δὲ γίγνεται δῆλον καὶ ποιεῖ τὸν ἴλιγγον εἰς ἓν μόριον ἀθρόον ἀποκλῖναν (διὸ καὶ τὰ ὠμὰ ⟨ᾠὰ⟩ οὐ δύναται δινεῖσθαι ἀλλὰ καταπίπτει)· κινουμένων δὲ τὸ ὑγρὸν ὁμοίως ἔχει; ἀνίστανται μὲν οὖν ἠρεμήσαντες ὅτε οὕτω διάκεινται· καθιζάνουσι δὲ ἐν κινήσει γενόμενοι ὅτε ὁμαλῶς ἔχει τὸ ὑγρὸν καὶ ἐσκέδασται.*

2.50. *Διὰ τί τῶν μὲν ἄλλων πνευμάτων αἱ διέξοδοι οἷον φύσης καὶ ἐρευγμοῦ οὐχ ἱεραί, ἡ δὲ τοῦ πταρμοῦ μόνη; πότερον ὅτι τριῶν ὄντων*

ὅτι ὅ τε] *ὅτε* tantum Oil || *τῇ* om. OSbil 3 *πνεύματι* T || *τὸ πνεῦμα τὸ γινόμενον* coni. Buss. 4 *στέγειν* T Buss. Us.: *στέργειν* MOPRSTbeilm

2.48 (MOPRSTbeilm*p* Buss. Us.)

1 *ἀναπνέουσιν ἐλαττονάκις* codd. Us.: *ἀναπνέομεν ἐλαττονάκις p*: *ἔλαττον οἱ ἄνεμοι ἀναπνέουσιν* M Buss. || *μὲν* om. P 2 *ἀλέαν* Buss. || *ἱκανὸς* T || *καταψύχειν* T: *καταψύχεται* coni. Buss. 3 *τοῦτο*] *τούτου* M || *τὸ περὶ* m 5 *γὰρ* om. T || *οὐ* om. T

2.49 (BLMOPRSTbeilm*p* Buss. Us.; [Ar.] 6.4; cf. Thphr. *Vert.* 2, 13)

1 *μᾶλλον* om. BLOSbil || *καθέζουσιν* BL: *καθιζάνουσι* [Ar.] codd. Louis 2 *αὐτοῖς* om. [Ar.] || *κινουμένης* T: *κενουμένοις p* 2-3 *συνίσταται – ἀποκλῖναν*] *εἰς ἓν μόριον ἀθρόον ἀποκλίνει* tantum [Ar.] 3 *ἀποκλίνας* BL 4 *ᾠὰ* [Ar.] Us., cf. Thphr.: om. codd. Buss. || *δινεῖσθαι* [Ar.] Us. Thphr.(cod. B[2] et Tiph.): *κινεῖσθαι* codd. Buss.: *διανονεῖσθαι* sic Thphr.(cod. A): *διανοεῖσθαι* Thphr.(codd. plerique) || *κινουμένων*] *κινούμενον* [Ar.]: *κινουμένῳ* Gaza ad [Ar.] 6 *καθιζάνουσι*] *καθέζουσι* BL Buss. || *ἐν κινήσει* Buss. Us.: *κινήσει* [Ar.] codd.: *ἡ κίνησις* codd. || *γενόμενοι* [Ar.] Buss. Us.: *γιγνομένοις* codd. || *ὅτε* Buss. Us.: *ὅθ'* [Ar.]: *τὲ* codd. || *ἔχειν* B

2.50 (BLMOPRSTbeilm*p* Buss. Us.; [Ar.] 33.9)

1 *διέξοδοι*] *ἔξοδοι* [Ar.] || post *διέξοδοι* add. *δ'* BL || *φύσις* BLPe 1-2 *οἷον φύσης καὶ ἐρευγμοῦ* add. supra lin. l 2 *ἐρευγμοῦ*] *ἐρεγμοῦ* BL: *ῥυγμοῦ* T: *ἐρυγῆς p*: *ἐρυγμοῦ* [Ar.] || *οὐχ ἱεραὶ* PTm [Ar.] Buss. Us.: post *διέξοδοι* (1) habent BLMORSbeil || *τοῦ*

straw, by the movement, and that when the moist is heated *pneuma* is produced and ejected more quickly (for an equal place cannot contain the bulk when it becomes greater), and the ejection is called a sneeze?

2.48. Why do people breathe less often in the winter? Is it that in summer the air which is breathed in does not cool equally because of the heat, and then [something] to cause cooling is needed again, and for this reason [the person] breathes quickly?[288] This is also the reason why the dying breathe quickly; this happens because of the weakness of the inhalation. For a small amount [of air] inhaled does not cause cooling, and a small amount is inhaled because of the weakness of the drawing in. In the winter the opposite: for the season contributes to the cooling.

2.49.[289] Why does dizziness occur in those who stand up more than in those who sit down? Is it because when they are stationary the moisture gathers in the head more, and when they move this becomes apparent and causes the dizziness when it all inclines together to one part (and this is why raw <eggs>[290] cannot be spun, but fall over),[291] while when they are moving the moisture is even? They stand up, then, having been stationary, when they are disposed in such a way; but they sit down when they have been in motion,[292] at a time when the moisture is even and spread out.

2.50.[293] Why are other escapes of *pneuma*, such as breaking of wind and belching, not sacred, but only that in a sneeze?[294] Is it that of three regions,

[288] The explanation of breathing as cooling internal heat is orthodox Aristotelian doctrine: *De respiratione* 8 474b20, 11 476a16, *PA* 3.6 668b34 ff.

[289] = [Ar.], *Probl.* 6.4; a fuller version of Theophrastus, *Vert.* 13.

[290] Added by Usener from [Aristotle], and supported by Theophrastus.

[291] Cf. Theophrastus, *Vert.* 2.

[292] So [Aristotle]. The MSS of our text have "The movement is for those who come to sit down", but this gives a less good sense.

[293] = [Ar.], *Probl.* 33.9, cf. 33.7 (a compressed version of the same explanation).

[294] As when we say "bless you" to someone who has sneezed.

τόπων, κεφαλῆς καὶ θώρακος καὶ τῆς κάτω κοιλίας, ἡ κεφαλὴ θειότατον, (ἔστι δὲ φῦσα μὲν ἀπὸ τῆς κάτω κοιλίας πνεῦμα, ἐρευγμὸς δὲ τῆς ἄνω, ὁ δὲ πταρμὸς τῆς κεφαλῆς), διὰ δὲ τὸ ἱερώτατον εἶναι τόπον καὶ τὸ πνεῦμα τὸ ἐντεῦθεν ὡς ἱερὸν προσκυνοῦσιν; ἢ ὅτι τὰ πνεύματα σημαίνει τοὺς εἰρημένους τόπους βέλτιον ἔχειν, ὡς ἐπὶ τὸ πολὺ μὴ διαχωρούντων (κουφίζει γὰρ τὸ πνεῦμα διεξιόν), ὥστε καὶ ὁ πταρμὸς τὸν περὶ τὴν κεφαλὴν τόπον ὅτι ὑγιαίνει καὶ δύναται πέττειν; ὅταν γὰρ κρατήσῃ ἡ ἐν τῇ κεφαλῇ θερμότης τὴν ὑγρότητα, τὸ πνεῦμα τότε γίνεται πταρμός· διὸ καὶ τοὺς ἐκθνήσκοντας κρίνουσι πταρμικῷ, ὡς ἂν μὴ τούτῳ δύνωνται πάσχειν, ἀσώτους εἶναι· ὥστε σημεῖον ὑπὲρ τοῦ ἀρίστου τόπου καὶ ἱερωτάτου προσκυνοῦσιν ὡς ἱερόν, καὶ φήμην ἀγαθὴν ποιοῦνται.

2.51. *Διὰ τί ὁ ἄνθρωπος μᾶλλον πτάρνυται τῶν ἀλόγων ζῴων; πότερον ὅτι τοὺς πόρους εὐρεῖς ἔχει δι' ὧν τὸ πνεῦμα καὶ ὀσμὴ εἰσέρχεται; τούτοις γὰρ πληρουμένοις πνεύματος πτάρνυται· ὅτι δ' εὐρεῖς, σημεῖον ὅτι ἥκιστα ὀσφραντικὸν τῶν ἄλλων ζῴων· ἀκριβέστεροι γὰρ οἱ λεπτόποροι; ⟨εἰ οὖν⟩ εἰς μὲν τοὺς εὐρεῖς πλέον καὶ πλεονάκις εἰσέρχεται τὸ ὑγρόν, οὗ*

---

om. Oil || *μόνη*] *ἱερά* [Ar.] 3 *τόπων* MOPRSbeilm [Ar.] Buss. Us.: *τῶν τόπων* T: om. BL || *τῆς κεφαλῆς* BL || *καὶ*[1] om. BL || *κάτω* om. m || *θειοτάτη* M 3-4 *καὶ θώρακος – ἔστι* add. l[1] in marg. (v. Wegehaupt 1918, 472) 4 *δὲ*[1] Me [Ar.] Buss. Us.: om. cett. || *φῦσα*] *φύσει* L || *μὲν* om. e || *κοιλίας* om. BL || *δὲ* om. BL || *τῆς*[2]] *τοῖς* T 5 *δὲ*[2] MPRTem Us.: *γοῦν* cett.: om. [Ar.] || *τὸν ἱερώτατον* L: *τὸν ἱερότατον* B: *τοῦ ἱερώτατον* T 6 *τὰ πνεύματα*] *τὰ πνευμάτια* T: *ἅπαντα τὰ πνεύματα* [Ar.] 7 *ἔχει* BL 7-8 *γὰρ* post *διαχωρούντων* [Ar.] 8 *τὸ* om. Buss. 9 *ὑγιαίνει* [Ar.] Us.: *πνεῖ* codd. Buss. 10 *τὸ πνεῦμα* om. m 11 *ἐκθνήσκοντας* [Ar.] Buss. Us.: *ἔξω θνήσκοντας* codd. || *κρίνουσι*] *κίνουσιν* e: *κινοῦσι* [Ar.] || *πταρμικῷ*] *πταρμικᾶ* Oil: *πταρμικοῦς* P: *πταρνικῶν* m: *πταρτικῷ* *p* || *τούτῳ* M [Ar.] Us.: *τοῦτο* cett. Buss. || *δύνονται* B: *δίνονται* T 12 *πάσχει* BL || *ἀσώτους* [Ar.] Buss. (var. lect.) Us.: *ἀσώους* PTm: *σώους* cett. || *εἶναι*] *ὄντας* [Ar.] || *ὥστε*] *ὥστε ὡς* Buss.: *ὥσθ' ὡς* [Ar.] || *ὑπὲρ*] *ὑγείας* [Ar.] || *ἀρίστου*] *ἀορίστου* P: *ἀρρώστου* [Ar.] codd. 12-13 *τόπου* post *ἱερωτάτου* [Ar.] 13 *ἱερωτάτου* MPRTe[ac]m Buss. Us.: *ἱερώτατον* OSbe[pc]il : *ἱερότατον* BL

2.51 (CMOPRSTbeilm*p* Buss. Us.; [Ar.] 10.18, 33.10)

1 *ὁ ἄνθρωπος μᾶλλον πτάρνυται*] *πτάρνυται ὁ ἄνθρωπος μᾶλλον* P: *ἄνθρωπος πτάρνυται μάλιστα* (om. *ὁ*) [Ar.] || *ἀλόγων*] *ἄλλων* Pe [Ar.] 33.10: om. [Ar.] 10.18 || *πότερον* om. PT 2 *τοὺς* om. P || *εὐρεῖς*] *εὐχερεῖς* m [Ar.] 10.18 cod. Y[a1] (*εὐχρεῖς* cod. C[a]) || *ὀσμὴ* [Ar.] 10.18: *ἡ ῥύμη* codd. [Ar.] 33.10 Buss. Us. (hoc loco: sed *ὀσμὴ* coni. ipse Buss. in [Ar.] 33.10) || *τούτοις*] *τούτου* P 3 *πληρουμένου* P || *πνεύματος πληρουμένοις* [Ar.] 10.18 4 *ἄλλων*] *ἀλόγων* m: om. [Ar.] 33.10 || *γὰρ*] *δ'* M Us. || *λεπτοπόροι*] *λεπτοὶ πόροι* [Ar.] 33.10: *λεπτότεροι* [Ar.] 10.18 4-5 *εἰ οὖν εἰς μὲν* [Ar.] Us.: *εἰ μὲν οὖν εἰς* Buss.: *εἰς μὲν οὖν* T Wegehaupt: *εἰς μὲν* cett.

the head and the chest and the lower abdomen, the head is most divine: breaking of wind is *pneuma* from the lower abdomen, belching from the upper, but a sneeze from the head, and because it is the most holy place they also reverence the *pneuma* from there as holy? Or is it that the *pneuma* indicates that the afore-mentioned places are in a better condition, for the most part without passing [of excrement] (for *pneuma* brings relief when it passes out),[295] so that a sneeze [indicates] that the region around the head is healthy[296] and able to effect concoction? For when the heat in the head prevails over the moisture, it is then that the *pneuma* becomes a sneeze. And for this reason they test those who are dying with what causes sneezing, since if they cannot be affected by this, they are beyond recovery. So they reverence as holy the sign concerning the best and most holy place, and make it a good omen.

2.51.[297] Why does man sneeze more than irrational animals? Is it because in him the passages through which the *pneuma* and the scent[298] enter are wide? For it is when these are filled with *pneuma* that he sneezes. That they are wide is shown by the fact that he is less able to smell than the other animals; for those with slender passages are more accurate. <So, if> moisture enters into wide [passages] in greater quantity and more often, and it is when this is

---

[295] The text of [Aristotle] gives a better sense: "are in a better condition, for the most part; for without passing [of excrement] *pneuma* brings relief when it passes out."

[296] So Usener from [Aristotle]; the MSS have rather "breathes".

[297] = [Ar.], *Probl.* 10.18 = 33.10; cf. 10.54 (a different and much fuller explanation in terms of the upright posture of human beings and the rising of heat in the body), [Alex.], *Probl.* i.144 Ideler (a different explanation again, in terms of diet).

[298] Reading *osmê* with [Ar.] *Probl.* 10.18 against *rhumê* "inrush" with id. 33.10 (itself emended to *osmê* by Bussemaker) and the MSS of our text.

πνευματουμένου πταρμὸς γίνεται, τοιούτους δὲ μάλιστα τῶν ζῴων οἱ ἄνθρωποι ἔχουσι, πλειστάκις ἂν πτάρνυντο εἰκότως. ἢ ὅτι ἐλάχιστοι οἱ μυκτῆρες, ὥστε τὸ θερμανθὲν ὑγρὸν ταχὺ ἐξιέναι δύνασθαι πνεῦμα γενόμενον, ἐν δὲ τοῖς ἄλλοις διὰ μῆκος καταψύχεται πρότερον;

2.52. *Διὰ τί οἱ πρεσβῦται χαλεπῶς πτάρνυνται; πότερον ὅτι οἱ πόροι συμπεπτώκασι δι' ὧν τὸ πνεῦμα ἐκκρίνεται; ἢ τῷ ἀδυνατεῖν αἴρειν ῥᾳδίως, βίᾳ ἀφιᾶσι κάτω;*

2.53. *Διὰ τί ἐάν τις ἀπνευστιάζῃ, ἡ λὺγξ παύεται; ἢ διότι ἡ μὲν λὺγξ ὑπὸ καταψύξεως γίνεται διὸ καὶ οἱ φοβούμενοι καὶ οἱ ῥιγοῦντες λύζουσι, κατεχόμενον δὲ τὸ πνεῦμα ἐκθερμαίνει τὸν ἐντὸς τόπον;*

2.54. *Διὰ τί τῶν τετελευτηκότων οἱ ὄνυχες καὶ αἱ τρίχες αὔξονται ἐπί τινα χρόνον, καὶ τῶν πρεσβυτῶν; ἢ διότι τῶν περὶ τὸ σῶμα ἀπαθέστερα*

---

5 *τὸ ὑγρὸν εἰσέρχεται* P 6 *πνευματομένου* C: *πνευματούμενον* m || *ὁ πταρμὸς* [Ar.] Us. 7 *ἔχουσι* MPRTe^{ac}m [Ar.] Buss. Us.: *ἔχοντες* COSbe^{pc}il Wegehaupt || *ἂν* om. P || *πτάρνοιντο* Pm: *πταρνύοιντο* vel *πτάροιντο* vel *ἐπτάρνυντο* coni. Buss. || *ἢ ὅτι*] *ὅσοις* [Ar.] 33.10 || *ἐλάχιστοι κατὰ τὸ μῆκος* [Ar.] 10.18 8 *τὸ* om. R || *θερμαθὲν* C || *δύναται* [Ar.] 33.10 8-9 *ταχὺ – γενόμενον*] *ταχὺ δύναται πνεῦμα γίνεσθαι* [Ar.] 10.18 9 *γινόμενον* em || *τὸ μῆκος* Buss. Us.: *τὸ* om. codd. [Ar.] Wegehaupt

2.52 (BLMOPRSTbeilm*p* Buss. Us.; [Ar.] 33.12)

2 *συμπτώκασι* BL || *ἐκκρίνεται* om. [Ar.] || *τῷ*] *διὰ τὸ* P || *αἴρειν* nos ex [Ar.]: *ἄνω βλέπειν* MPRTem*p* Buss. Us.: *ἀναβλέπειν* BLOSbil 2-3 *τῷ – ῥᾳδίως*] *ἢ ὅτι αἴρειν τὰ* (*αὐτὸ* Richards) *ἄνω οὐκέτι δυνάμενοι ῥᾳδίως, εἶτα* [Ar.]

2.53 (BLMOPRSTbeilm*p* Buss. Us.; [Ar.] 33.13)

1 *ἀπνευστιάζῃ* [Ar.] Us.: *ἀποπνευματίζῃ* codd. Buss. || *λὺγξ*] *λὺξ* (bis) BLP || *ἢ* om. BL || *λὺγξ*^{2} om. [Ar.] || *ὑπὸ* MPRTem*p* [Ar.]: *ἀπὸ* BLOSbil Buss. Us. 2 *καταψύξεως*] *ψύξεως* BL Buss. || *ῥιγοῦντες* BLMORSTbeilm Buss. [Ar.]: *ῥιγῶντες* P*p* Us. || *λύξουσι* BL 3 *ἐκθερμαίνει*] *θερμαίνει* *p* Buss. || *τὸν τὸν* l || *ἐντὸς*] *ἐκτὸς* PRTbe(sed *ἐντὸς* e^{mg})m*p*

2.54 (BLMOPRSTbeilm Buss. Us.)

1 *τελευτηκότων* BLRSe^{pc} || *αὐξάνονται* PTm 2 *τινων* P || *χρόνων* PT || *τῶν περὶ*] *τὸ πυρὶ* BL || *ἀπαθέστερα* BLMRSTbelm Buss.: *ἀπαθέστερον* POi: *ἀπαθέστατα*

converted into *pneuma* that a sneeze occurs, and human beings most of [all] animals have these [wide passages], it is reasonable that [man] should sneeze very often. Or is it that the nostrils are very small, so that the heated moisture can pass through swiftly when it becomes *pneuma*, while in other [creatures] it is cooled first because of [the] length [of the nostrils]?

2.52.[299] Why do old people have difficulty in sneezing? Is it because the passages through which the *pneuma* is ejected have collapsed? Or because they cannot easily raise [the breath] up,[300] and [when they have, they][301] emit it forcibly downwards?

2.53.[302] Why, if someone holds his breath,[303] do the hiccoughs stop? Is it because the hiccoughs are brought about by chilling (and this is why those who are afraid, and those who shiver from cold, hiccough), and the breath when it is held back heats the interior part?

[Nails and hair][304]

2.54.[305] Why do the nails and hair of those who have died grow for a time, and those of old people? Is it that, because these are the less affected things

299 = [Ar.], *Probl.* 33.12.

300 Reading *airein* from [Aristotle] who, as emended by Richards, has "because they can no longer raise [the breath] up". The MSS of our text have not *airein* but *anô blepein*, "because they cannot easily look up."

301 So [Aristotle].

302 = [Ar.], *Probl.* 33.13. See also above, 2.45; below, 2.111; and [Alex.], *Probl.* i.48 Ideler (we are grateful to Lou Filius for drawing our attention to the latter parallel).

303 So [Aristotle]; the MSS of our text have rather "breathes out".

304 Theophrastus wrote a treatise on hair: **328** no.13 FHS&G. Cf. Ar., *HA* 3.10-12; *GA* 5.3-6, and Sharples 2004.

305 2.74 below explains both the cases, that of the old and that of those who have died, by weakness and lack of heat causing nourishment to be directed to the less necessary parts, 2.59 the case of the old by the presence of more earthy residue in old age. These explanations are not necessarily inconsistent. [Alex.], *Probl.* i.145 gives two different explanations in the case of those who have died: that the growth is only an apparent effect of the flesh wasting away, and that it is because hair and nails are made up of residues.

ταῦτ᾽ ἐστίν (ὕστερον οὖν τὸ πάθος τοῦ συμπτώματος), ἐπί τινα χρόνον αὔξεται;

2.55. *Διὰ τί ποτε ἐν τῷ πώγωνι τρίχες βραδύτερον παραγινόμεναι θᾶττον φθείρονται; ἢ διότι ὡς ἐπὶ τὸ πολὺ αἱ ἐν τῷ πώγωνι τρίχες πυρρότεραι τῶν ἐν τῇ κεφαλῇ γίνονται; ἐλάττων οὖν ἡ μετάβασις ἐκ τῆς πυρρᾶς τριχὸς εἰς τὴν λευκὴν ἢ ἐκ τῆς μελαίνης. ἢ ὅτι τὰ ἐκ τῆς κεφαλῆς ῥεύματα πάντα εἰς τὸ στόμα φέρεται καὶ εἰς τοῦτον τὸν τόπον; διὸ ἀσθενεστέρα ἡ θρὶξ γίνεται· διὰ τὴν ἀσθένειαν οὖν μᾶλλον πολιοῦται. εἶτα τριχὸς οὐχ αὕτη ἐστι διαφορὰ μόνον, ἀλλὰ καὶ τὸ ἐκρεῖν τὴν κεφαλὴν μᾶλλον ἢ ἐν τῷ πώγωνι ἐγγίνεται· φαλακροὶ γὰρ τὴν κεφαλὴν τελέως γίγνονται;*

2.56. *Διὰ τί ὁ πώγων ὡς ἐπὶ τὸ πολὺ καὶ μᾶλλον καὶ πρότερον πολιοῦται τῆς κεφαλῆς ὕστερος γιγνόμενος; ἢ ὅτι ἡ πολιὰ δι᾽ ἀσθένειαν παραγίνεται τόπου καὶ τροφῆς, ὥστε ὅπου τοιαῦται αἱ τρίχες, καὶ ὅστις μάλιστα τοιοῦτος τόπος, οὗτος ἂν μάλιστα [πλεῖστα] πολιοῖτο· αἵ δ᾽ ἐν τῷ γενείῳ τρίχες τοιαῦται, καὶ ὁ τόπος τοιοῦτος; σημεῖον δὲ καὶ τὸ θῆλυ· ἐνταῦθα γὰρ μόνον τρίχας ἁπλῶς οὐκ ἔχει, ἐπὶ δὲ τῆς ἥβης καὶ ἐν ταῖς μασχάλαις ἔχει· καὶ οἱ τὴν γένεσιν διαφθειρόμενοι μάλιστα κυρίως τοῦτο ἐπιτηροῦνται·*

Us. 3 *οὖν*] *δὲ* P || *ἐπὶ*] *δ᾽ ἐπὶ* Tm

2.55 (MOPRSTbeilm*p* Buss. Us.)

1 *αἱ τρίχες p* Us.: *αἱ* om. codd.: ante *ἐν* add. Buss. (in var. lect.) 2 *φθείρονται*] *λευκαίνονται* Me[mg] Buss. 3 *ἐν τῇ κεφαλῇ*] *ἐκ τῆς κεφαλῆς* M || *ἢ ἐλάττων* e 4 *τὴν*] *τὸ* P || *λευκὸν* P || *τῆς*[2] om. Sbil Us. 5 *τὸν* om. S || *ἡ θρὶξ τῆς κεφαλῆς* m 6 *οὐχ αὕτη* Us.: *οὐχ ἡ αὐτὴ* MPRSTbem*p* Buss.: *οὐχὶ αὐτὴ* Oil 7 *τὸ* Buss. (in var. lect.) Us.: *ἐν τῷ* MPRSTbem Buss.: *ἐκ τῶ* Oil || *ἐκρεῖν*] *ἐκρρεῖν* Oil: *ἐκκλίνειν* P || *τὴν κεφαλὴν*] *ἐν τῇ κεφαλῇ* Buss. (in var. lect.) Us. 8 *τελείως* M

2.56 (DGMOPRSTYbeilmo*p* Buss. Us.)

1 *καὶ*[1] om. DOil || *μᾶλλον* om. D || *καὶ*[2] om. D 1-2 *διὰ τί – ἢ*] *ὕστερον πολιοῦται ὁ πώγων τῆς κεφαλῆς p* 2 *ὕστερον* Oil Wegehaupt || *γενόμενος* T[ac] || *ἢ* om. DY Buss. || *ἡ* om. OPRSTbeilm || *διασθένειαν* Ol || *ἀσθένειαν τινὰ* Go || *παραγίνεται*] *γίνεται* DGYo*p* Buss. 3 *τόπου* DGYe Buss. Us.: *πότου* MOPRSTbilo: *ποτοῦ* m*p* || *αἱ* om. DY || *ὅστις*] *ὅπου* D 4 *οὗτος*] *οὕτως* m || *ἂν* om. D || *πλεῖστα* om. DY, del. Us.: *καὶ πλεῖστα* Buss. || *πολιοῖτο*] *πολιοῦται* DGYo || *αἱ δ᾽*] *αἱ τε οὖν* D || *γενείω*] *γένει* T: *πώγωνι* D 4-9 *αἱ δ᾽ ἐν τῷ γενείῳ – πάσχει* om. o 5 *τρίχες τοιαῦται* D Buss. Us. || *καὶ*[2]] *ὅτι καὶ* DY || *γὰρ* om. D 6 *ἁπλῶς* om. DY: trasp. post *ἔχει* m || *ἔχει ἐπὶ*] *ἔχει ἐν ἄλλοις δὲ μέρεσι ἐπὶ* D || *δὲ*] *τὲ* D || *ταῖς* om. GOPSTYbilm || *ἔχει*] *ἐστὶ* D 7 *γένεσιν*] *γέννησιν* G: *γένει* Y || *διαφθειρόμενος* e: *μάλιστα – ἐπιτηροῦνται*] *δι᾽ ἀσθένειαν παρατηροῦνται τοῦτο πάσχειν* D Buss. || *μάλιστα κυρίως*] *κυρίως μάλιστα* Y: om. Buss. 8 *καὶ ὅσοι* D

in the body (so the affection [occurs] after the mishap),[306] they grow for a time?

2.55. Why do hairs in the beard come more slowly but go grey[307] more quickly? Is it because for the most part the hairs in the beard are redder than those on the head? So there is a lesser change from red hair to white than there is from black. Or is it that all the fluxes from the head travel to the mouth and this region, and for this reason the hair becomes weaker? So it goes grey more because of the weakness.[308] And then this is not the only difference in the hair,[309] but also the head sheds [hair] more than happens in the beard; for [people] become completely bald on their heads.

2.56. Why does the beard for the most part go grey more than and earlier than [the hair on] the head, when it grows later? Is it because the greyness comes because of a weakness in the place and the nourishment,[310] so that where the hairs are like this, and where the region is most like this, this will go grey most, and the [hairs] on the chin are like this, and the place is like this? This is indicated by the female [sex]; for here alone [this] absolutely does not have hairs, but has them in the genital region and the armpits; and males who have become sterile[311] are most observed [to have] this as a

[306] i.e.: it takes time for nails and hair to be affected in their growth by the death of the whole body.

[307] Literally "corrupt". Cf. below, 2.64, 2.78.

[308] See below, 2.56; decline in heat, 2.75; insufficient concoction of nourishment, 2.64, 2.78, 2.81.

[309] The Greek might rather suggest "this is not a difference only in the hair", but that seems to fit the sequel less well, unless indeed the contrast is between the nature of the hair itself on the one hand and its shedding on the other.

[310] See above, 2.55, and below, 2.64, 2.75, 2.78, 2.81.

[311] "who have become sterile through weakness", Buss.

ὥστε ὅσοι πολιοῦνται, διὰ τοῦτο τὸν πώγωνα πολιώτατοι γίνονται· πρῶτος γὰρ δι' ἀσθένειαν μάλιστα πάσχει.

2.57. *Διὰ τί ἐν τῇ ἥβῃ τρίχες βραδύτεραί εἰσι καὶ ἧττον αὔξονται τῶν ἐν τῷ πώγωνι πρότερον γινόμεναι καὶ ἐν ὑγροτέρῳ τόπῳ οὖσαι; διότι ὁ τόπος ἀραιὸς ὁ περὶ τὴν ἥβην ἐστὶ καὶ ὑγρός, ὥστε ἀεὶ μὲν ἔχει τροφὴν (διὸ οὐκ ἐλλείπει) ἀλλ' ὀλίγην· οὐ γὰρ ὑπομένει ὡς αὐξητικόν· ὅτι δ' ἀραιὸς ὁ τόπος, δηλοῖ ὁ γινόμενος ὄγκος τῶν αἰδοίων· ὁ δὲ τοῦ γενείου ἧττον τοιοῦτος.*

2.58. *Διὰ τί ἐν τοῖς ἐπὶ τῶν ποδῶν ὄνυξι τὰ λευκὰ σημεῖα οὐκ ἐγγίνεται ὥσπερ ἐπὶ τοῖς τῶν χειρῶν, ἃ καλοῦσιν οἱ μὲν ἐραστάς, οἱ δὲ ψεύδη; ἢ διότι τῷ βραδέως αὐξάνεσθαι τοὺς ἐπὶ τῶν ποδῶν ὄνυχας ταχὺ μελαίνεται πρὶν κινηθῆναι καὶ αὐξηθῆναι [ἐπεὶ δῆλον] τὸ ἐπὸν ἐπὶ τοῦ δέρματος, τοῦτο δ' ἐστὶ τὰ λευκὰ σημεῖα, ἅπερ ἐν τοῖς ἄνω ὄνυξι [τὰ λευκὰ] φαίνονται· οἱ δ' ἐπὶ τῶν χειρῶν ταχὺ αὔξονται.*

2.59. *Διὰ τί οἱ τῶν γερόντων ὄνυχες σκληρότεροι ἢ οἱ τῶν νεωτέρων εἰσίν, αὔξονται δὲ θᾶττον; ἢ σκληρότεροι μὲν διότι καὶ τὸ δέρμα σκληρότερον καὶ τὰ ὀστᾶ καὶ ὅλα τὰ γεώδη διὰ τὸ ἐξανηλῶσθαι τὸ ὑγρόν, ὁ δὲ ὄνυξ τοιούτου γένους ἐστί· θᾶττον δὲ αὐξάνεται, διότι πλείονα τροφὴν λαμβάνει· πλεῖον μὲν γὰρ τό γε περίττωμα τὸ γεῶδες ἐν τῷ γήρᾳ, ἐκ δὲ τῶν τοιούτων περιττωμάτων τρέφονται τρίχες καὶ ὄνυχες. ἢ ὅτι καὶ τὰ μὲν*

|| *πολλὰ πολιοῦνται* D || *τὸν πώγωνα διατοῦτο* 1[ac] || *διὰ – γίνονται*] *καὶ λευκότατον ἔχουσι τὸν πώγωνα* D || *πρώτως* G 8-9 *πρῶτος γὰρ* om. D 9 *ἀσθένεια* D || *μάλιστα πάσχει*] *πάσχει μάλιστα* Oil: *μάλιστα τοῦτο πάσχουσι* D

2.57 (MOPRSTbeilm*p* Buss. Us.)
αἱ τρίχες p Us. || *βραχύτεραί* e[ac] Buss. (in var. lect.) 2 *πρότερον*] *πρότεραι* Tm 3 *περὶ*] *παρὰ* M 4 *ἐλλείπει*] *ἐλλιπῇ* P: *ἐκλείπει* Buss. || *ὀλίγον* M Buss. || *αὐξητικόν* MPRTem Buss.: *αὐξητέον* OSbe[mg]il Us. || *ἀρεὸς* T 5 *τοῦτο δηλοῖ* P Us. || *δηλοῖ*] *δῆλον* e[ac] || *ὁ*[1]] *ὃ* T || *τῶν αἰδίων ὄγκος* m 5-6 *ὁ δὲ – τοιοῦτος* om. T

2.58 (MOPRSTbeilm*p* Buss. Us.)
2 *τοῖς* om. OSbil || *ἃ*] *ὃ* PTm || *οἱ δὲ*] *τινὲς δὲ* m 3 *τῷ*] *τὸ* OTil 4 *ἐπεὶ δῆλον* del. Us.: *οὐκ ἐπίδηλον* Buss. (in var. lect.) || *ἐπὸν*] *ἔπον* MRTe Buss.: *λεπτὸν* m 5 *ἅπερ* P Us.: *ὅπερ* cett.: *διόπερ* coni. Buss. || *τὰ λευκὰ* del. Us. 6 *ταχὺ*] *τὰ μεταξὺ* T

2.59 (DGMOPRSTYbeilmo*p* Buss. Us.)
1 *οἱ*[1] om. o || *ἢ* om. o || *νεωτέρων*] *νέων* DGTYo 2 *εἰσὶν* om. DY: post *σκληρότεροι* o || *αὔξονται δὲ*] *αὐξῶνται δὲ* G: *δὲ αὔξονται* o || *ἢ* om. D || *διότι*] *ὅτι* T 3 *καὶ* om. DGYo || *ὅλα*] *ὅλως* Buss. || *ἐξανηλῶσθαι*] *ἐξαναλῶσθαι* G: *ἐξαναλωθῆναι* DYo 4 *πλείονα*] *πλείω* DGTYmo: *πλείων* P 5 *λαμβάνεται* P || *πλεῖον*] *πλείω* GTmo || *γε* Us.: om. D: *τε* cett. Buss. || *ἐκ*] *ἐν* o 6 *περιττωμάτων*

proper [feature]. So it is for this reason that those who go grey become most grey in their beard; for because of its weakness it undergoes this first most of all.

2.57. Why are hairs in the genital region slower [to grow] and grow less than those in the beard, [although] they occur earlier and are in a moister region? Because the region around the genitals is of loose texture and moist,[312] so that it always has nourishment (and that is why [this hair] does not stop), but not much of it; for it does not remain, so as to cause growth.[313] That the region is of loose texture is shown by the size to which the genitals grow; the [size] of the chin is less like this.

2.58.[314] Why do the white marks which some people call "lovers" and others "lies" not occur in the toenails in the same way that they do in the fingernails? Is it because the toenails grow slowly and so go dark before what is on the skin is moved and grows, and this is the white marks which appear in the upper part of the nails, while the finger[nails] grow swiftly?

2.59. Why are the nails of old men harder than those of younger ones, but grow faster? Is it that they are harder because the skin is harder and the bones and all the earthy parts, because the moisture has been exhausted, and the nail is of this sort, while it grows more swiftly because it receives more nourishment? For there is more earthy residue in old age, and it is from such residues that hairs and nails are nourished.[315] Or is it because some bones

---

[312] See below, 2.81, and n.331. The connection of *araios*, "of loose texture", with the growth of pubic hair seems to come not from the discussion of hair in *GA* 5, but from *GA* 1.20 728b26, where it is the fact that such air starts growing *at all* at puberty that is explained by the claim that the region (not specifically the skin) becomes more *araios* then. (This is the only occurrence of *araios* in *GA*).

[313] Buss.'s *auxêtikon* is easier than Usener's *auxêteon*.

[314] Cf. [Alex.], *Probl.* i.146, which also discusses why the marks on the fingernails occur more in children, and has a different explanation for both points; they are due to phlegm, and in the feet this is dispersed by exercise.

[315] See below, 2.74; Ar., *GA* 5.3 782b1.

ὀστᾶ μέχρι τινὸς αὐξάνεται, τὰ δὲ οὔ;

2.60. *Διὰ τί οἱ τοῦ ἀνθρώπου ὄνυχες μαλακώτατοι; πότερον ὅτι καὶ τὸ δέρμα, οἱ δ' ὄνυχες ἐκ τοῦ δέρματος γίνονται; ἢ ὅτι τοῖς ἄλλοις ἄλλοι ὄνυχες· ὅσοις οὖν σκληρότεροι, τούτοις χρήσιμοί εἰσι μᾶλλον;*

2.61. *Διὰ τί οἱ ἐκ γενετῆς τυφλοὶ οὐ γίνονται φαλακροί; πότερον ὅτι μεθέστηκεν ἡ θερμότης πηρωθέντων τῶν ὀμμάτων εἰς τὸν πλησίον τόπον· οὐκοῦν οὐ καταψύχεται ὁ τόπος, ἀλλ' ἀεὶ ἔχει τροφήν· ἡ δὲ φαλακρότης ἀπὸ ψύξεως; ἢ διότι ποτὲ μὲν ὁ τόπος ἔχει θερμότητα ὥστε ἀναλίσκειν τὴν τροφὴν καὶ οὐ γίνεται ἐν τῷ προσθεν περίττωμα· πηρωθέντων δὲ κατέψυκται, ὥστε πολὺ τὸ περίττωμα ἐν τῷ τόπῳ, αἱ δὲ τρίχες γίνονται ἀπὸ περιττώματος;*

2.62. *Διὰ τί ὁ μὲν πώγων γίνεται πυρρὸς ἄνευ τῆς κεφαλῆς, ἡ δὲ κεφαλὴ πυρρὰ ἄνευ τοῦ πώγωνος οὐ γίνεται; ἢ διότι αἵ τε πυρραὶ τρίχες δι' ἀσθένειαν γίνονται, καὶ ἐν τῷ πώγωνί εἰσι τρίχες ἀσθενέστεραι φύσει; σημεῖον δέ· ἀγένειοι γὰρ ⟨οὐ⟩ γίνονται ἀλλ' ἢ ὀψὲ φαλακροί· δῆλον οὖν ὡς ὁ τόπος οὐ δύναται φέρειν· διὸ ἐὰν μὲν καὶ ὁ ἰσχυρότερος πυρρὰς φέρῃ, καὶ τὸν ἀσθενέστερον ἀνάγκη· ἐὰν δὲ οὗτος, οὔπω ἀνάγκη καὶ τὸν ἰσχυρότερον.*

2.63. *Διὰ τί αἱ μακραὶ τρίχες ῥέουσιν; ἢ διότι μοχλὸς μὲν ἡ θρίξ, ὑπομόχλιον δὲ τὸ δέρμα γίνεται, εὐκίνητα δὲ τὰ πλέον ἀπέχοντα τοῦ ὑπομοχλίου· καὶ ὅλως τοῦ ἠρεμοῦντος, δῆλα δὲ ταῦτα ἐν τοῖς* Μηχανικοῖς.

---

post *ὄνυχες* habent OPRSTbilm || *καί*^{2} om. DGYo 7 *αὐξάνεται*] *αὐξάνονται* OSbilm: *αὔξονται* Do: *αὐξῶνται* GY

2.60 (KMOPRSTbeilm*p* Buss. Us.; *ν* 2.2)

1 *πρότερον* e 2 *οἱ δ'*] *οἱ γὰρ* Se^{mg}*ν* || *ὄνυχες χρήσιμοι* K*ν* || *τοῖς* om. K 3 *τούτου* P

2.61 (KMOPRSTbeilm*p* Buss. Us.; *ν* 2.3)

4 *ποτὲ*] *ὁπότε* K 5 *γίνεται*] *γίνεσθαι* Us. || *πρόσθεν*] *ἔμπροσθεν* P Us. || *πηρωθέντων*] *πυρωθέντων* e: *πηρωθεὶς* K: luminibus ... orbatus *ν* 7 *περιττωμάτων* P

2.62 (KMOPRSTbeilm*p* Buss. Us.; *ν* 2.4)

2 *πυρὰ* S || *τε* om. Oi*p* 3 *καὶ ⟨αἱ⟩* Buss. (in var. lect.) Us. 4 *ἀγγενίοι* T: *οἱ γὰρ ἀγένειοι* Buss. (in var. lect.) || *οὐ* add. Buss. (in var. lect.) Us. || *ἀλλ' ἢ ὀψὲ φαλακροὶ*] *φαλακροὶ δ' οὐ ἀλλ' ἢ ὀψὲ* K*ν* 6 *τὸν ν* Buss. (in var. lect.) Us.: *τὸ* codd. || *δὲ*] *τε* T || *τὸν ν* Buss. (in var. lect.) Us.: *τὸ* codd.

2.63 (BKLMOPRSTbeilm*p* Buss. Us.; *ν* 2.5)

1 *αἱ μακραὶ*] *ῥεῦμα καὶ* BL 2 *δέρμα*] *ῥεῦμα* L: *ῥεῦγμα* B || *τὸ πλέον* BL || *ἀπέχοντα*] *ὑπάρχοντα* BL 3 *δὲ* om. BLOil

grow up to a certain point, some not?

2.60. Why are man's nails very soft? Is it because the skin [is] too, and the nails come from the skin? Or is it because different [creatures] have different nails, so those who have harder ones have more useful ones?

2.61. Why do those who are blind from birth not become bald? Is it because when the eyes were impaired the heat moved to the neighbouring region, so the region is not chilled, but always has nourishment, and baldness comes from chilling? Or is it because [normally] the region possesses heat, so that the nourishment is used up, and there is no residue in the forward part, but when [the eyes] are impaired [the region] is chilled, so that there is much residue in it, and the hairs come from the residue?

2.62. Why does the beard become red without the head [doing so], but not the head without the beard? Is it because red hairs result from weakness,[316] and in the beard hairs are weaker by nature?[317] An indication [of this]: men without beards do <not> become bald except late in life.[318] It is clear then that the region cannot produce [hair]; so if the stronger region produces red [hairs], the weaker one must [do so] too, but if this does, it is not therefore necessary for the stronger one [to do so] too.

2.63. Why do long hairs fall out? Is it because the hair is a lever and the skin the fulcrum, and those things are more easily moved which are further away from the fulcrum and in general from what is at rest? This is [made] clear in the *Mechanics*.[319]

---

316 Cf. Ar., *GA* 5.5 785a20 (in the context of explaining why red hair goes grey faster than black); below, 2.72.

317 See above, 2.55.

318 The Greek (as emended by Usener whom we follow here) could also mean "men do not lose their beards, apart from bald men who do so late in life", but that is neither true nor to the point. The argument as given in the translation is relevant – if the hair on the head was weaker than the beard, beardless men would go bald quickly – though hardly a good one (many men who *do* have beards retain them after going bald). The reading in K, followed by Valla, *could* mean "there are men who have no beards, but none who are bald except late in life".

319 [Ar.], *Mechanica* 3 850b2 (for the general point concerning the lever, with no specific reference to hair). The explanation may seem bizarre, but such enthusiasm for the principles of mechanics is not unknown in the Bekker *Problems* (cf. 2.23, where this is admittedly only one explanation of the mechanics of erections).

2.64. Διὰ τί ἐν ταῖς βλεφαρίσι τρίχες οὐ γίνονται πολιαί; πότερον δι' ἰδιότητα τῆς τροφῆς τῆς περὶ τοὺς ὀφθαλμούς; σημεῖον δέ· πίων ἡ σὰρξ γίνεται· ἀπὸ δὲ τῆς τοιαύτης τροφῆς εὐλόγως οὐκ αὔξεται ἐπὶ πολὺ ἡ θρὶξ αὐταῖς (ἐκ γὰρ τῶν γεωδῶν γίνεται περιττωμάτων) οὔτε πολιοῦται· φθορᾷ γάρ τινι καὶ φαυλότητι τῆς τροφῆς πολιοῦσθαι συμβαίνει.

2.65. Διὰ τί οἱ ἐπὶ τῶν χειρῶν ὄνυχες θᾶττον αὐξάνονται ἢ οἱ ἐπὶ τῶν ποδῶν; ἢ ὅτι εἰς τὰ ἄνω ἡ αὔξησις μᾶλλον χωρεῖ; διὸ καὶ τρίχες μᾶλλον αὐξάνονται ἄνωθεν ἐπὶ τοῦ γενείου ἢ ἐπὶ τῆς ἥβης. ἢ διότι ἐν ἀλέᾳ μὲν αἱ χεῖρες, οἱ δὲ ἐν ψύξει, αὔξεται δὲ τὰ ἀλεαινόμενα μᾶλλον πάντα; ἢ ὅτι θερμότεραι αἱ χεῖρες καὶ ὑγρότεραι, οἱ δὲ πόδες ψυχρότατοι τοῦ σώματος, καὶ τὰ τοιαῦτα μέρη τῶν ἄλλων, λέγω δ' οἷον τρίχες τε καὶ ὄνυχες; διὰ τοῦτο οὖν αὐτὸ τὸ καὶ ψυχροὶ καὶ ἐν ψυχρῷ τόπῳ εἶναι, δυσαυξεῖς εἰσιν· οἱ δ' ἐπὶ τῶν χειρῶν τοὐναντίον.

2.66. Διὰ τί αἱ τρίχες οὖλαι γίνονται; ἢ ὅτι δύο φορὰς φέρονται κρατουμένης δ' ἔσωθεν φορᾶς * * * ⟨ἢ⟩ ὑπὸ τῆς ἐν τῷ περιέχοντι θερμότητος; διὸ καὶ οἱ Αἰθίοπες τοιοῦτοι.

2.67. Διὰ τί ἐν τοῖς πυρικαύστοις τρίχες οὐ φύονται; πότερον ὅτι πυκνὸν τὸ δέρμα γίνεται, δεῖ δὲ ἀραιὸν εἶναι, ᾗπερ ἐκφύεται θρίξ;

2.64 (KMOPRSTbeilm*p* Buss. Us.; *v* 2.6; 1 Petrus ap. Olivieri 1988, 197)
1 *αἱ ἐν* KMObi Buss. Us. || *τοῖς* m^{pc}K || *βλεφάροις* K || *πότερον* K*v* Us.: om. cett. Buss. 1-2 *δι' ἰδιότητα* K*v*: *ἐκ ποιότητος* Buss. Us.: om. cett. 2 *τῆς τροφῆς* *v* Buss. Us.: post *ὀφθαλμοὺς* exhib. K: om. cett. || *τῆς*^{2}] *ταῖς* Obil || *ὅτι πίων* K*pv* 3 *οὐκ*] *οὔτε* K || *ἐπὶ πολὺ*] *ἐπὶ πολλὰς* P 4 *αὐταῖς* K Us.: *αὐτῆς* cett.: *αὐτή* vel *αὖθις* coni. Buss. || *γεωδῶν* Us.: *τοιούτων* codd.: *τοιούτων* ⟨*οὐ*⟩ Buss. || *γίνονται* OSbe^{mg}il 5 *φαυλότητα* e
2.65 (KMOPRSTbeilm*p* Buss. Us.; *v* 2.7)
1 *ὄνυχες* K^{pc}MR^{pc}e^{pc}*v* Buss. Us.: om. cett. || *θᾶττον*] *μᾶλλον* m*p* || *οἱ*^{2} om. RTe Buss. 2 *τὰ*] *τὸ* KOPSbil || *αἱ τρίχες* *pv* Buss. (var. lect.) Us. 3 *τοῦ* om. OPSTbilm || *ὅτι* K || *ἀλέα* Buss. 4 *οἱ δὲ*] *οἱ δὲ πόδες* K*v* Buss. (var. lect.) || *ἐν ψύχει* KOil || *ἀλεαινόμενα* Buss.: *ἀλαινόμενα* MRe 5 *ψυχρότατον* Tm 6 *μέρη*] *τὰ μέρη* T 7 *οὖν αὐτὸ τὸ καὶ* Us., ita quoque (vel *οὖν αὖ τὸ καὶ*) coni. Buss.: *τὸ αὐτὸ οὖν καὶ* m: *αὐτοὶ οὖν καὶ* P: *αὐτὸ οὖν καὶ* T: *οὖν αὐτὸ καὶ* cett. Wegehaupt || *εἶναι*] *ὄντες* KOSbil*v* Wegehaupt
2.66 (ACKMOPRSTbeilm*p* Buss. Us.; *v* 2.8)
1-2 *Διὰ τί – περιέχοντι* amissa in A 1 *οὐλαὶ* Pm*p* || *φορᾶς* P 2 *δ'*] *δὲ τῆς* K || *ἔσωθεν*] *introrsum* *v*, ex quo *ἔσω* Buss. (in var. lect.) || lacunam post *φορᾶς* stat. Us. || *ἢ* suppl. Us. || *ὑπὸ* K Buss. (in var. lect.) Us.: om. cett.
2.67 (ADGKMOPRSTYbeilm*p* Buss. Us.; *v* 2.9)
1 *πυρικαύτοις* RTYem Buss. || *πότερον* om. D || *πυκνὸν*] *πικρὸν* m^{ac} 2 *ᾗπερ*] *ἧπερ* G: *ἤπερ* m: *εἴπερ* AY: *ὅθεν* D || *ἐκφύονται* DGY || *θρὶξ*] *ἡ θρὶξ* m: *θριξὶ* S: *τρίχες* DGY

2.64. Why do hairs on the eyelids not become grey? Is it from a particular quality of the nourishment around the eyes? An indication [of this]; the flesh becomes fatty. It is reasonable that from such nourishment the hair on [the eyelids] does not grow much (for [hair] grows from earthy residues),[320] nor go grey; for it is by some corruption and inferiority of the nourishment that going grey occurs.[321]

2.65. Why do the fingernails grow faster than the toenails?[322] Is it because growth tends upwards? And for this reason [hairs][323] grow more above, on the chin, than in the genital region. Or is it because the hands are in warmth, the [feet] subject to chilling, and all things grow more if they are heated? Or is it because the hands are hotter and moister, the feet the coldest parts of the body, and such parts are [colder] than the others, I mean hairs and nails? So that for this very reason, that they are both cold and in a cold region, [the toenails] grow with difficulty; the opposite for the finger[nails].

2.66.[324] Why do hairs become curly? Is it because they move with two movements, and when the movement inwards prevails ... ? <Or>[325] is it [brought about] by the heat in the surrounding environment? That is why the Ethiopians are like this.

2.67. Why do hairs not grow on [places] that have been burned by fire? Is it because the skin becomes dense, and it needs to be rare where hair grows?[326]

---

[320] Adopting Usener's conjecture. Bussemaker conjectures "[hair] does <not> grow from such residues".

[321] Above, 2.55, 2.56; below, 2.75, 2.78, 2.81.

[322] An explicit reference to nails is present in only two MSS, M and e, and in Valla's translation; the other MSS (including the one Valla himself used, K) have simply "those on the hands" and "those on the feet". It looks very much as if this problem originally followed another referring to nails (e.g. 2.60), so that the reference could easily be supplied from the context, and that only M and e have added, as an emendation, the term now required.

[323] Added in the text itself by Psellus, Bussemaker and Usener.

[324] Cf. Ar., *GA* 5.3 782b19-783a2, which gives the two explanations here in much fuller detail; either the tension between the heat and the earthiness in the air twists it, or drying due to heat causes the hair to contract and twist.

[325] Usener's addition of "Or" is justified by the two alternative explanations in Ar. loc.cit.; for the second explanation see also cf. 2.73 below.

[326] 2.57 above at first sight seems in conflict with this, as does 2.71 below. But see below, n.331.

2.68. Διὰ τί αἱ παχεῖαι τρίχες οὐ γίνονται μακραί; ἢ ὅτι ἡ εἰς τὸ μῆκος αὔξησις εἰς τὸ πάχος ἀναλίσκεται; ἢ ὅτι αἱ παχεῖαι σκληραί εἰσι, τὸ δὲ σκληρὸν δυσκίνητον ἐπὶ τὴν αὔξησιν, ἐπεὶ [σκληρὰς] διὰ τὸ γεῶδες συμβαίνει, ἡ δὲ γεώδης τροφὴ οὐ θέλει ἄνω πέμπεσθαι· διὰ βραχύτητα οὖν τῆς κινήσεως τῷ αὔξοντι θερμῷ φθάνει καταλαμβανομένη ἡ ἑτέρα ὑπὸ τῆς ἑτέρας, ὥστε εἰς βάθος καὶ πλάτος αὔξεσθαι καὶ οὐκ εἰς μῆκος;

2.69. Διὰ τί δὲ πυκνότεραι τῶν λεπτῶν αἱ παχεῖαι; πότερον ὅτι φαίνονται διὰ τὸ πάχος; ἔλαττον γὰρ τῶν ἴσων τὸ μεταξὺ φαίνεται. ἢ διὰ τὸ βραδέως εἰς μῆκος αὔξεσθαι κατὰ πλείους τόπους ἡ τροφὴ ἀνατέλλει ἐπιβιαζομένη ἐντός, οὐ ταχὺ ὑπεξαγούσης τῆς ἔξω εἰς τὸ ἄνω;

2.70. Διὰ τί αἱ σκολιαὶ τρίχες, ὅταν πάθωσιν, ἂν εἰς τοὐναντίον ἔλθῃ, τοσοῦτον μᾶλλον εἰς τὴν φύσιν καθίστανται; ὅτι ἐκείνων ἡ φύσις σκολιά, τοῦ δ' εὐθὺ ἐναντίον.

2.71. Διὰ τί οὐκ ἐν τοῖς ὀστώδεσι λεπτότεραι αἱ τρίχες; ἢ ὅτι παχὺ τὸ δέρμα ταύτῃ, ἐκ δὲ τοῦ παχέος παχύτεραι γίνονται;

2.68 (AKMOPRSTbeilm*p* Buss. Us.; *v* 2.10. de m cf. Introd. 2.1.7)
1 *ἢ* om. P || *ἡ* om. Oi 2 *αἱ παχεῖαι σκληραὶ εἰσι*] *καὶ σκληραὶ καὶ παχύαι* sic A || *δὲ*] *οὖν* A 3 *ἐπὶ*] *περὶ* A || *ἐπεὶ [σκληρὰς]* Us.: *τὸ δὲ σκληρὸν* A Buss.: *καὶ σκληρὰς* MOPRSTbeil Wegehaupt: *ἡ γὰρ σκληρότης* K*v* (*siquidem* = *γάρ*; cf. e.g. 2.76.4) 3-6 *ἐπεὶ ... μῆκος* om. m 4 *ἄνω πέμπεσθαι*] *ἀναπέμπεσθαι* A Buss. || *βραδυτῆτα* K*v* 5 *τοῦ αὔξοντος θερμοῦ* coni. Buss. || *θερμὸν* A || *φθάνεται* Buss. || *καταλαβομένη* A Buss. || *ἡ* A Buss. Us.: om. cett. 6 *ὥστε*] *ὅστε* A ut saepissime || *αὔξεσθαι*] *αὔξασθαι* Oil: om. A || *καὶ οὐκ εἰς μῆκος* Me Buss. Us.: *οὐκ εἰς μῆκος* Kv: *καὶ εἰς μῆκος* OPRSTbil: *καὶ μήκος* A
2.69 (AKMOPRSTbeil*p* Buss. Us.; *v* 2.11)
1 *δὲ* APT Us. Buss.: om. cett. Wegehaupt || *πότερον* om. P || *ὅτι* P Buss. Us.: *γὰρ* A: om. cett. 2 *πάχος*] *πάθος* P || *τῶ μεταξὺ* P || *ἢ*] *ἢ ὅτι* K*v*: om. A || *διὰ τὸ* Buss. Us.: *τὸ διὰ* MORSTbeil: *τὸ διὰ τὸ* P: *διὰ* K*v*: *τὸ* A || *βραδέως*] *βραχέος* KOil*v* 3 *αὔξεται* K*v* || *πλείους οὖν* K*v*: *πολλοὺς* A || *τόπους* M*p* Us.: *τρόπους* cett. || *ἐπιβιβαζομένη* P Us. 4 *οὐ παχὺ* e^{ac} || *ἔξω* Us.: *ἔξεως* sic T: *ἕξεως* cett. || *τὸ* M*p* Buss. Us.: *τὰ* cett.
2.70 (APTl^{2} Buss. Us.: *διατί* tantum habet e: 2.69* Buss.)
1 *ὅταν πάθωσιν*] *ὅσον* Buss. || *ἔλθῃ*] *ἔλθωσιν*, supra *ω*: *η*, l^{2} 2 *ὅτι* om. PTl^{2} || *ἐκείνων δ'* PTl^{2} 2-3 *ἐ|κείνων*: *κείνων – ἐναντίον* in marg. l^{2} 3 *τοῦ*] *τὸ* Buss.
2.71 (AKMOPRSTbeil*p* Buss. Us.; *v* 2.12; 2.70 Buss.)
1 *διὰ τί*] *διότι* e || *οὐκ* om. *v* || *ἢ* om. A Buss. || *διότι* T 1-2 *τὸ δέρμα παχὺ* KOil: *pellis crassa* non *est* v 2 *ταύτῃ* Buss. Us.: *ταυτὶ* APSTbe: *τουτὶ* MR: om. KOil*v*

2.68. Why do thick hairs not grow long? Is it because the [potential] for growth in length is used up in thickness? Or is it because thick ones are hard, and what is hard is moved to grow with difficulty, since it results from earthiness, and earthy nourishment does not wish to be sent upwards; so on account of the slowness of the movement for the heat that produces growth one is overtaken by the other, so that it grows in breadth and depth but not in length?

2.69. Why are thick [hairs] more closely packed than thin ones? Is it that they seem [to be so] because of the thickness? For what is between appears less when they are equal [in number].[327] Or is it that because they grow slowly in length the nourishment rises up in more places as it is forced[328] within, since what is outside does not draw it swiftly upwards?

2.70.[329] Why do crooked hairs, when they are affected,[330] if they go in the opposite direction, go back so much the more to their original condition? Because their nature is crooked, and opposite to the straight?

2.71. Why are hairs not thinner in bony [parts]? Because the skin is dense there, and from what is dense they grow more thickly.[331]

---

327 I.e., for hairs at a given spacing, the thicker the hairs the less the space between them.

328 Usener adopts the reading of P: "applied".

329 This problem is present in only four MSS, APTl. See Introduction, §§ 2.1.1 and 2.1.13.

330 So the MSS. Bussemaker reads "... hairs, to the extent that they go in the opposite direction ...".

331 At first sight this seems inconsistent with 2.67 above (and it is presumably for this reason that Valla omitted "not" in the first sentence but added it in the second: "the skin is not dense there"). However, Aristotle's discussion of the issue in *GA* 5.3 speaks both of skin being thick (*pakhu*) or thin (*lepton*) *and* of its being rare (*manon*) or dense (*puknon*) (782a24-26). The first pair of alternatives relates to the quantity of skin present, the second to the size of the passages through which the residue that in Aristotle's view forms hair is excreted; and both factors play a part (782b1-2). Thickness of hair goes with thickness of skin (*ta pakhudermotera pakhutrikhotera*, 782b6: for if there is more skin there is more earthy matter to produce hair), but not always, because the size of the passages plays a part (783a2). However, thick skin plus fine passages does not inhibit the growth of hair altogether, rather causing it to be fine (*leptotrikha*) but plentiful, as in sheep (ibid.) Both the present text and 2.67 are consistent with the Aristotelian doctrine; in the case of 2.67, even if the

2.72. Διὰ τί αἱ πυρραὶ τρίχες μελάντεραι γίνονται πυρούμεναι; ἢ διότι πέττεται μᾶλλον ἐν ὀλίγῳ τὸ ὑγρὸν γινόμενον ὑπὸ τῆς ἐνούσης ἐν τῇ ἀρχῇ θερμότητος· ἔστι δ' ἡ μὲν πυρρὰ ἐκ νοσούσης, ἡ δὲ μέλαινα ἐξ ὑγιοῦς τροφῆς;

2.73. Διὰ τί Αἰθίοπες μὲν οὐλότριχες, Σκύθαι δὲ καὶ οἱ πρὸς ἄρκτον εὐθύτριχες; ἢ διότι ξηραινομένη ἡ θρὶξ ὑπὸ τοῦ ἡλίου συγκάμπτεται ὥσπερ καὶ τὰ ἄλλα, οἱ δὲ Αἰθίοπες πρὸς τῷ ἡλίῳ, καὶ ἐν ὑπερβολῇ θερμοῦ διατελοῦσιν· οἱ δὲ Σκύθαι ἐν ψυχρῷ τόπῳ, οὗπερ τὸ ὑγρὸν τὸ ἐν ταῖς θριξὶν οὐκ ἐξατμίζει ὁ ἥλιος, ὥστε οὐδὲ συγκάμπτονται; ἐπεὶ οὖν ἀνάγκη κεκλασμένας εἶναι ἢ εὐθείας, εἰ μὴ κεκλασμέναι, εὐθεῖαι ἔσονται.

2.74. Διὰ τί τοῖς τε ἀσθενοῦσι καὶ τοῖς πρεσβύταις τὸ μὲν ἄλλο σῶμα φθίνει, οἱ δὲ ὄνυχες καὶ αἱ τρίχες αὔξονται; ἢ ἐπειδὴ τῆς τροφῆς ἡ μὲν εἰς σάρκας, ἡ δὲ εἰς χολήν, ἡ δὲ εἰς τὰ λεπτομερῆ διαδίδωσιν· ὅσα μὲν οὖν θερμὰ καὶ ἰσχυρὰ σώματα, εἰς χολὴν δύνανται καὶ εἰς αἷμα διδόναι τὴν τροφήν, τὰ δ' ἀσθενῆ εἰς ταῦτα μὲν οὔ; εἰς δὲ τὰ ἐλαχίστης δεόμενα τροφῆς οἷον ὄνυχάς τε καὶ τρίχας; ὥσπερ καὶ τοὺς Ἰλλυριοὺς λέγουσιν ἐξυρημένων ἐν ταῖς μάχαις καὶ ἀποθανόντας ἐν ὀλίγαις ἡμέραις δασείας

2.72 (ABDGKLMOPRSTYbeil*p* Buss. Us.; *v* 2.13; 2.71 Buss.)
1 *μελάντεραι γίνονται*] *μελαίνονται* D ‖ *πυρρούμεναι* MRSbe: *πυρούμεναι μᾶλλον* D: om. *pv* ‖ *ἢ* om. D ‖ *διότι*] *ὅτι* DGY 2 *μᾶλλον*] *πλέον* D ‖ *ἐν ὀλίγῳ τὸ ὑγρὸν γινόμενον*] *ἐν ὀλίγῳ γινόμενον τὸ ὑγρὸν* Y: *τὸ ἐν αὐτοῖς ὀλίγον ὑγρὸν* D ‖ *ἐν τῇ ἀρχῇ* om. D 3 *δ' ἡ*] *δεὶ* A: *δὲ ἡ* BDGLY ‖ *πυρρὰ*] *λιπαρὰ τῆς* BL: *πυρρὰ θρὶξ* D ‖ *ὑγιοῦς*] *ὑγιᾶς* R: *ὑγροῦς* BL
2.73 (ADGKMOPRSTYbeil*p* Buss. Us.; *v* 2.14; 2.72 Buss.)
1 *δὲ*] *τε* GY ‖ *ἄρκτον ἄλλοι* D 2 *ἢ* om. DGY ‖ *διότι*] *ὅτι* DGKY ‖ *ἡ* om. AKORSTbeil 3 *ἄλλα αὐτῶν μέλη μελαίνει* D ‖ *δὲ*] *γὰρ* DY ‖ *πρὸς τῷ ἡλίῳ*] *ἐγγὺς τοῦ ἡλίου* D: *περισῶ ἡλίω* sic A 4 *σκύθαι καί οἱ λοιποὶ* D ‖ *τόπῳ*] *οἰκοῦσι τόπῳ τῷ ἀρκτικῷ* D: *τόπῳ διατελοῦσιν* Y ‖ *οὗπερ* Us.: *ὅπουπερ* MR Buss.: *ὅπου* Ke: *ὅπερ* GOPSTbe[mg]il: *ὦπερ* sic A: *ὅθεν* D: *ὅτι* *p*: om. Y ‖ *τὸ γὰρ ὑγρὸν* Y ‖ *τοῖς* T 4-5 *τὸ ὑγρὸν τὸ ἐν ταῖς θριξὶν οὐκ*] *οὐδὲ τὸ ἐν ταῖς θριξὶν αὐτῶν ὑγρὸν* D 5 *ἐξατμίζει*] *ἀτμίζει* GY ‖ *ὥστε οὐδὲ*] *διατοῦτο οὐδὲ* Y: *καὶ διατοῦτο δὲ* D ‖ *συγκάμπτονται* DY*v* Buss. Us.: *συγκάπτεται* cett. ‖ *οὖν* Buss. Us.: *οὐκ* MOPRSTbeil: *οὐκ* A: *δὲ* K*v* ‖ *ἀνάγκη ἢ* Us. 5-6 *ἐπεὶ – ἔσονται* om. DGY 6 *εἶναι ἢ εὐθείας*] *ἢ εὐθείας εἶναι* KOil*v* ‖ *εἰ*] *ἢ* OPRSbeil: *cum v*
2.74 (AGKMOPRSTbeilo*p* Buss. Us.; *v* 2.15; 2.73 Buss.)
1 *τε* om. o 2 *αὐξῶνται* G ‖ *ἢ* om. Go 3 *τὰς σάρκας* e ‖ *διαδίδωσιν*] *δίδωσιν* M: *δίδοται* coni. Buss. 3-4 *ἡ δὲ εἰς τὰ – χολὴν* om. A 4 *εἰς* om. O ‖ *διδόναι*] *διδόσθαι* Oil: *διαδόσθαι* coni. Wegehaupt: *δέχεσθαι* K 5 *τὰ δ' ἀσθενῆ – οὔ* om. Go 6-9 *ὥσπερ – διέδωκεν* om. Go 6 *ὄνυχές τε καὶ τρίχες* o ‖ *τε* om. P ‖ *ἰλλυρικοὺς* e[ac] 7

2.72. Why do red hairs become blacker when they are burned? Is it because the moisture, produced by the heat that was in them originally, is concocted more in a short [time], and red [hair] comes from unhealthy nourishment,[332] black from healthy?

2.73.[333] Why do Ethiopians have curly hair, Scythians and those in the north straight hair? Is it because the hair, when it is dried by the sun, is bent together, like other things, and the Ethiopians dwell towards the sun and in an excess of heat, while the Scythians dwell in a cold place, where the sun does not steam out the moisture in the hairs, so that they are not bent together? Since then it is necessary for them to be bent or straight, if they are not bent, they will be straight.

2.74.[334] Why in those who are weak and in the old does the rest of the body waste away, but the nails and hair grow? Is it because part of the nourishment is distributed to the flesh, part to bile and part to the rarer parts; bodies that are hot and strong can distribute the nourishment to bile and blood, while the weak cannot [distribute it] to these, but to the [parts] that are least in need of nourishment, like nails and hair? As they say also that the Illyrians who are scalped in battles and die have bushy heads within a

---

fineness of pores causes thinner rather than less hair in the case of sheep, in the limiting case it must prevent the growth of hair altogether. And 2.57 above is compatible with Aristotle's view too, provided we take the point of *araios* to be not that there are many pores, but that there is less earthy matter present; whether that can be the case without there being more pores depends on the view of the structure of matter that one adopts, and indeed in Aristotle's own account it would seem that two aspects of thickness or density, one relating specifically to pores and the other not, are both in play (unless indeed "thick" and "thin" are to be understood purely in terms of the dimension from outer surface to inner surface, which does not seem to be the case in 2.57 at least).

332 See above, 2.62.

333 Cf. 2.66 above, and Ar., *GA* 5.3 782b24ff.

334 Cf. 2.54, 2.59.

τὰς κεφαλὰς ἔχειν, ὡς φθίνοντα τὰ σώματα εἰς ταῦτα τὴν τροφὴν διέδωκεν.

2.75. *Διὰ τί αἱ ἐν τοῖς ποσὶ τρίχες ἥκιστα πολιοῦνται ἢ ὀψὲ λευκαίνονται τοῖς πολλὰ ἔτη ζῶσι; πότερον ὅτι αἱ ἐν τῷ προσώπῳ πρότεραι πολιοῦνται ἤπερ αἱ ἐν τῇ κεφαλῇ ἢ ἐν τῷ γενείῳ, ἐναντιώτατος δὲ ὁ τόπος ὁ ἄνω τῷ κάτω, ἐν δὲ τοῖς ἐναντίοις εὔλογον ἐναντίον συμβαίνειν; ἢ διότι ἡ πολιὰ δοκεῖ γήρως εἶναι σημεῖον, τὸ δὲ γῆρας φθίσις τις καὶ ἔκλειψις θερμότητος; ὁ μὲν οὖν τρόπος, διότι γίνονται πολιαί, ἄλλου λόγου· ἀλλὰ τὸ γένος τῆς αἰτίας τοῦ πάθους τούτου συμβαίνει πρεσβυτέρων γινομένων καὶ τῷ τὸ θερμὸν ὑπονοστεῖν· διὸ καὶ ῥυτιδοῦται τὰ πρόσωπα καὶ τὰ αἵματα παύεται [τὰ] ἐκ τῶν ῥινῶν ῥέοντα, αἱ δὲ κνῆμαι ἐξικμεῖς τῶν πρεσβυτῶν καὶ αἱμορροΐδες γίνονται· εἰ οὖν ἡ μὲν πολιὰ ἐνδείᾳ γίνεται ὑγρᾶς θερμότητος, εἰς δὲ τοὺς πόδας κατέρχεται ἡ τοιαύτη δύναμις, ὥσπερ καὶ ἐν τῇ ἥβῃ ὕστερον πολιοῦνται τῶν ἄνω, ἔτι καὶ αἱ ἐν τοῖς ποσὶ μᾶλλον τούτων· ὅταν δὲ ἐκ παντὸς ἐξικμασθῇ ἡ ὑγρὰ θερμότης διὰ τὸ πλῆθος τῶν ἐτῶν, τότε καὶ ἐν τοῖς ποσὶ τρίχες γίνονται λευκαί.*

2.76. *Διὰ τί τὰ μὲν ἔρια τῷ ἐλαίῳ ῥαινόμενα, λευκότερα γίνεται αἱ δὲ τρίχες τῶν ἀλειφομένων μελάντεραί εἰσί (καὶ γὰρ τὰ ἔρια τρίχες); ἢ ὅτι τὰ μὲν ῥυπαρὰ ξανθότερα γίνεται, αἱ δὲ τρίχες ξηραινόμεναι καὶ ἀπιόντος τοῦ ἐν αὐταῖς ὑγροῦ λευκαίνονται; τὸ οὖν ἔλαιον κωλύει· ἥκιστα γὰρ*

ἐξυρημένων] *ἐξηριμένων* sic A: *ἐξηρημένων* M Buss.: *ἐξευρημένων* T || *ἐν ταῖς μάχαις* om. Buss. || *ἀποθανόντας* P Us.: *ἀποθανόντες* T: *ἀποθανόνταις* sic A: *ἀποθανόντων* cett. Buss. 8 *ὡς* A*pv* Buss. Us.: *καὶ* cett. || *φθίνοντας* ATe[ac] 9 *διέδωκεν*] *διαδίδοσθαι* K*v*

2.75 (AKMOPRSTbeil*p* Buss. Us.; *v* 2.16; 2.74 Buss.)

1 *αἱ* KMP*p* Buss. Us.: om. cett. || *τοῖς*] *ταῖς* e: om. KOSbil 2 *πρότεραι* O[pc]Sbe[mg]il Buss. (in var. lect.) Us.: *πρότερον* K*v*: *πρῶται* AMPRTe 3 *αἱ* om. T || *ἢ αἱ* M Us. || *ἢ ἐν τῷ γενείῳ* om. Buss. || *ὁ τόπος ὁ ἄνω*] *ὁ ἄνω τόπος* Buss. 4 *εὐλόγων* A || *ἐναντίως*] *ἐναντίον* KMOil Us: *contraria v* || *συμβαίνειν* AR Buss. Us.: *συμβαίνει* cett. 5 *τις* om. P 7 *τούτου* KOSbil*v* Us.: *τοῦτο* AMPRTe Buss. || *πρεσβύτερον* T || *γινομένων*] *γενομένων* P Buss. Us.: *γιγνόμενον* T || *καὶ* om. K, del. Buss. in var. lect. 8 *καὶ ῥυτιδοῦται τὰ πρόσωπα* Us.: *καὶ τὰ πρόσωπα ῥυτιδοῦται* Buss.: *καὶ τὰ πρόσωπα ῥυσσὰ γίνεται* K*v*: *καὶ πρόσωπα* A: *καὶ προσώπῳ* T: *καὶ πρόσωπα* cett. || *παύονται* MORSbeil 9 *τὰ* del. Us. || *ἐξικμεῖς*] *ἐξικμάζουσι* K 11 *δὲ* om. e || *ἡ* om. S || *αἱ ἐν τῇ ἥβῃ* M Buss. Us.: *τὰ ἐν τῇ ἥβῃ* K 12 *ἔτι*] *οὕτω* K*v* || *αἱ ἐν* M Buss. Us. 13 *ἡ* om. Oil 14 *αἱ ἐν* M Buss. Us.

2.76 (AKMOPRSTbeil*p* Buss. Us.; *v* 2.17; 2.75 Buss.)

2-3 *ἢ ὅτι τὰ μὲν – τρίχες* om. T 4 *λευκαίνονται*] *λευκαὶ γίνονται* A Buss. || *οὖν*]

few days, since the bodies as they decayed directed the nourishment to these.

2.75.[335] Why do the hairs on the feet go grey least, or turn white late in the case of those who live for many years? Is it because those on the face go grey before those on the head or those on the chin, and the region above is most opposite to the region below, and it is reasonable that in opposite [regions] things should happen in opposite ways? Or is it because greyness seems to be a sign of old age, and old age is a wasting and decline in heat? The way in which they become grey belongs to another account;[336] but the kind of reason [there is] for this effect comes about when they are growing older and because the heat subsides. And for this reason also their faces become wrinkled and they cease having nose-bleeds, and the shins of older people are dried up and liable to bleeding. So, if greyness comes about through lack of moist heat, and this sort of power descends into the feet, then as [hairs] in the genital region too go grey later than those above, this applies on the feet even more than to these; but when the moist heat is dried out of the whole [body] because of the number of years, then hairs become white on the feet, too.

2.76. Why does wool sprinkled with oil become whiter, while the hair of those who are anointed is blacker (for wool too is hair)? Is it because greasy things become more yellow, but the hairs, when they are dried and the moisture in them departs, become white? The oil then prevents [this]; for

335 Cf. [Alex.], *Probl.* i.147, where the explanation is that the feet are in constant movement and therefore the phlegm in them is dispersed and does not make the hairs there grey; for the genital region however the same explanation in terms of heat is given there as here.

336 Cf. 2.55, 2.56, 2.64, 2.78, 2.81.

ξηραίνονται· ἐπὶ δὲ τούτῳ ἀνομοίως συμβαίνει ἐχούσας προβάλλειν τὰς τρίχας, ἀλλὰ τὰς μὲν ἔτι ζώσας ἐν τῷ σώματι, τὰ δ' ἔρια ἀφηρημένα.

2.77. *Διὰ τί τῶν μὲν προβάτων μαλακώτεραι αἱ τρίχες ἀναφύονται, τιλλομένων, τῶν δ' ἀνθρώπων σκληρότεραι; ἢ ὅτι αἱ μὲν παχύνονται ὅσῳ ἂν μακρότεραι γένωνται, ἀποψύχει γὰρ δι' ἀσθένειαν τοῦ θερμοῦ τοῦ ἐν αὐταῖς ὁ ἀὴρ ὁ κύκλῳ, καὶ μᾶλλον ὅσῳ ἂν πορρωτέρω τοῦ οἰκείου θερμοῦ γίνωνται ὥστε ἐπιτελοῦνται· αὐξανόμεναι δὲ διὰ μὲν παχύτητα πλέον ἔχουσι τὸ θερμόν, διὰ δὲ τὸ λίαν ἀποξηραίνεσθαι ῥέοντα εἰ μή τι κωλύει, λεπτότεραι ἔτι διὰ τὴν αὔξησιν;*

2.78. *Διὰ τὶ τοὺς κροτάφους πολιοῦνται μὲν τάχιστα, φαλακροὶ δὲ ἥκιστα γίνονται, καίτοι μηδέτερον ἢ ἄμφω ἔδει; ἢ ὅτι ἡ μὲν πολιὰ γίνεται τῷ ποιάν τινα εἶναι τὴν τροφήν τὴν εἰς τὰς τρίχας διδομένην, ἡ δὲ φαλακρότης τῷ ποσήν; ὅταν γὰρ ψυχθῇ ἡ τροφή, οὐκ αὔξεται, ἀλλ' ὥσπερ τὰ δένδρα τοῦ χειμῶνος, καὶ ὁ τόπος ἕκαστος φυλλορροεῖ ψυχρᾶς γενομένης καὶ οἷον παγείσης· ὀλίγη παροῦσα ἀδύνατος τρέφειν· ἡ δὲ πολιὰ ἐκ σαπρᾶς τροφῆς ἐστι· σήπεται δὲ, ὅταν ἀπεπτότερα ἕλκῃ· οὔκουν ἡ*

γοῦν T 5 *ἐπὶ*] *ἐπεὶ* A: *ἔτι* Buss. || *τούτῳ*] *τοῦτο* APT Buss. || *ἀνομοίως* Us.: *οὐχ ὁμοίως* K*v*: *ὁμοίως* cett. Buss. || *ἐχούσας* Us.: *ἔχειν* K*v*: *ἐχούσαις* cett. || *προβάλλειν τὰς* Us.: *comparemus v*: *προσβάλλοντας* PTe: *προβάλλοντας* cett. 6 *ἀλλὰ* Us.: *ἅμα* codd. Buss. || *τῷ* om. P

2.77 (AKMOPRSTbeil*p* Buss. Us.; *v* 2.18; 2.76 Buss.)

1 *ἀναφύονται* KM Us.: *ἀναφέρονται* AOPRSTbeil Buss.: *renascunt v* 2 post *αἱ μὲν* (vel post *σκληρότεραι*) plura fortasse exciderunt 3 *γίνωνται* KMRe*p* Buss.: *γί(γ)νονται* cett.: *γένωνται* Us. 4 *αὐταῖς*] *αὐτοῖς* Buss. 5 *γίνονται* AMPT || *ὥστε ἐπιτελοῦνται*] *unde exoriuntur v* 7 *ἔτι*] *εἰσι* A Buss. || *λεπτότεραι – αὔξησιν* om. T

2.78 (AKMOPRSTbeil*p* Buss. Us.; = *v* 2.19; 2.77 Buss.)

1 *τὰ μὲν τάχιστα* O 2 *καίτοι*] *καὶ ἔτι* T || *μὲν* om. R 3 *ποῖαν* M || *τὴν*[2]] *τῆς* APRT: *τὴ* Sb || *διδομένης* APRT 4 *τῷ ποσὴν*] *τόπος ἣν* A || *καὶ οὐκ* P || *αὔξονται* Buss. (in var. lect.) || *ἀλλ'* om. K 5 *καὶ ὁ τόπος* om. K*v* || *ἕκαστος*] *ἐκεῖνος* Us.: om. K*v* || *φυλλοροεῖ* AMORe: *φυλορροεῖ* K 6 *παχείσης* A || *ὀλίγη*] *ὀλίγους* M: *ὀλίγης παρούσης* Buss. (in var. lect.) || post *ὀλίγη* habent *γὰρ* K Us.: *autem* v: *δὲ* Buss. (in var. lect.). || *ἀδύνατος* P Us.: *ἀδυνατεῖ* K: *ἀδύνατον* cett. 7 *ἕλκει* A: *ἀπεπτοτέρα ἕλχθῃ* Buss. (in var. lect.) || *οὔκουν* KM*v* Buss. Us.: *οὐκοῦν* cett. 9 *τὸ*

they are dried least.[337] In addition, the hairs proposed [in the problem] are not in a similar condition, but [the hairs] are still alive on the body, while the wool has been removed.

2.77.[338] Why do the hairs of sheep grow softer when they are plucked, but those of humans harder?[339] Is it because the former grow thicker to the extent that they become longer, for the surrounding air chills them on account of the weakness of the heat in them, and the more so the further they get from the proper heat, so that they are completed? But while they are growing they have more heat because of their thickness, but because they are excessively dried they fall out if nothing prevents it, but they are still finer because of the growth?

2.78.[340] Why do people go grey most quickly on their temples, but become bald [there] least, although either neither should happen, or both? Is it because greyness comes about because the nourishment that is distributed to the hairs is of a certain sort, baldness because [it is] of a certain quantity? For when the nourishment is chilled, [the hair] does not grow, but like trees in winter each place sheds its leaves when [the nourishment] becomes cold and as it were congealed. Being present in a small amount it is unable to provide nourishment. But greyness comes from nourishment that has rotted, and it rots when it attracts what is less concocted.[341] So the reason of both is

---

337 Athenaeus 15 692b = Aristotle fr. 235 Rose³ cites "Aristotle in the natural problems" as answering the question why anointing *with myrrh* makes the hair go grey by saying that myrrh has a drying effect. Cf. Prantl 1851, 369 and n.80; Rose 1863, 220.

338 Cf. [Ar.], *Probl.* 10.22.

339 The explanation which [Aristotle] goes on to give is in terms of human hair being more deeply rooted than sheep's wool. If what follows here is indeed still part of the same problem and there is not a lacuna in the text, it appears that we have only the part of the explanation that relates to sheep (different from that in [Aristotle], who argues that plucking opens the pores so that subsequent growth is well-nourished), not that relating to humans.

340 Cf. [Alex.], *Probl.* i.1, with the same explanation as here for the greyness; Ar., *HA* 3.11 518a17, *GA* 5.3 782a15 (simply noting the phenomenon), 5.4 784b35 (the same explanation as here for the greyness, with less detail). Bussemaker 1857, xix regards *GA* 5.4 as referring to our text; Prantl 1858, 206 rightly rejects any "direct reference" between the two passages. Cf. Rose 1863, 220.

341 See above, 2.64; also 2.55, 2.56, 2.75, 2.81.

αὐτὴ αἰτία ἐστὶν ἀμφοῖν, διὸ οὔτε ἅμα οὔτε τοῖς αὐτοῖς παραγίνεται τόποις [τὸ δὲ σκέμμα ὅλως φυσικώτερον ἢ προβληματικώτερον]. οἱ δὲ κρόταφοι ἐπὶ φλεβῶν εἰσι πολλῶν καὶ πυκνῶν, ὥστ᾽ ἔχουσι θερμότητα καὶ οὐχ ὑπολείπει ἡ τροφή, ἀλλὰ μᾶλλον διὰ τὸ πολλὴν ἅμα εἶναι προϊούσης τῆς ἡλικίας οὐ δύναται πέττεσθαι, ἀλλὰ σήπεται· ἐκ τοιαύτης δὲ γίνεται πολιά, ὥσπερ καὶ οἱ μύκητες ἐξ εὐρωτιώσης ἰκμάδος· οὐ γὰρ πάντα μελαίνει τὰ σηπόμενα, ἀλλ᾽ ὁ εὐρὼς λευκός, καὶ ὁ σκώψας κωμῳδοποιός εὖ ἔτυχεν εἰπών·

εὐρῶτι γήρως τὰς τρίχας βεβαμμένος.

2.79. *Διὰ τί οὐκ ἐν ταῖς νόσοις, ἀλλ᾽ ἐκ τῶν νόσων αἱ τρίχες ἐκρέουσιν; ἢ ὅτι τότε μὲν ἔχουσι τροφήν (ἐκ τοῦ περιττώματος γὰρ γίνονται), μετὰ τὰς νόσους δὲ ὀλίγον τοῦτο διὰ τὸ τρέφεσθαι τὸ σῶμα· ἡ δὲ εἰσιοῦσα τροφὴ εἰς τὸ κατὰ φύσιν οὐκ αὔξει αὐτάς, ἀλλὰ διὰ τὴν κένωσιν προκαταναλίσκεται.*

2.80. *Διὰ τί τῆς κεφαλῆς τὸ ἔμπροσθεν φαλακροὶ γίνονται, τὰ δ᾽ ὄπισθεν οὐ γίνονται ἢ ἧττον; ἢ ὅτι οὐκ ἔνεστιν ἐν τῷ ὄπισθεν ἡ περίττωσις (κενὸς γὰρ καὶ ξηρὸς ὁ τόπος) τὸ δὲ ἔμπροσθεν διὰ τὸν ἐγκέφαλον πλῆρες; ὅσοις οὖν ἐνταῦθα ψυχρὰ συλλέγεται περίττωσις διὰ τὸ κρατεῖν τῆς θερμότητος καὶ μὴ πέττεσθαι οὐ τρέφει τὰς τρίχας, ὥστε ἀποπίπτουσι· τὸ δ᾽ ὄπισθεν οὐκ ἔχει τοιαύτην ἀπόκρισιν, ἢ ἧττον διὰ τὴν εἰρημένην αἰτίαν.*

δὲ σκέμμα – προβληματικώτερον del. Us. || *ἢ*] *ἡ* A 10 *φλεβῶν φυσικώτερον* A || *πολλῶν καὶ πυκνῶν*] *πυκνῶν ἢ πολλῶν* P 11 *τὸ μᾶλλον* T || *ἅμα* om. APT Buss. 12 *ἡ πολιά* Buss. (in var. lect.) 13 *μύκητες* l² in marg. (Wegehaupt) Us.: *μυκτῆρες* cett. Buss.: *in naribus muci v* || *μελαίνει τὰ*] *μελαίνεται* K 14 *ὁ*] *ὅ* T || *εὐρὼς λευκὸς* Us. Buss.: *εὐρὼς λευκὸν* P: *εὐχερῶς λευκὸν* AMRTe: om. KOSbil*v* || *ὁ σκώψας*] *ὠσκύνψας* A: *ὁ κώψας* T 15 *εὐρῶτι γήρως* K*v* Us.: *εὐρωτογῆρας* Buss.: *εὔρω τὸ γῆρας* AMPRTe*p*: *εὐρῶτα γῆρας* OSbil || *βεβαμμένος* Us.: *βεβαμμένας* KMOPRSTbeil*p*: *βεβλαμμένας* A Buss.: *addictos v*

2.79 (ABGKLMOPRSTbeilo Buss. Us.; *v* 2.20; 2.78 Buss.)

1 *ταῖς*] *τοῖς* BLo || *ἐκ τῶν νόσων*] *μετὰ τὰς νόσους* Me (sed *ἐκ τῶν νόσων* e[mg]) 2 *τότε*] *τὸ* AMPRe || *τὴν τροφὴν* Go || *γί(γ)νονται* APT Us.: *γίνεται* Go Buss.: om. cett. *v* || *τὰς*] *τοὺς* A 3 *δὲ*[1]] *γὰρ* BL || *τρέφεσθαι*] *ali ... opus v* 4 *τὸ κατὰ φύσιν*] *id quod naturam sublevat v* || *αὐτὰ* o || *ἀλλ᾽ ἢ* o || *προκαταναλίσκεται*] *προμεταναλίσκεται* o

2.80 (AKMOPRSTbeil*p* Buss. Us.; *v* 2.21; 2.79 Buss.)

1 *τὸ*] *τὰ* P*p* || *φαλακρὰ* P: *φαλακρόν p* 2 *οὐ* om. P || *γίνονται*] *γίνεται* Buss. || *ἔνεστιν* AMPRTe Buss. Us.: *ἔστιν* KOSbil*v* || *περίττωσις*] *περίπτωσις* AMT[pc]e: *παράπτωσις* P || post *περίπτωσις* add. *διὰ τὸ κρατεῖν τῆς θερμότητος* A 3 *ξηρὸς*] *σκληρὸς* A || *τὸ δὲ*] *καὶ τὸ* M: *τὰ δὲ* Oi: *ὁ δ᾽* Buss. || *τὸ δὲ – πλῆρες* delendum coni. Wegehaupt || *πλῆρες*] *πλήρις* sic A: *πλήρεις* T: *πλήρης* Buss. 4 *περίττωσις*] *περίπτωσις* A || *τὸ κρατεῖν τῆς θερμότητος*] *ibidem a calore detinetur v*

not the same, and for this reason they do not happen together or in the same regions. (In general the enquiry is more to do with nature than with the discussion of problems.)[342] The temples have many thick veins beneath them, so that they possess heat and the nourishment does not fail, but rather, because there is a lot of it all together, as age advances it cannot be concocted, but rots; and from such [nourishment] greyness comes, as mushrooms do from moisture that decays; for not everything that rots goes black, but decayed moisture[343] is white, and the comic poet happened to get it right when he said in jest "old age which has dyed its hair with mould."[344]

2.79.[345] Why does hair fall out not during disease, but after it? Is it because then [the hairs] have nourishment (for they grow from residue),[346] but after diseases there is little of this because the body is being nourished? The nourishment that comes into what is in a natural state does not cause them to grow, but is used up first because of the emptiness [caused by the previous disease].

2.80.[347] Why do people become bald on the front of the head, but on the back they do not, or [do so] less? Is it because residue is not produced at the back (for the region is empty and dry), but the front is full because of the brain? So in those in whom cold residue gathers together here, because it overpowers the heat and is not concocted it does not nourish the hairs, so that they fall out, while the back does not have a shedding of this sort, or less so, for the reason that has been stated.

---

342 Rightly bracketed by Usener as a marginal comment.

343 I.e. mould. Cf. Ar., *GA* 5.4 784b13 for this explanation of greyness.

344 Fr. 53 in Demianczuk 1912; the poet is unidentified.

345 Cf. [Alex.], *Probl.* ii.19 (the same explanation as here, but specifying fever as the disease).

346 Above, 2.59.

347 Cf. Ar., *GA* 5.3 783b37 ff. (explanation through coldness, as here, in the context of why sexual activity produces baldness); [Alex.], *Probl.* i.2 (explained rather through rareness and *dryness* of front part of head), ii.13 (explained through dryness as in i.2, but contrasting top of head and temples, with an analogy to plants growing in valleys rather than hills); Macrobius, *Sat.* 7.10.5 (unnatural dryness, with the same connection as in [Alex.] i.2 between dryness and curly hair); Adelard, *Quaest. Nat.* 20, pp.128-130 Burnett 1998 (heat given off from the brain requires the pores at the front of the head to be larger, and the hair therefore falls out more easily).

2.81. Διὰ τί ἐν ταῖς μασχάλαις καὶ ἐν τῇ ἥβῃ τρίχες ἧττον καὶ αὔξονται καὶ πολιοῦνται τῶν ἐν τῇ κεφαλῇ; ἢ ὅτι μέγισται καὶ οἰκεῖαι αἱ ἔγγιστα τῆς κεφαλῆς τῶν ἄλλων; διὰ γὰρ τὴν γειτνίασιν τῆς τροφῆς μάλιστα ἀπολαύουσιν· εἰκότως οὖν τῶν λοιπῶν ἐν τῷ γενείῳ μέγισται. ἢ οὐ τοῦτό ἐστι τὸ αἴτιον, ἀλλὰ καθάπερ ἡ ἐλαία καί ἡ ἄμπελος καὶ ὁ στάχυς, καὶ ἄλλο ὁτιοῦν τῶν τοιούτων ὡρισμένον τι μέγεθος ἔχει ὃ οὐχ ὑπερβάλλει, καθάπερ οὐδ' ὁ ἄνθρωπος οὐδ' ἵππος οὐδὲ τὰ ἄλλα ζῷα; τὰ μὲν οὖν αὐτῶν μείζω καὶ ἐλάττω γίνεται καὶ παρὰ τοὺς τόπους καὶ ⟨παρὰ⟩ τὰς τροφάς, εἰς μέντοι τὰ ἀλλήλων οὐκ αὔξεται μεγέθη. τὰ μὲν οὖν εἰρημένα ῥᾴδιόν ἐστι διαγνῶναι, ὅτι ἕτερα τῷ εἴδει (μεγάλαι γὰρ αἱ διαφοραί), ἐνίων δ' οὐκέτι οἷον τὴν τοῦ πυροῦ καὶ κριθῆς καλάμην ἑτέραν οὖσαν τῷ εἴδει· ὁμοίως δὲ οὐδὲ τὰς τρίχας οὐ διαγινώσκομεν διὰ τὴν ὁμοιότητα οὐσῶν ἑτέρων τῷ εἴδει. οὐδὲν οὖν θαυμαστὸν τὸ μὴ γίνεσθαι ἴσας ⟨οὐχ ὁμοίας⟩ οὔσας τῷ εἴδει ἀλλὰ τὰς μὲν μικράς, τὰς δὲ μεγάλας. τοῦ μὲν οὖν μὴ γίνεσθαι μεγάλας αὕτη αἰτία. τοῦ δὲ μὴ ὁμοίως πολιοῦσθαι νῦν λεχθήσεται· αἱ γὰρ τρίχες οὐκ ἔχουσιν οὔτε τὸ μέλαν οὔτε τὸ λευκὸν οἰκεῖον χρῶμα, ἀλλ' ἀπὸ τῆς τροφῆς ἴσχουσι, καὶ μέλαιναι μέν εἰσιν ὅταν πολλὴν ἕλκωσι τροφήν· συναναςπῶσι γὰρ τὸ γεῶδες τοῦτο δὲ

2.81 (A(BKLMO)P(RS)T(bei)l(*p*) Buss. Us.; *ν* 2.22; 2.80 Buss.)
1 ⟨*αἱ*⟩ *ἐν* (bis) Buss. Us.: om. codd. || *ἐν*^{2} om. A || *αἱ τρίχες* BKL 2 *τῶν*] *τὸν* A || *οἰκεῖαι*] *precipui ν* || *αἱ* om. ABST 2-4 *καὶ οἰκεῖαι – μέγισται* om. P 3 *ἔγγισται* L 4 *ἀπολαύουσιν* Buss. Us.: *ἀπολύουσιν* BKLOSbil*ν*: *ἀπολλύουσιν* AMRTe || *ἐν τῷ* ABLORSTbeil: *αἱ ἐν τῷ* KLM Buss. Us. || *μέγισται*] *αὔξονται* K 4-33 *ἢ οὐ τοῦτό – πολιοῦνται* APT Buss. Us.: add. in marg. l^2: om. BKLMORSbeil^1*pν* 5 *τοῦτό ἔστι* l^2 Buss. Us.: *τοῦτέστιν* A: *τοῦτ' ἐστὶ* PT || *ἢ*^{2} PT Us.: om. Al^2 Buss. 6 *τί* PT 7 *οὐδ'*^{1}] *οὔθ'* ATl^2 || *οὐδ'ἵππος* PTl^2 Buss.: *οὐδὲ ἱππος* sic A: *οὐδ' ⟨ὁ⟩ ἵππος* Us. 8 *αὐτῶν* A || *μεῖζων* sic A || *παρὰ*^{1}] *περὶ* P || ⟨*παρὰ*⟩ add. Buss. Us. 9 *μεγέθει* P 11 *δ'*] *τ'* P || *πυρροῦ* P 12 *δὲ οὖσαν* A: *δ' οὖσαν* Buss.: *γ' οὖσαν* Us. || *οὖσαν τῷ εἴδει* PTl^2 Us.: *οὖσαν. τῷ δ' οὐκέτι τῷ εἴδει* A: *τὸ δ' οὐκέτι τῷ εἴδει* Buss. (in var. lect.) || *ὁμοίως δὲ οὐδὲ*] *ὁμοίως οὐδὲ* A: *ὁμοίας εἶναι* Buss. (in var. lect.) || *διαγινῶσκωμεν* A: *διαγιγνόσκομεν* T 13 *οὐσῶν* Us.: *ὅσον* A Buss.: *ο[.]σον* A: *ὅσων* PTl^2 || *ἑτέρων*] *ὅσων ἑτέρων* T || *τὸ*] *τῶ* PTl^2 14 *ἴσας* om. A || *οὐχ ὁμοίας* add. Buss. Us. || *οὔσας* om. l^2 || *ἀλλὰ τὰς μὲν* Us. Wegehaupt: *ἀλλὰ ⟨τοῦ⟩ τὰς μὲν* Buss.: *διἄλλα τὰς μὲν* sic A: *δι' ἄλλα* PTl^2 15 *τοῦ – γίνεσθαι* om. A Buss. || *μεγάλας* om. A || *αὕτη*] *αὔτη* T: *αὕτη <ἡ>* Buss. Us. Wegehaupt || *τοῦ δὲ μὴ ὁμοίως*] *το μὴ ὁμοίος δὲ* sic A: *τοῦ μὴ ὁμοίως δὲ* Buss. 16 *λεγχθήσεται* T 16-17 *οὖτε τὸ λευκὸν οὖτε τὸ μέλαν* sic A 18 *πολλὴν*

2.81.[348] Why do hairs both grow less and go grey less in the armpits and in the genital region than those on the head? Is it because those on the head are largest and proper compared with the others? For because of their proximity to the nourishment they derive most benefit from it? And reasonably, of the rest [those] on the chin are largest. Or[349] is this not the cause, but just as the olive tree and the vine and the ear of corn and anything else like this has a definite size which it does not exceed, just as a human being and a horse and other animals do not? So some of them grow larger and others smaller as a result both of the regions and of their nourishment, but they do not grow to each others' size. Well, the stated [examples] are easy to distinguish, because they are different in kind (for the differences are great), but that is no longer [so] in some cases, for example the stalk of wheat and of barley which is different in kind. Similarly we do not distinguish the hairs, because of their similarity, although they are different in kind. So it is in no way surprising that they do not grow equally when they are <not similar> in kind but some small, others large. So, this is the reason why they do not grow large. Now the reason for their not going grey in a similar way will be stated. The hairs do not have either black or white as their proper colour, but they derive it from their nourishment, and on the one hand they are black when they attract much nourishment, for they draw up what is earthy with it,

---

348 Cf. [Alex.], *Probl.* i.10 (referring to looseness of pores, as at the end of this text, but also to heat).

349 The remainder of this problem is found only in three MSS (APT) and as a marginal addition in one more (l). See Introduction, §§ 2.1.1 and 2.1.13.

*μελαίνει * * *. καθάπερ ⟨γὰρ⟩ ἀπὸ τοῦ πυρὸς γῆ ἀνωθουμένη, ὅπου ἂν προσκαθίζηται, μελαίνει προσμένουσα ἄνω, τὸν αὐτὸν τρόπον ὑπολαβεῖν δεῖ καὶ ἐν τῷ σώματι, ὅταν ἀνασπασθῇ εἰς τὸν ἀνωτάτω τόπον, ὄντος ἐν ἡμῖν θερμοῦ ὑπὸ τούτου μελαίνεσθαι· ὅταν δὲ ἢ δι᾽ ἔκθλιψιν τροφῆς ἢ δι᾽ ἀσθένειαν τὴν τροφὴν παύσηται ἀνάγον, πολιοῦται αὐτοῦ τοῦ ὑγροῦ ἀναγομένου. ὑποκειμένου δὲ τοῦ οὕτως πολιοῦσθαι, ἡ κεφαλὴ καὶ ὁ πώγων καὶ διὰ τὴν πυκνότητα καὶ τὸ πλῆθος τῶν τριχῶν καὶ τὴν ἀσαρκίαν ὅταν μὴ δύνωνται τρέφειν ὡς εἴρηται, λευκαίνονται. αἱ γὰρ τρίχες πολλαὶ οὖσαι καὶ κειρόμεναι προσέλκουσιν εἰς αὑτὰς τροφήν· ὅταν δὲ μὴ ἕλκωσιν [ὅθεν ἐξέλκουσιν] τὴν ξηράν, καὶ τὴν ὑγρὰν ἀνέλκωσι, λευκαίνονται. ἡ δ᾽ ἥβη καὶ αἱ μασχάλαι βραχείας ἔχουσαι τῇ φύσει καὶ ἀραιὰς ῥᾳδίως μελαίνονται· συναναςπῶσι γὰρ καθάπερ προείρηται. ἔτι ὁ περὶ τὰ αἰδοῖα τόπος ἐστὶ τοιοῦτος οἷος τρέφειν ἄριστα· καὶ γὰρ ἀραιὸς καὶ χαῦνος, ἔτι δὲ τὸ σπέρμα κατὰ τοῦτον τὸν τόπον ἐξέρχεται, ὥστε πολλὴν εἶναι δεῖ τροφήν. εἰκότως οὖν καὶ ἐν τούτοις τοῖς τόποις ἧττον πολιοῦνται.*

---

om. A Buss.: *πολὴν* T ‖ *συναναςπῶσι* Tl^2: *συναναςπῶσαι* P Us.: *συναναςπῶ* A: *συνασπῶ⟨σι δὲ θερμότητα⟩* Buss. (in var. lect.) 18-19 *γὰρ τὸ γεῶδες* (*τὸ γεῶδες* Wegehaupt: *τοὺς γεώδεις* T: *τοὺς γαιώδεις* l^2) *τοῦτο δὲ μελαίνει* Tl^2: om. AP Buss. Us. 19 lacunam statuit Us. ‖ *γὰρ* add. Us. ‖ *ἀπὸ*] *ὑπὸ* Us. 20 *μελαίνει*] *μελαίνη* T 21 *εἰς τὸν ἀνώτατον τόπον* l^2 Us.: *κάτω* Buss. ‖ *ὑπὸ τοῦ ὄντος* Us. 22 *ὑπὸ τούτου* om. Us. ‖ *ἔκθλιψιν*] an *ἔκλειψιν* legendum? ‖ *τροφεῖς* A 23 *ἀνάγον*] *ἀνάγων* A Buss. ‖ *πολιοῦνται* P Us. Wegehaupt ‖ *αὐτοῦ* A 24 *οὕτως*] *οὗτος* sic A 25 *καὶ*^{1} del Buss. (in var. lect.) ‖ *τὸ*] *τῷ* A ‖ *καὶ διὰ τὸ πλῆθος καὶ τὴν πυκνότητα* PTl^2 25 *διὰ* ante *τὴν ἀσαρκίαν* add. Us. 26 *δύνονται* A ‖ *τρέφεσθαι* coni. Buss. (in var. lect.) 27 *καὶ κειρόμεναι* T^{ac} Buss. Us.: *και κυρῶμέναι* A: *κεκηρωμέναι* PT^{pc}l^2 ‖ *πρὸς ἕλκουσιν* T ‖ *αὐτὰς* Buss. 28 *ὅθεν ἐξέλκουσι* l^2 Buss. (in var. lect.): *ὅθεν ἐξ ἕλκουσι* T: *ὅθεν ἐξέλκουσαι* A: *ὅταν ἐξέλκωσιν* P: del. Us. ‖ *ἢ τὴν ὑγρὰν* PT Buss.: *ἡ τὴν ὑγρὰν* A: *ἢ τὴν ὑ* | (post litteram *ὑ* in fine folii deficit) l^2: *καὶ τὴν ὑγρὰν* Us. ‖ *ἀνέλκουσι* PT: *ἀνέλκωσι* Us.: *ἀνέλκουσαι* A Buss. 29 *βραχεῖας* T ‖ *ἔχουσι* A 30 *ἀρεὰς* T ‖ *ῥάδιος* A ‖ *τε γὰρ* T Buss. 30-31 *ἔτι ὁ περὶ* Buss. (in var. lect.): *ἐνι *. ὅ τε* Us.: *ἐνίοτε* codd. 32 *τόπον*] *τρόπον* P 32-33 *ὥστε – τροφὴν* om. Buss.

and this causes blackness ... [350] <For> just as earth which has been pushed up away from a fire, wherever it settles down, is black when it remains above, in the same way one must suppose that in the body too [the earthy stuff], when it is drawn up into the topmost region, since there is heat in us, is made black by this; but when, because the nourishment is squeezed out[351] or on account of weakness it stops drawing up the nourishment, it goes grey because the moisture itself is drawn up. Now it has been established that this is how going grey [comes about], the head and the beard, both because of the thickness and quantity of hair and [because of] their lack of flesh, become grey when they[352] cannot nourish [the hair] in the way described.[353] For the hairs, being many in number and being shorn, attract nourishment to themselves; and when they do not attract dry [nourishment][354] and draw up moist [nourishment], they become white. The genital region and the armpits, having [hairs] which are short by nature and rare, become black easily; for they draw up [the dry nourishment], as has been said. Moreover, the region around the genital organs is such as to nourish [hairs] best; for it is both loose-textured and porous,[355] and moreover the seed is emitted in that region, so there needs to be much nourishment. So it is reasonable also that in these regions [the hairs] go grey less.

---

350 "for they draw up ... blackness" is based on the reading found in MSS Tl; the other MSS just have "drawing up along with ...". Bussemaker and Usener mark a lacuna, and this is needed even if we incorporate the reading of Tl, as the second half of the antithesis ("they are white when they do not draw up nourishment") is missing.

351 For "is squeezed out" (*ekthlipsin*) one might expect rather "fails" (*ekleipsin*).

352 That is "the head and the beard", or rather the parts of the head from which hair and beard grow.

353 Cf. 2.55, 2.56, 2.65, 2.75, 2.78.

354 The MSS add here "from the place whence they draw it out" or, in one case (P), "when they draw it out". The latter is redundant, and "from the place whence" is attested in two branches of the tradition, which makes it less likely that it is a correction of "when" resulting from repetition. One might be tempted to emend *helkôsin* to *ekhôsin*, "when they do not have a place to draw it out from", but that combines *hothen,* which appears in all the MSS except P, with the subjunctive which appears only in P.

355 See above, 2.57. What is said here is not necessarily inconsistent with what is said there about the hair around the genitals growing less than the beard.

2.82. *Διὰ τί ἐν τῇ κεφαλῇ μάλιστα γίνονται φθεῖρες; ἢ ὅτι ζῷα μὲν πέφυκε γίνεσθαι, οὗ ἂν περίττωμα ᾖ ὑγρότητος καὶ θερμότητος, ἔτι δ' ἔχῃ τόπον· ἡ δὲ κεφαλὴ ὑγρὰ καὶ θερμὴ ἡ τῶν νέων, καὶ οἱ πόροι οἱ ἐν αὐτῇ εὐρύτατοι; σημεῖον δέ· αἱ τρίχες παχύταται ἐνταῦθα, ὥστε ἔχει καὶ ἐξ οὗ συσταίη ζῷον καὶ ἐν ᾧ.*

2.83. *Διὰ τί ἐκ τῶν πότων καὶ τῶν ἐμέτων καὶ ἐν τοῖς ψύχεσι φθέγγονται βαρύτερον; ἢ διὰ τὴν ἔμφραξιν τοῦ λάρυγγος ὑπὸ τοῦ φλέγματος; ἐπικατασπᾷ γὰρ ῥευμάτιον τοῖς μὲν ὁ ἔμετος ἢ τὸ ποτόν, τοῖς δὲ ἡ ὥρα, ὥστε βραδυτέρα ἡ φορὰ τοῦ πνεύματός ἐστιν· ἡ δὲ βραδεῖα λίαν καὶ βαρεῖά ἐστιν.*

2.84. *Διὰ τί ποτε ἐγγύθεν τῆς βαρυτέρας φωνῆς μᾶλλον ἐξακούομεν, πόρρωθεν δὲ ἧττον; ἢ διότι [καὶ] ἡ βαρυτέρα φωνὴ πλείω ἀέρα κινεῖ, ἐπ' ἔλαττον δὲ τοῦτο ποιεῖ; πόρρωθεν μὲν οὖν ἧττον ἀκούομεν, διότι ἐπ' ἔλαττον ἀφικνεῖται· ἐγγύθεν δὲ μᾶλλον, διότι πλείων τοῦ ἀέρος κίνησίς ἐστιν.*

2.82 (ABKLMOPRSTbeil*p* Buss. Us.; v 2.23; 2.81 Buss.)
1 *γίνονται φθεῖρες*] *γίνονται φθεῖραις* sic A: *γίνονται φθεῖραι* OSbi: *φθεῖραι γίνονται* BL || *ζῷα*] *ζῶσιν* Oi 2 *γίνεσθαι* om. BL || *οὗ ἂν περίττωμα* om. T || *ἂν*] *ἐὰν* Sbe[mg] || *ᾖ*] *ἢ* T || *ὑγρότητος*] *ὑγρότης* BL: *ὑγρότατον* Buss. || *θερμότητος*] *θερμότης* BL: *θερμότατον* Buss. || *ἔχῃ* Buss. (var. lect.) Us.: *ἔχει* codd. (om. BL) 3 *τόπον*] *πόρους* K: *spiramenta v* || post *θερμὴ* in fine folii deficit l || *οἱ*[2] BLMR Buss. Us.: om. cett. 4 *παχύτατες* BL 5 *συσταίη*] *συσταί* sic A || *ζῷον*] *ζώων* ABL: *ζώου* OSei
2.83 (ABKLMOPRSTbei*p* Buss. Us.; *v* 2.24; [Ar.] 11.18; 2.82 Buss.)
1 *ποτῶν* AT*p* || *τῶν*[2] om. BL || *ἐν τοῖς ψύχεσι*] *ἐν τῷ ψύχει* A Buss.: *τοῖς ψύχεσι* BL 1-2 *βαρύτερον φθέγγονται* [Ar.] 2 *ἔμφραξιν* BKLOSbe[mg]i [Ar.] Buss. Us.: *ἔκφρυξιν* MRe[ac]: *ἔκφριξιν* APT*p* || *λάρυγγος*] *φάρυγγος* [Ar.] || *τὴν γινομένην ὑπὸ* M [Ar.] Us. 3 *ῥευμάτιον εἰς αὐτὸν· ⟨καὶ⟩* [Ar.] (suppl. Sylburg) || *ἢ τὸ ποτόν*] *ἐπὶ ποτὸν* BL: *κατὰ ποτὸν* T: *τὸ πότον* e: *τὸν τόπον* M: *ἢ ὁ πότος* [Ar.] Us. || *ἡ ὥρα*] *αἱ ὥρα* B: *ἡ αι ὥρα* sic L: *ἡ ὥρα, καὶ τὸ συμπλήρωμα στενώτερον ποιεῖ τὸν φάρυγγα* [Ar.] 4 *βραδυτέρα* Us.: *βραδυτέρα γίνεται* [Ar.]: *βαρυτέρα* codd. Buss. || *πνεύματος*] *ῥεύματος* P || *ἐστιν* om. [Ar] || *βραδεῖα* [Ar.] Us.: *βαρεῖα* codd. Buss. || *λίαν*] *καὶ λεῖα* BL || *καὶ* om. ABPRSTbe || *βαρεῖά* Us. cf. [Ar.]: *βραδεῖα* codd. Buss. (*βραδεῖαν* L) 4-5 *λίαν – ἐστιν*] *φορὰ* [*καὶ* del. Bekker] *βαρεῖαν* (*βραδεῖαν* Ap.) *ποιεῖ φωνήν* [Ar.]
2.84 (AFGKMOPRSTbeio*p* Buss. Us.: *v* 2.25; [Ar.] 11.19; 2.83 Buss.)
1 *ποτε* om. F [Ar.] || *τῆς – ἐξακούομεν*] *μὲν* (om. C[a]) *βαρυτέρα μᾶλλον ἐξακούεται* [Ar.] || *μᾶλλον* om. F 2 *διότι*] *ὅτι* Go || *καὶ* om. Go*v* [Ar.], del. Buss. (in var. lect.) Us. || *ἡ* om. AFOSTbei || *πλείω μὲν* [Ar.] || *ἀέρι* Oi 2-3 *ἐπ' ἔλαττον – ποιεῖ*] *οὐκ εἰς μῆκος δέ* [Ar.] 3 *οὖν* om. Go || *διότι* M [Ar.] Buss. Us.: *διὸ καὶ* o: *διὸ* cett. 4 *ἀφικνεῖται*] *κινεῖται* [Ar.] || *δὲ*] *μὲν* A || *ἐγγύθεν*] *ἐκεῖθεν* K: *proxime v* || *δὲ*] *καὶ* A || *ἐγγύθεν δὲ*] *μᾶλλον ἐγγύθεν* o || *πλεῖον* A: *πλείω* GTo || *ἡ τοῦ ἀέρος* K || *τοῦ – ἐστιν*] *ἡμῖν ἀὴρ πρὸς τὸ αἰσθητήριον προσπίπτει* [Ar.]

2.82. Why do lice occur most on the head?[356] Is it because living creatures are naturally produced wherever there is a residue of moisture and heat, and where there is room, and the head of young people is moist and hot,[357] and the pores in it are very wide? An indication [of this]; the hairs are thickest here, so there is both matter for the composition of a living creature and a place [for it].

[Voice and sound]

2.83.[358] Why do people speak with a deeper voice after drinking and vomiting and when it is cold? Is it because of the blocking of the larynx by phlegm?[359] For a flux is drawn in afterwards by the vomiting or the drink in some cases and by the season in others, so that the breath travels more slowly; and when it is excessively slow it is deep.[360]

2.84.[361] Why do we hear a deep voice better from close at hand, but less from a distance? Is it because the deeper voice sets more air in motion, but does this to a lesser extent? So from a distance we hear less, because less [air] reaches [us]; but from close at hand more, because there is more movement of the air.

---

356 The solution would seem rather to be to the question why lice occur most on the heads *of young people*, as Aristotle notes at *HA* 5.31 557a7. Lice were believed to be spontaneously generated from the body, ibid. 556b28.

357 Cf. 2.75.

358 = [Ar.], *Probl.* 11.18.

359 Cf. 2.95.

360 Cf. 2.85, 2.89, 2.93, 2.95, 2.96.

361 = the first part of [Ar.], *Probl.* 11.19 (which goes on to suggest also that shrill sound involves quicker movement).

2.85. Διὰ τί ἐν τῷ χειμῶνι ὀξύτερον φθέγγονται, θέρους δὲ καὶ μεθύοντες βαρύτερον; ὅτι ὀξυτέρα μέν ἐστιν ἡ ταχυτέρα ἢ ἀπὸ συντεταμένου φωνή· τῶν δὲ νηφόντων καὶ ἐν τῷ χειμῶνι τὰ σώματα συνέστηκε μᾶλλον ἢ μεθυόντων καὶ ἐν τῷ θέρει· τὸ γὰρ θερμὸν καὶ αἱ ἀλέαι διαλύουσι τὰ σώματα.

2.86. Διὰ τί τὰ ἐννεάμηνα οὐ φθέγγεται, τὰ δὲ δεκάμηνα φθέγγεται εὐθὺς γεννώμενα; ὅτι ὀψιαίτατα τελειοῦται ἡ φωνὴ τοῖς ἀνθρώποις· σημεῖον δέ· ἤδη γὰρ ηὐξημένοι καὶ ψελλίζουσι καὶ τραυλίζουσιν· ἢ ὅτι τὰ μὲν ἐννεάμηνα ⟨οὐ⟩ τελειογόνα * * * ὥστε τὰ μὲν ἀπειληφότα ἤδη γίγνεται τὴν φύσιν, τὰ δὲ οὔ· τὸ οὖν τελευταῖον ἐπιτελούμενον τοῖς μὲν ἀτελέσιν οὐχ οἷόν τε ὑπάρχειν ὡς ἐπὶ τὸ πολύ, τοῖς δὲ ἐπιτετελεσμένοις ὑπάρξει· τὰ δὲ ἑπτάμηνα ἔτι ἧττον διὰ τὴν αὐτὴν αἰτίαν· οὕτω γὰρ ἀτελῆ σφόδρα καὶ ἀσθενῆ γίνεται, ὥστε καὶ σπαργανοῦται ἐρίοις.

2.87. Διὰ τί ἡ φωνὴ ὕστατον τελειοῦται τοῖς ἀνθρώποις τῶν φθεγγομένων; ἢ διότι πλείστας ἔχει διαφορὰς καὶ εἴδη, τὰ δ' ἄλλα ζῷα ἢ οὐδὲν γράμμασιν ἢ ὀλίγα διαλέγεται· τὸ δὲ ποικιλώτατον καὶ πλείστας

2.85 (AFGKMOPRSTYbei*p* Buss. Us.; [Ar.] 11.56; 2.84 Buss.)
1 *ἐν τῷ χειμῶνι*] *τοῦ μὲν χειμῶνος* [Ar.] ‖ *τῷ* om. GY ‖ *ὀξυτέρα* G: *ὀξύτερα* P: *ὀξυτέρω* Y ‖ *φθέγγονται καὶ νήφοντες* [Ar.] ‖ *θέρος* P ‖ *μεθύονται* A 2 *βαρυτέρων* A: *βαρύτερα* Y ‖ *ἢ ὅτι* K [Ar.] Us. ‖ *ὅτι*] *διότι* FP ‖ *ἡ ταχυτέρα* [Ar.] Buss. Us.: *ἡ παχυτέρα* AGPY Ap.: *ἢ παχυτέρα* FMORSTbei: *ἡ ταχυτέρα φωνὴ* K: *ταχυτέρα, ταχυτέρα δ' ἐστὶν* [Ar.] ‖ *ἢ ἀπὸ* KMOi Buss. Us.: *ἡ ἀπὸ* FGPRSTYbe [Ar.]: *ἡ ἀπὸ* sic A ‖ *συντεταγμένου* M 3 *μᾶλλον* S [Ar.] Buss. (in var. lect.) Us.: *μάλλιστα* sic A: *μάλιστα* cett. 4 *αἱ ἀλέαι*] *ἀλέα* Y: *ἀλέα* G
2.86 (ABFKLMOPRSTbei*p* Buss. Us.; *v* 2.26; 2.85 Buss.)
1 *ἐννεάμηνα*] *ἑπτάμηνα* K*v* ‖ *φθέγγεται*[1]] *φέγγονται* sic A ‖ *δὲ δεκάμηνα*] *δεκάμηνα* F: *δὲκάμηνα* sic A: *δὲ ἐννεάμηνα* K*v* 2 *γεννώμενα*] *γενόμενα* BL: *γεννηθέντα* A Buss. ‖ *ἢ ὅτι* K*v* 3 *γὰρ* om. BL ‖ *καὶ*[1] om. K*v* ‖ *ψιλλίζουσιν* Oi ‖ *ἢ* K*v* Buss. Us.: om. cett. ‖ *τὰ* om. BL 4 *οὐ* add. Buss. ‖ post *τελειογόνα* lacunam statuit Us. 6 *οἷόν τε*] *οἴονται* T ‖ *ἐπιτελεσμένοις* AT 6-8 *τὰ δὲ ἑπτάμηνα – ἐρίοις* del. Us. 7 *ἔτι* coni. Buss., prob. Us.: *ἐστιν* codd. ‖ *ἧττον* L Buss. (coni.) Us.: *ἥττονα* cett. ‖ *διὰ τὴν αὐτὴν αἰτίαν* om. Buss. ‖ *αὔτὴν* A ‖ *οὕτω*] *ὅτι* K*v*: *οὔτε* BL ‖ *ἀτελεῖ* AR 8 *ἀσθενεῖ* A: *ἀσθενεῖν* P ‖ *ὥστε*] *σημεῖον ὅτι* K*v* ‖ *σπαργανοῦσθαι* T
2.87 (AFKMOPRSTbei*p* Buss. Us.; *v* 2.27; [Ar.] 11.57; 2.86 Buss.)
1-2 *τοῖς φθεγγομένοις τῶν ἀνθρώπων* Se[pc] 2 *δ'*] *γὰρ* [Ar.] 3 *οὐδὲν γράμμασιν* FKPTe*v*: *οὐδ' ἐν γράμμασιν* MORSbi: *οὐδ' ἐν γρᾶ* sic A: *οὐθὲν γρᾶμμα* [Ar.]: *οὐδὲν γράμμα* Buss. Us. ‖ *ἢ* K*v* [Ar.] Buss. Us.: om. cett. ‖ *διαλέγονται* [Ar.] 4

2.85.[362] Why do people speak with a higher voice in winter, with a deeper one in summer and when drunk? Is it because the voice is higher when it is swifter or from what is taut,[363] and in the case of those who are sober and in winter bodies are more compacted than in the case of those who are drunk and in summer? For heat and hot conditions relax bodies.

2.86. Why do nine-month children not make a sound, but ten-month children do as soon as they are born? Because the voice is the last thing to be perfected in human beings. An indication; when they are have already grown they speak inarticulately and lisp. Is it because nine-month children are <not>[364] born complete, so that the former are born already having received their nature, the latter not. So what is perfected last cannot for the most part be possessed by those that are imperfect, but it will be possessed by those that have been perfected. Seven-month children [make a sound] even less for the same reason; for they are so very incomplete and weak that they are swaddled in wool.[365]

2.87.[366] Why is voice perfected last in human beings of [all creatures] that make a sound? Is it because it has very many differences and kinds, while other creatures utter either nothing in letters, or few, and what is most

---

362 = [Ar.], *Probl.* 11.56.

363 cf. 2.83, 2.89, 2.93, 2.95, 2.96.

364 Supplied by Bussemaker. For ten months as the normal term of human pregnancy cf. Ar., *HA* 7.4 584a37; see also *GA* 4.4 772b9.

365 Usener brackets the last sentence; but this is not necessary if we follow Bussemaker in what precedes – see the previous note.

366 = [Ar.], *Probl.* 11.57; cf. [Alex.], *Probl.* i.148, with a similar answer, adding that the sounds made by animals are more due to nature than to soul.

ἔχον διαφορὰς ἀνάγκη οὖν ἐν πλείστῳ χρόνῳ ἀποτελεῖσθαι.

2.88. Διὰ τί οὐ τῇ αὐτῇ φωνῇ πάντες διαλέγονται; ἢ ὅτι πρῶτον ἀπὸ τῆς μιμήσεως τῶν ἀλόγων αἱ φωναὶ εὑρέθησαν οἷον τὰ παιδία τὸν βοῦν ἀπὸ τοῦ μυκᾶσθαι καὶ τὸν κύνα ἀπὸ τοῦ ὑλακτεῖν; ἔχοντος δὲ τοῦ αὐτοῦ ζῴου πολλὰς μιμήσεις, ἀφ᾽ ὧν ἐκλέξεταί τις τὴν ὁμοίωσιν (καὶ γὰρ αὐτὸ κέρατα ἔχει, καὶ τετράπουν καὶ μακρόν), οὐκ ἐπὶ ταὐτὸν ἕκαστος ὁρμήσας ταὐτὸν μὲν ἐσήμανε τὸν βοῦν, οὐ μὴν τῇ αὐτῇ φωνῇ· ὁ δὲ τὸ αὐτὸ τῆς πρώτης παραλλαξάσης ἤδη, εἰκότως ἐκ τούτων παραλλάττων μικρόν, οἷον οἱ μὲν Διόνυσον, οἱ δὲ Ζώνυσον, καὶ οἱ μὲν Ἡραγόρην, οἱ δ᾽ Ἡραγόραν.

2.89. Διὰ τί οἱ μὲν ἀγωνιῶντες βαρὺ φθέγγονται, οἱ δὲ φοβούμενοι ὀξύ; ἢ ὅτι τοῖς μὲν φοβουμένοις καταψύχεται ὁ τόπος ὁ περὶ τὴν καρδίαν, κάτω ὁρμῶντος τοῦ θερμοῦ, ὥστ᾽ ὀλίγον ἀέρα κινοῦσιν (ἡ γὰρ ἰσχὺς ἐν τῷ θερμῷ)· τοῖς δὲ ἄνω φέρεται τὸ θερμόν, ὥσπερ τοῖς αἰσχυνομένοις; δι᾽ αἰσχύνης γὰρ ἀγωνιῶσι, τοῖς δὲ αἰσχυνομένοις ἄνω φέρεται πρὸς τὸ πρόσωπον· σημεῖον δὲ ὅτι ἐξέρυθροι γίνονται μᾶλλον· συντήκει οὖν καὶ παχὺν ποιεῖ τὸν ἀέρα ᾧ φθέγγεται, ὁ δὲ τοιοῦτος βραδέως ὠθεῖται, τὸ δὲ βραδὺ ἐν φωνῇ βαρύ ἐστιν.

ἔχον om. Oi: post *διαφορὰς* exhib. K || *οὖν* om. Kp*v* [Ar.] Buss. || *ἀποτελεῖσθαι* APT*p* [Ar.] Buss. Us.: *ἀποτελεῖται* F: *διατελεῖσθαι* KMORSbei: *consumari* (sic, i.e. *consummari*) *v*

2.88 (AFKMOPRSTbei*p* Buss. Us.; 2.87 Buss.)
1 *οὐ*] *ἐν p* || *πάντες* FMRSTb*p* Buss. Us.: *πάνταις* A: *πάντα* KOPei || *τῆς* om. Buss. 2 *τὸν τὸν* P 3 *μυκᾶσθαι*] *μηκᾶσθαι* T || *τὸν κύνα* AKOSbi Buss. Us.: *τὴν κῦνα* FMPRTe || *ἔχοντος* Buss. Us.: *ἔχοντες* codd. 4 *τὸ αὐτὸ* K 5 *κέρας* F || *κέρατα – μακρόν*] *τετράπουν ἐστὶ καὶ μακρὸν καὶ κέρατα ἔχει* K || *καὶ*[1]] *ὡς* F || *ἕκαστος* FKP Buss. Us.: *ἕκαστον* MORSTbei 5-7 *μακρὸν – παραλλάττων* om. A 6 *ἐσήμανε* KORSbi Us.: *ἐσήμαινε* FMPe Buss.: *ἐσήμενε* T 7 *παραλλάξας* Buss. (in var. lect.) || *εἴδη* P || *παραλλάττων* KP Us.: *παραλλάττον* cett. Buss. || *μικρὰν* P 8 *δὲ*[1] om. Oi || *ζώνυσον* AFP Buss. Us.: *ζόνυσον* MORSTbei: *ὑζόνυσον* K || *Ἡραγόρην* M Us. Buss.: *οιραγόρην* sic A: *ἡραγώραν* P: *κραγόρην* cett. FKMORSTbei || *Ἡραγόραν*] *ἡραγόραν* K: *ηραγόραν* T: *ἡραγόρην* M: *ἡραγωρᾶν* P: *κραγόραν* F

2.89 (ABFGKLMOPRSTYbeij*p* Buss. Us.; *v* 2.28; [Ar.] 11.32; 2.88 Buss.)
2 *καταψύχεται*] *καταμάχεται* Yj 2-3 *κάτω ὁρμῶντος*] *καταορμῶντος* BL: *κατὰ ὁρμοῦντος* j 4-5 *δι᾽ αἰσχύνης – αἰσχυνομένοις* om. Y 5 *αἰσχύνης* AFGMPRTej Buss. Us.: *αἰσχύνην* BKLOSbi [Ar.] || *καὶ* (om. Y[a]C[a]) *ἀγωνιῶσι* [Ar.] 6 *δὲ*] *οὖν* GYj || *γίνωνται* j 7 *παχὺ* BGLYj || *ᾧ* PT [Ar.] Buss. Us.: *ὃ* GYj: *ὁ* K*v*: *ὡς* BLMORSbei: *ὃς* sic A: *καὶ* F || *φθέγγεται*] *φθεγγόμενος* K*v*: *φθέγγονται* [Ar.] || *δὲ*[1] om. Yj || *ὠθεῖται*] *ὠθῆται* A: *φθέγγεται* GYj 8 *βαρὺν* Gj: om. BL

complex and has most differences necessarily is perfected in the longest time?

2.88. Why do all not speak in the same language? Is it because sounds were discovered first from imitation of irrational animals, as children [name] the cow[367] from its mooing and the dog from its barking? But since there are many ways of imitating the same animal, and one will select the resemblance from among these (for the same [creature] has horns, and has four feet and is tall), each does not go after the same feature and indicate the cow as the same thing or by the same sound. And another [will indicate] the same thing with the first sound having already changed, reasonably changing it a little, so that some say Dionysus and others Zonysus, and some Heragores and some Heragoras.[368]

2.89.[369] Why do those who are struggling in pain speak with a deep voice, those who are fearful with a shrill one? Is it because in those who are fearful the region around the heart is chilled, the heat tending downwards, so that they set a small amount of air in motion (for strength is in heat), while in the others the heat travels upwards, as in those who are ashamed? For they are pained by the shame, and in those who are ashamed [the heat] is carried upwards to the face. An indication is that they become more red. So [the heat] condenses and thickens the air with which they speak, and when it is like this it is propelled slowly, and in voice slow is deep.[370]

---

[367] In the Greek, the bull. But it is of a moo-*cow* that English children speak.

[368] The forms "Zonysus" and "Heragores" occur only here in the Greek literature in the TLG (on-line version, checked March 2006). "Heragoras" is used as an example to explain vowel-quantity in derived words (in this case from "Hera") by Georgius Choeroboscus, *Prol. et schol. in Theodosii Alexandrini canones*, *Grammatici Graeci* 4.1 186.22.

[369] = [Ar.], *Probl.* 11.32; cf. id. 11.53, a fuller version of the same solution.

[370] Cf. 2.83, 2.85, 2.93, 2.95, 2.96.

2.90. Διὰ τί τῆς φωνῆς ἐπειδὴ ῥύσίς τίς ἐστι φύσιν ἐχούσης ἄνω φέρεσθαι, μᾶλλόν ἐστιν εὐήκοα τὰ ἄνωθεν κάτω ἢ τὰ κάτωθεν ἄνω; ἢ διότι ἡ φωνὴ ἀήρ τίς ἐστι μεθ' ὑγροῦ; βαρὺς οὖν ὢν οὗτος ὑπὸ τοῦ ὑγροῦ φέρεται κάτω καὶ οὐκ ἄνω, τοῦ γὰρ ὑγροῦ κατὰ φύσιν ἡ κάτω φορά· διόπερ τοῖς κάτω μᾶλλον ἀκούεται· ἢ τοῦτο μὲν ἐπὶ τῆς τοῦ ζῴου φωνῆς μόνον γίγνεται· (αὔτη γὰρ μεθ' ὑγροῦ), τὸ ⟨δὲ⟩ συμβαῖνόν ἐστι καὶ ἐπὶ τῶν ἄλλων ψόφων; καθάπερ οὖν ἄποψις ἂν μὲν ἄνωθεν κάτω προσπέσῃ, τὴν ἀνάκλασιν ἄνω ποιεῖται, κάτωθεν δ' ἄνω προσπεσοῦσα ἀνεκλάσθη κάτω, τὸν αὐτὸν δὴ τρόπον ἡ φωνὴ φύσιν ἔχουσα ἄνω φέρεσθαι, προσκόψασα τῷ ἐξ ἐναντίας ἀέρι, βιάζεσθαι μὲν οὐ δύναται πλεῖον ὄντα καὶ βαρύτερον, ἀνακλασθεὶς δὲ ὁ κινηθεὶς ἀὴρ ἐπὶ τοὐναντίον ἠνέχθη κάτω, διὸ καταχεόμενος ἀκούεται μᾶλλον κάτω. τοιοῦτο δὲ καὶ τὸ περὶ τὴν ἠχὼ συμβαῖνον· ἔστι γὰρ ἀνάκλασις τῆς φωνῆς ἐπὶ τοὐναντίον.

2.91. Διὰ τί μεθυόντων μᾶλλον ἀπορρήγνυται ἡ φωνή, ἢ νηφόντων; ἢ διότι τῷ πεπληρῶσθαι ταχέως ἀπορρηγνῦσι τὴν φωνήν; σημεῖον δὲ τούτου· οὔτε γὰρ οἱ χοροὶ μελετῶσιν ἐξ ἀρίστου οὐθ' ὑποκριταί, ἀλλὰ νήστεις ὄντες· ἐν δὲ τῇ μέθῃ πληρέστεροι ὄντες εὐλόγως μᾶλλον

2.90 (AFGKMOPRSTYbei*p* Buss. Us.; *v* 2.29; [Ar.] 11.45; 2.89 Buss.)
1 *φύσιν*] *φύσις* F || *ἔχουσα* K*v* 2 *εὐήκοα τὰ ἄνωθεν κάτω*] *εὐήκοα τὸ ἄνωθεν κάτω* G: *τὰ ἄνωθεν κάτω εὐήκοα* K: *εὐήκοος ἄνωθεν κάτω* [Ar.] || *τὰ*² om. [Ar.] || *ἢ*² om. GY 3 *διότι*] *ὅτι* [Ar.] || *βαρὺς*] *βαρυνόμενος* [Ar.] || *οὖν* om. GY || *ὢν οὗτος*] *οὗτος ὢν* P: *οὗτος* [Ar.] || *ὑπὸ*] *ἀπὸ* GY 5 *διόπερ* KOSbi*v* Us.: *διότι* cett.: *διὸ* [Ar.] Buss. (in var. lect.) || *ἐπὶ*] *ὑπὸ* [Ar.] || *τῆς* om. e 5-13 *ἢ τοῦτο μὲν – τοὐναντίον* om. GY 6 *γίγνεται*] *μίγνυται* e[ac] || *αὔτη* e [Ar.] Buss. Us.: *αὐτῇ* A: *αὐτῇ* K*v*: *αὐτὴ* cett. || *δὲ* ex. [Ar.] suppl. Buss. Us.: om. codd. || *συμβαίνων* A || *καὶ ἐπὶ*] *οὐκ ἐπὶ* K*v* 7 *ἄποψις*] *ἄμπωτις* Buss.: *ἄμμπω*[...] A: *ἡ ὄψις* [Ar.] 8 *προσπεσοῦσα*] *scandens v* || *ἀνακλασθῇ* AFOPR[ac]STbi: *ἀνεκλύσθη* M 9 *αὐτὸν* A || *δὴ* KOSbi Buss. (in var. lect.) Us.: *δὲ* AFMPRTe: om. [Ar.] || *ἡ φύσιν* [Ar.] || *ἔχουσα*] *φέρουσα* A || *προσκόψασα*] *προκόψασα* T: *presecta v* 10 *μὲν οὖν* Oi Us. || *πλεῖον* ORSbei Us.: *πλέον* FT: *πλείω* K: *πλέονα* A Buss.: *πλείονα* MP [Ar.] Us. || *καὶ* K*v* [Ar.] Us. Buss.: om. AFMOPRSTbei 11 *δὲ*] *οὖν* T 11-12 *διὸ – κάτω* om. KMORSbei*pv* 12 *καταχεόμενος* A [Ar.] Us.: *κατεχόμενος* FPT Buss || *μᾶλλον κάτω* FP Us.: *κάτω μᾶλλον* AT [Ar.] Buss. 13 *γὰρ* om. [Ar.]
2.91 (AFKMOPRSTbei Buss. Us.; *v* 2.30; [Ar.] 11.46; 2.90 Buss.; paraphrasin tantum habet *p*)
2 *διότι*] *διὰ* [Ar.] || *τῷ*] *τὸ* T [Ar.] || *ἀπορρηγνῦσι* Us.: *ἀπορρήγνυσι* codd. et [Ar.] codd. plerique: *ἀπορρηγνύασι* Buss.: *ἀπορρηγνύει* [Ar.] Y[a] 3 *μελετούσιν* sic A || *οὐθ' οἱ ὑποκριταὶ* K Us. 4 *νῆστις* Fbi et [Ar.] cod. Y[a2]: *νῆσεις* O || *μέθει* F || *πληρέστεροι* AFPRT [Ar.] Buss. Us.: *πληρέστατοι* KOSbei: *πληρέστερα* M:

2.90.[371] Why, when there is a flowing of voice which naturally travels upwards, are things above more easily heard from below than things below from above? Is it because voice is a certain air accompanied by moisture? So being heavy from the moisture it travels downwards and not upwards, for what is moist naturally travels downwards, and for this reason it is heard more by those below. Or does this happen only in the case of the sound of a living creature (for this is accompanied by moisture), but what happens applies to other sounds too? So just as sight if it strikes something below from above is reflected upwards, but if it strikes something above from below is reflected downwards, in the same way sound which naturally travels upward, striking the air opposite, cannot force a way through it because it is greater in quantity and heavier, and the air that has been set in motion is reflected in the opposite direction and travels downwards, and for this reason, being poured down, it is heard better below. Like this too is what happens with the echo; for it is a reflection of the sound in the opposite direction.

2.91.[372] Why is the voice of those who are drunk more fractured than of those who are sober? Is it because through being filled they break off their utterance swiftly? An indication of this: choirs and actors do not practice after lunch, but on an empty stomach; in drink, being more full, it is

[371] = [Ar.], *Probl.* 11.45.
[372] = [Ar.], *Probl.* 11.46.

ἀπορρήγνυνται τὰς φωνάς.

2.92. Διὰ τί ἡ σῦριγξ καὶ ἡ ὀξεῖα φωνὴ ἁπλῶς ὥσπερ ἐρημίαν ποιεῖ φαίνεσθαι; ἢ ὅτι ἡ ὀξεῖα φωνὴ μακρὰν βαδίζει καὶ οὐκ ἐπὶ πολὺ διαχεῖται, καθάπερ αἱ ἄλλαι φωναί;

2.93. Διὰ τί ποτε ἡ φωνὴ ὀξυτέρα φαίνεται τοῖς μακρότερον ἀφεστηκόσι, τοῦ ὀξέος ὄντος ἐν τῷ ταχέως φέρεσθαι, τὸ δὲ μακρότερον βραδύτερον κινεῖται; ἢ ὅτι ἡ ὀξύτης οὐ μόνον ἐστὶν ἐν τῷ ταχέως κινεῖσθαι, ἀλλὰ καὶ ἐν τῷ ἐλάττω τὸν ψόφον γίνεσθαι, τοῖς δὲ μακρότερον ἀφεστηκόσιν ἀεὶ ἐλάττων ἡ φωνὴ ἀφικνεῖται διὰ τὴν ὀλιγότητα τοῦ ἀέρος τοῦ κινουμένου;

2.94. Διὰ τί, εἴπερ ἡ φωνὴ ἀήρ τίς ἐστιν ἐσχηματισμένος καὶ φερομένη διαλύεται πολλάκις τὸ σχῆμα, ἡ ἠχὼ γίνεται πληγέντος τοῦ τοιούτου πρός τι στερεὸν καὶ οὐ διαλύεται ἀλλὰ σαφῶς ἀκούεται; ἢ διότι ἀνάκλασις, οὐ κατάκλασίς ἐστι, τοῦτο δ' ὅλον ἀφ' ὅλου; εἶτα τὸ πάθος ἀφ' ὁμοίου· ἀπὸ γὰρ τοῦ ἀέρος ἀνακλᾶται τοῦ ἐν τῷ κοίλῳ οὐκ ἀπὸ τοῦ κοίλου.

2.95. Διὰ τί χειμῶνος ὄντος αἱ φωναὶ βαρύτεραι; ἢ ὅτι παχύτερος ὁ ἀήρ, παχυτέρου δὲ ὄντος, βραδυτέρα ἡ κίνησις, ὥσθ' ἡ φωνὴ βαρυτέρα; ἢ ὅτι

---

saturi tantum *v* || *ὄντα* M 4-5 *ἀπορρήγνυνται εὐλόγως μᾶλλον* P 5 *φωνὰς*] *φύσεις* F

2.92 (AFKMOPRSTbei*p* Buss. Us.; *v* 2.31; 2.91 Buss.)

1 *σῦριγξ* Us.: *σύρηξ* T: *σύριγξις* F: *σύριγξ* cett. Buss.

2.93 (AFKMOPRSTbei*p* Buss. Us.; *v* 2.32; [Ar.] 11.20; 2.92 Buss.)

1 *ποτε* om. [Ar.] || *τοῖς μακροτέρων ἀφεστικόσιν ἀεὶ ἐλάττον* (sic) A 2 *ὀξέως* AT et [Ar.] Y[a] ante corr. || *μακρότερον*] *μακρότερον φερόμενον* [Ar.] Y[a]C[a]: *μακροτέρω φερόμενον* [Ar.] cett. 3 *ἢ ὅτι* K*v* [Ar.] Us.: om. cett. || *ἡ* om. M || *ὀξύτης τῆς φωνῆς* [Ar.] || *ἐστὶν* om. KOSbi 4 *ἐλάττω τὸν*] *λεπτότατον* [Ar.] || *μακρότερον*] *μακροτέραν* [Ar.] codd. plerique: *μακρότερον* [Ar.] C[a]: *μακροτέρῳ* [Ar.] Ap.C[a] 5 *ἐλάττων*] *λεπτοτέρα* [Ar.] || *ἡ φωνὴ* om. K*v* || *ἀφικνεῖται* om. F || *τοῦ*[2] AKMOSbei*v*, [Ar.] Buss. Us: *οὐ* FPT: om. Re[ac]

2.94 (AFKMOPRSTbei*p* Buss. Us.: *v* 2.33; [Ar.] 11.51 et (1-4) 11.23; 2.93 Buss.)

1 *ἐστιν* om. KORSTbei: ante *ἀήρ* exhib. [Ar.] 11.23: post *ἐσχηματισμένος* [Ar.] 11.51 || *φερόμενος* P [Ar.] 11.23 2 *ἡ ἠχὼ*] *ἢ ἠχὼ* F: *ἡ δὲ* (*δὲ* om. Y[a]C[a]) *ἠχώ, ἢ* [Ar.] 11.51 3 *καὶ* om. [Ar.] || *διαλύεται αὕτη* [Ar.] 11.23 || *σαφῶς*] *σοφῶς* F || *ἀκούεται*] *ἀκούομεν* [Ar.] 11.51: *ἀκούομεν αὐτῆς* [Ar.] 11.23 || *διότι*] *ὅτι* R[ac]e[ac] [Ar.] 11.23 3-4 *οὐ κατάκλασίς* om. A 4 *ἐστι* post *ἀνάκλασις* (3) coll. [Ar.] || *δ' ὅλον*] *δὲ ὅλον* M [Ar.] Us.: *εὖ ὅλον* O: *εὐ ὅλον* sic i || *εἶτα* M [Ar.] Us.: *εἰς* cett. Buss. || *ἀφ' ὁμοίου* [Ar.] Us.: *ἀφομοιοῦται* cett. Buss. 5 *ἀνακλᾶται*] *ἀνακᾶται* O || *τοῦ*[2] MR Buss. Us.: *καὶ τοῦ* cett.: om. [Ar.]

2.95 (ADFGKMOPRSTYbei*p* Buss. Us.; *v* 2.34; [Ar.] 11.61; 2.94 Buss.)

1 *χειμῶνος ὄντος*] *τοῦ χειμῶνος* [Ar.] || *ἢ* om. D || *ὁ* om. Y 2 *δὲ* om. A || *ἡ κίνησις – βαρυτέρα*] *ἡ κίνησις γίνεται τῆς φωνῆς* D || *ὥσθ'*[Ar.] Us.: *ὡς* cett.

reasonable that they break off their utterances more.

2.92. Why do the pan-pipes and a shrill voice give an impression of an empty place? Is it because a shrill voice travels a long way[373] and does not spread out much, as other sounds do?

2.93.[374] Why does the voice seem more shrill to those who are standing a long way off, when shrillness depends on swift travel,[375] but movement over a longer distance is slower? Is it because shrillness depends not only on swift movement, but also in the sound being less, and for those who are standing further away correspondingly less sound arrives because of the small amount of the air that is set in motion?

2.94.[376] Why, if sound is air that has been given a shape, and as it travels its shape is often distorted, does an echo occur when what is like this strikes against some solid object, and it is not distorted but heard clearly? Is it because this is reflection, rather than breaking up,[377] and this is of a whole from a whole? Moreover what happens to it is from something similar;[378] for it is reflected from the air in the hollow thing, not from the hollow thing itself.

2.95.[379] Why are voices deeper in winter? Is it because the air is thicker, and when it is thicker the movement is slower, so the voice is deeper?[380] Or is it

---

373 See the next problem. The first sentence here may perhaps allude to musical effects in theatrical performances.

374 = the first part of [Ar.], *Probl.* 11.20, which then adds a comparison between the diminution of sound and the diminution of number terminating in the unit.

375 Cf. 2.83, 2.85, 2.89, 2.93, 2.95, 2.96.

376 = [Ar.], *Probl.* 11.51; cf. id. 11.23, which omits the last sentence here and replaces it by the observation that there are then two parts, the initial sound and the reflected part.

377 The Greek words are similar to each other: *anaklasis* rather than *kataklasis.*

378 So Usener following [Aristotle]. The MSS of our text have "... from a whole, that it is made like what happens to it". But the reading of [Aristotle] gives a better sense in the light of what follows.

379 = [Ar.], *Probl.* 11.61; cf. id. 11.17, which gives a fuller version of the first answer here, but substitutes for the second that we are sleepier in winter. That voices are deeper in winter contradicts 2.85 above, but the contradiction is present also in the [Aristotle] *Problems* to which 2.85 and 2.95 correspond; cf. Prantl 1851, 367.

380 Cf. 2.83, 2.85, 2.89, 2.93, 2.96.

διὰ τῶν στενῶν βραδύτερον χωρεῖ ὁ ἀήρ, συμφράττονται δὲ τὰ περὶ τὸν φάρυγγα τῇ ποιότητι τοῦ ψυχροῦ καὶ τοῦ ἐπιρρέοντος φλέγματος;

2.96. *Διὰ τί μεγαλόφωνοι οἱ θερμοὶ τὴν φύσιν ὄντες; ἢ ὅτι ἀνάγκη καὶ ἀέρα ἐν τούτοις πολὺν ψυχρὸν εἶναι; ἕλκει γὰρ τὸ πνεῦμα τὸ θερμὸν καὶ ποιεῖ πλείω· ἡ δὲ μεγάλη φωνὴ γίνεται ἐν τῷ πολὺν ἀέρα κινεῖν, ὀξεῖα δὲ ἐν τῷ ταχέως, βαρεῖα δὲ ἐν τῷ βραδέως.*

2.97. *Διὰ τί ἀκούουσι μᾶλλον κατέχοντες τὸ πνεῦμα ἢ ἐκπνέοντες (διὸ ἐν ταῖς θήραις παραγγέλλουσιν ἑαυτοῖς μὴ πνευστιᾶν); πότερον ὅτι τὸ αἰσθητικὸν ἄνω ἔρχεται τῶν φλεβῶν συναιρομένων [ἢ ἀραιουμένων]; καθευδόντων δὲ κάτω, διὸ καὶ μᾶλλον ἐκπνέουσιν. ἢ καὶ τὸ αἷμα ἀπέρχεται κάτω καταπεπνευκότων, ὥστε κενοῦται τὰ ἄνω, ἀκούουσι δὲ τῷ κενῷ; ἢ ὅτι φύσει ὁ ἀσθμὸς ψόφος τίς, οὗτος δ' ἐν τῷ πνεῖν γινόμενος κωλύει ἀκούειν;*

Buss. || *ὅτι*] *διότι* [Ar.] 3 *στενῶν*] *στερνῶν* [Ar.] || *συμφράττονται*] *συμφράττεται* [Ar.] || *τὰ*] *τὸ* [Ar.] 4 *τῇ ποιότητι τοῦ ψυχροῦ*] *ὑπό τε τοῦ ψύχους* D: *ὑπό τε τοῦ ψυχροῦ* [Ar.]

2.96 (ADFGKMOPRSTYbeio*p* Buss. Us.; *v* 2.35; [Ar.] 11.3; 2.95 Buss.)

1 *οἱ θερμοὶ*] *πάντες εἰσὶν οἱ θερμοὶ* [Ar.] || *ὄντες* om. [Ar.] || *ἢ* om. D 2 *ἐν τούτοις ἀέρα* KMORSbei || *ἐν τούτοις* post *ψυχρὸν* [Ar.] || *πολὺν* post *εἶναι* P || <*καὶ*> *ψυχρὸν* Us. ex [Ar.] || *ψυχρὸν* om. D || *τὸ πνεῦμα τὸ θερμὸν*] *τὸ θερμὸν καὶ πνεῦμα πρὸς ἑαυτὸ καὶ ἀέρα* [Ar.] || *πνεῦμα*] *φλέγμα* A 3 *ποιεῖ πλείω*] *ποιεῖ πλεῖον* A Buss.: *πλείω τὸ πλεῖον* [Ar.] || *δὲ*[1] om. MRe || *ὀξεῖα δὲ* DFKY*v* Buss. Us.: *ὀξεῖαν δὲ* MOPRSTbei: *ὀξείαν δὲ* A: *ὀξεία δὲ* G: *ὀξέα δὲ* o: *καὶ ὀξεῖα* [Ar.] 4 *βαρεῖα*] *βαρεία* G: *βαρείαν* P: *βαρεῖαν* AFRSTb: *βραδεῖα* DKY: *βραδεῖαν* Oi || *δὲ ἡ* [Ar.] || *βαρέως* e[mg]

2.97 (AFKMOPRSTbei*p* Buss. Us.; *v* 2.36; [Ar.] 11.41; 2.96 Buss.)

1 *κατέχονται* P || *ἢ*] *οἱ* T || *ἐκπνέονται* P: *ἐκπίνοντες* S 1-2 *ἐν ταῖς*] *ἡ ἐν ταῖς* A: *καὶ ἐν ταῖς* K*v* [Ar.] Us. 2 *ἑαυτὸν* P || *μὴ*] *οὐ* Buss. 3 *αἰσθητικὸν*] *αἰσθητὸν* M || *τῶν φλεβῶν – ἀραιουμένων*] *αἰρομένων τῶν φλεβῶν* [Ar.] || *συναιρομένων* nos ducente [Ar.]: *συναιρουμένων* codd. Buss. Us. || *ἢ ἀραιουμένων* MRe Buss.: del. Us.: *καὶ ἀραιουμένων* K*v*: *ἀναιρουμένων* F: *ἀραιουμένων* cett. 4 *δὲ*] *γὰρ* [Ar.] || *καὶ μᾶλλον*] *μᾶλλον* F: *καὶ μᾶλλόν τε* [Ar.] || *ἐκπνέουσιν* K*pv* Buss. Us. *ἔκπνευσιν* cett.: *ἐκπνέουσι καθεύδοντες ἢ εἰσπνέουσι, καὶ ἀνήκοοί εἰσιν* [Ar.] || *καὶ*[2] om. FKOSbei 5 *ἀπέρχεται*] *ἔρχεται* T: *ἀνέρχεται ἐκπεπνευκότος* [Ar.] || *ἄνω*] *κάτω* Septali ad [Ar.] || *ἀκούουσι*] *exaudiunt proclivius v* 6 *κενῷ*] *καινῷ* T || *φύσει ὁ ἀσθμὸς*] *φύσει ὁ ἀτμὸς* T Buss.: *ὁ φυσιασμὸς* [Ar.] || *τίς*] *τίς ἐστιν* [Ar.] Buss. Us. || *ἐν* om. A Buss. Us. || *πνεῖν*] *ἐκπνεῖν* [Ar.] || *γινόμενος*] *γενόμενος* KMORSbei*pv* et [Ar.] cod. C[a]

because the air travels more slowly through narrow places,[381] and the places around the throat are blocked by the quality of the cold and by the phlegm that flows in?[382]

2.96.[383] Why do those who are hot by nature have loud voices? Is it because there is also necessarily much cold air in them? For heat draws in the *pneuma* and increases it; a loud voice involves moving much air, a shrill voice [moving it] quickly, a deep one slowly.[384]

2.97.[385] Why do people hear better when holding their breath than when breathing out (which is why when hunting they urge each other not to pant)? Is it because the perceptive [faculty] rises upwards when the veins are inflated[386] [or rarefied],[387] but downwards in sleep, which is why they breathe out more? Or is it because the blood departs downwards when we breathe [in], so that the upper parts are emptied,[388] and we hear with what is empty? Or is it because gasping is naturally a noise, and this, coming about when we breathe, prevents us hearing?

---

381 [Aristotle] has "through the chest" (a difference of one letter in the Greek). Both make good sense.

382 Cf. 2.83.

383 = [Ar.], *Probl.* 11.3.

384 Cf. 2.83, 2.85, 2.89, 2.93, 2.95.

385 = [Ar.], *Probl.* 11.41; cf. id. 48, which repeats the last solution given here; cf., with Flashar 1962b, 546, [Aristotle] *Probl.* 11.44 (noise and breath in the ears when yawning) and Ar., *GA* 5.2 781a30 (which argues that breathing sets the principle of hearing in movement and fills the ears with *pneuma*).

386 Following [Aristotle]; the MSS here have rather "are contracted".

387 Bracketed by Usener.

388 [Aristotle] has "Or is it because the blood rises up when we breathe out, so that the upper parts are emptied", emended by Septali to "... so that the lower parts are emptied"; but the point is rather that the upper parts become full. In the version here *katapepneukotôn* seems to mean just "breathe" rather than "breathe in", and what is required is "breathe in *and hold our breath*".

2.98. Διὰ τί οἱ τοῦ γράσου ὄζοντες, ὅταν ἀλείψωνται μύρῳ, δυσωδέστεροί εἰσιν; ἢ διότι ἐπὶ πολλῶν τοῦτο γίνεται ⟨οἷον⟩ ὀξύ * * *; ἢ ὅτι πάλιν ἱδρώσαντες, δυσωδέστεροί εἰσι· τὸ δὲ μύρον θερμαντικόν ἐστιν, ἱδρῶτας οὖν παρασκευάζει;

2.99. Διὰ τί ἡ μασχάλη δυσωδεστάτη τῶν ἄλλων τόπων; πότερον ὅτι ἀπνούστατός ἐστιν, ἡ δὲ δυσωδία μάλιστα ἐν τοῖς τοιούτοις διὰ τὸ σῆψιν γίνεσθαι ὑπὸ τῆς ἡσυχίας τῆς ὑγρότητος, διότι ἀκίνητος καὶ ἀγύμναστος;

2.100. Διὰ τί τὰ δυσώδη ἥκιστα φαίνεται δυσώδη τοῖς ἐδηδοκόσιν; ἢ διὰ τὸ συντετρῆσθαι τὴν ὄσφρησιν τῷ στόματι κατὰ τὸν οὐρανὸν πλήρης ἡ αἴσθησις γίνεται, καὶ τῆς τε δυσωδίας οὐκέτι ὁμοίως αἰσθάνεται (τὸ γὰρ πρῶτον αἰσθάνονται πάντες, ὅταν δὲ ἅψηται, οὐκέτι, ὥσπερ συμφύουσα)· καὶ ἡ ἔξωθεν ὁμοία ἀφανίζεται ὑπὸ τῆς ἔσω.

2.101. Διὰ τί χειμῶνος ὄντος ἧττον ὀσφραινόμεθα καὶ ἐν τοῖς πάγοις; ἢ ὅτι

2.98 (AFGKMOPRSTYbei*p* Buss. Us.; *v* 2.37; [Ar.] 13.9; 2.97 Buss.)
1 τοῦ om. GKY || γράσου om. GY: *situm et fetorem olentes v* || ὅταν] ὅτε GY || ἀλείψονται GY: ἀλείφωνται [Ar.] || μῦρον P 2 εἴσιν] γίνονται [Ar.] || ἢ διότι ἐπὶ – ὀξὺ om. GY: ἢ διότι ἐπὶ – εἰσι (3) om. P || τοῦτο ἐπὶ πολλῶν [Ar.] || post γίνεται add. οἷον Us.: οἷον ὀξὺ καὶ γλυκὺ συμμιχθὲν τὸ ὅλον γλυκύτερον ἐγένετο [Ar.] || post ὀξύ lacunam statuit Us. || ὅτι] διότι Buss. Us. || πάλιν] πᾶσιν Y: πολλοὶ K: *multi v* || ἢ ὅτι πάλιν] εἶτα [Ar.] 3 ἱδρώσαντες] υδρῶσαντες sic A || δυσωδέστεροί om. G: πάντες δυσωδέστεροί [Ar.] || εἰσι om. G || μῦρον KPSTYb
2.99 (ADFGKMOPRSTYbeio*p* Buss. Us.; *v* 2.38; [Ar.] 13.8; 2.98 Buss.)
1. δυσωδεστάτη DGMPYeo: δυσωδέστατος FKORSTbi*v*: δυσωδεστέρα A Buss. Us.: δυσωδέστατον [Ar.] || τῶν] ἐστὶ τῶν Ae Buss. Us. || τῶν ἄλλων τόπων trasp. post ἡ μασχάλη D: ἄλλων om. [Ar.] 1-2 πότερον ὅτι ἀπνούστατός ἐστιν] διὰ τὸ μὴ ἔχειν διάπνοιαν D 2 ἀπνούστατος] εὐπνούστατος o || ἡ δὲ] ἡ AFPT: ἡ γὰρ D: καὶ ἡ Yo: ἐστιν; ἔστι δὲ ἡ [Ar.] || μάλιστα post τοιούτοις [Ar.] || τοιούτοις γίνεται τόποις D || τὸ τὴν σῆψιν K 2-3 διὰ τὸ σῆψιν – ἀγύμναστος] διὰ τὴν ὑγρότητα καὶ τὴν σῆψιν καὶ τὴν ἀκινησίαν καὶ ἀγυμνασίαν D 3 γίνεσθαι] γίγνεσθαι T: γενέσθαι Y || ἡσυχίας] ἡσύχου [Ar.] familia *βγ* Marenghi || ὑγρότητος] ποιότητος [Ar.] familiae *αβγ* Marenghi: πιότητος [Ar.] fam. *δ* Marenghi, quod probavit Ruelle: et ὑγρότητος et πιότητος ideo introducta ut ποιότητος explicaretur censuit Marenghi 1991, 131 || ἢ διότι [Ar.]
2.100 (AFKMOPRSTbei*p* Buss. Us.; *v* 2.39; [Ar.] 13.2; 2.99 Buss.)
1 ἥκιστα – ἐδηδοκόσιν] τοῖς ἐδηδοκόσιν οὐ δοκεῖ ὄζειν [Ar.] || ἢ ὅτι coni. Buss. 3 γίνεται ταχύ [Ar.] || τῆς] τοῖς A et [Ar.] fam. *δ* Marenghi || τε delendum coni. Buss. || δυσωδίας] ἔσω [Ar.]: <ἔσω> δυσωδίας Us. || 4 ἅψωνται [Ar.] || συμφύουσα] συμφυοῦς [Ar.] Buss. Us. 5 ἡ ἔξωθεν ἡ ὁμοία [Ar.]
2.101 (AFKMOPRSTbei*p* Buss. Us.; *v* 2.40: [Ar.] 12.6; 2.100 Buss.)
1 χειμῶνος ὄντος] χειμῶνος KOi: τοῦ χειμῶνος [Ar.] || ὀσφραινόμεθα ἧττον A

[Odour]

2.98.[389] Why do those who smell like goats have a worse smell when they are anointed with unguents? Is it because in many cases this [the unguent] becomes sharp?[390] Or is it because when they sweat again they have a worse smell, and unguent[s] cause heating and so stimulate sweat?

2.99.[391] Why is the armpit the worst-smelling of [all][392] places? Is it because it is least exposed to air, and bad smell occurs most in such [places] because rotting is brought about by the stillness of the moisture,[393] since[394] it is not moved or exercised?

2.100.[395] Why do things with a bad smell seem to have a bad smell least to those who have eaten? Is it because the sense of smell is connected to the mouth through the roof of the mouth, [and that] for this reason it no longer perceives the bad smell[396] in the same way (for everyone perceives at first, but no longer when in contact, as it becomes united), and what is similar outside is obscured by what is inside?

2.101.[397] Why is our sense of smell weaker in winter and when it is frosty? Is

---

389 = [Ar.], *Probl.* 13.9. Cf., with Marenghi 1991, 132 n.3, ibid. 2.13 = 20.33, 13.11, and Theophrastus, *Sud.* 10, with Fortenbaugh 2003, 80-81.

390 Usener supplements the text from [Aristotle] to give "Is it because this happens in many cases, as <when> sharp <and sweet are mixed the whole becomes sweeter>?". Flashar 1962b, 558 emends the text of [Aristotle] from our *Problem* to give "... the whole becomes sharper". Cf. [Ar.], *Probl.* 13.11, and Marenghi 1991, 132.

391 = [Ar.], *Probl.* 13.8. Cf., with Marenghi 1991, ibid. 2.6 and Theophrastus, *Sud.* 9, with Fortenbaugh 2003, 80.

392 Literally "the worst-smelling of the other places": for the "compendious comparison" cf. Thucydides 1.10.3, "greatest of those before it". [Aristotle] simply has "is the worst-smelling of places".

393 For "moisture" most MSS of [Aristotle] have "quality"; one family, followed by Ruelle, has "fattiness". Marenghi 1991, 131 suggests that both "moisture" and "fattiness" originated as explanations of "quality".

394 [Aristotle] here has "Or is it since...".

395 = [Ar.], *Probl.* 13.2.

396 [Aristotle] has "the [bad smell] inside", adopted by Usener here.

397 A shorter version of [Ar.], *Probl.* 12.6. Cf., with Marenghi 1991, 98, Aristotle, *Sens.* 5 443b14-16 and Theophrastus, *CP* 6.17.5, where the explanation is that the cold interferes with the *production* of the odour.

ὁ ἀὴρ ἀκινητότερος ἐν τῷ ψύχει; οὐκοῦν οὐκ ἐφικνεῖται ἡ κίνησις ἡ ἀπ⟨ὸ τοῦ σώματος τοῦ τὴν⟩ ὀσμὴν ἔχοντος διὰ τὴν δυσκινησίαν;

2.102. Διὰ τί αἱ κύνες οὐκ ὀσφραίνονται τῶν ἰχνῶν ὅταν ἀποθάνῃ δασύπους, οὐδ' ἂν λάθῃ αὐτὰς ὁ τεθνεώς, οὐκέτι ὁμοίως εὑρίσκουσιν; ἢ ὅτι τεθνεῶτος μὲν οὐκ αἰσθάνονται διὰ τὸ ἄλλην ὀσμὴν γίνεσθαι, τὴν δ' οἰκείαν διαφθείρεσθαι, πάντων φθειρομένων, ὥσπερ φωνὴν καὶ χρῶμα καὶ χυμόν; πάντα γὰρ μεταβάλλεται ταῦτα, ζῶντος μὲν οὖν διὰ τὸ συνεχῆ εἶναι τὴν ὀσμὴν ἀπὸ τοῦ θηρίου αἰσθάνονται· τεθνεῶτος δὲ πέπαυται ῥέουσα· οὐ γὰρ καταλείπει, ὥσπερ Ἐμπεδοκλῆς, ὡς "... ἀπέλειπε ποδῶν ἀπαλὴν περὶ ποίην"· οὐχ οἷόν τε γὰρ διασεσῶσθαι τὴν ὀσμὴν οὐδὲ τὴν χρόαν, ἀλλ' ὅταν ἀποθάνῃ, κἀκεῖνα καὶ αὕτη ἔφθαρται.

2.103. Διὰ τί αἱ ὀδμαὶ ἐγγύθεν ἧττον εὐώδεις καὶ θυμιαμάτων καὶ ἀνθῶν; ἢ ὅτι τοῖς πλησίον συναναφέρεται τὸ γεῶδες, ὥστε κεραννύμενον ἀσθενῆ ποιεῖ τὴν δύναμιν· εἰ δὲ πόρρω, καταφέρεται ὀδμή; διὰ τοῦτο δὲ καὶ

Buss. || *πάγοις*] *πάγεσιν ἥκιστα* [Ar.] || *ἢ* om. Oi 2 *ὁ* om. Oi || *κινητότερος* sic A || *ψύχειν* sic A || *οὐκοῦν οὐκ ἐφικνεῖται*] *οὔκουν ἀφικνεῖται ὁμοίως* [Ar.] || *ἀφικνεῖται* M [Ar.] Buss. Us. || *ἡ κίνης* T 2-3 suppl. Us. ex [Ar.]: *ἡ ἀποσμὴν* M: *ἢ ἀποσμὴν* cett. (om. F) Buss. 3 *ἔχοντος* P [Ar.] Us.: *ἔχοντες* cett. Buss.: *aut redolentia non excipitur v* || *δυσκινησίαν*] *δυσκινησίαν τῆς ἀπορροῆς καὶ τοῦ ἀέρος ἐν ᾧ ἐστίν* [Ar.]

2.102 (AFKMOPRSTbei*p* Buss. Us.; 2.101 Buss.)

1 *αἱ*] *οἱ* A Buss. || *ὀσφραίνονται*] *αἰσθάνονται* M || *ἰχνῶν*] *ἰσχνῶν* KOi 2 *αὐτὰς*] *αὐτοὺς* FP Buss.: *αὐτοῦς* A: *αὐτοῖς* MRe[ac] || *ὁ* del. Buss. 3 *τεθνεῶτος* FMP Buss. Us.: *τεθνεῶτας* KOSbi: *τεθνεῶτες* Re: *τεθνεότος* T: *τεθνεὄταις* sic A 4 *φωνὴ* T 4-5 *φωνῆς καὶ χρώματος καὶ χυμοῦ* Buss. (in var. lect.) 5 *μεταφάλλεται* F || *συνεχῆ* AFKMPRTe Us.: *συνενεχθῆναι* OSbe[mg]i: *συνεχὲς* coni. Buss. 6 *εἶναι* om. Buss. || *ὀσμὴν* coni. Buss., accepit Us.: *νόσον* codd. || *τεθνεότος* T: *τεθνεὄτως* sic A || *πέπαυται*] *παύεται* T 7 *οὐ γὰρ καταλείπει* om. Buss. || *ὡς*] *ὅσσ'* Empedocli attribuit Nauck || *ἀπέλειπε* KMP (per emend. metri causa?) Buss. Us.: *ἀπέλιπε* AFORSTbei || *ποδῶν*] *παρὰ ποδῶν* e: *ποδῶν* cett. Buss. Us.: *πολῶν* coni. Usener 1859, xiii-xiv, Empedoclis fr. 31B101 DK conferens 8 *ἀπαλῇ* Diels: *ἀπαλὴ* FMPRSbe Buss.: *ἀπαλῆ* AT: *ἀπειλὴ* O: *ἀπειλὴ* Ki: *ἀπαλὴν* Us. || *περὶ ποίῃ* Diels: *περὶ ποία* MP: *περὶ ποίαν* e: *περίπνοια* Buss.: *περιποία* ART: *περίποια* F: *περιποιία* KOSbei: *περὶ ποίην* Us. || *οἷόν τε*] *οἷονται* sic A || *διασεσῶσθαι* Us.: *διασπᾶσθαι* codd. Buss. 9 ***κἀκείνη*** coni. Buss. || *αὕτη* P Us. Buss.: *αὐτῆ* sic A: *αὐτὴ* cett.

2.103 (AFKMOPRSTbei*p* Buss. Us.; de DGY v. infr.; *v* 2.41; 1-4 = [Ar.] 12.9; 4-5 = [Ar.] 12.4; 2.102 Buss.)

1 *ὀσμαὶ* [Ar.] 2 *τοῖς* om. [Ar.] || *συναναφέρεται*] *μὲν συναποφέρεται* [Ar.] || *ἀσθενῆ*] *ἀσθενέστεραν* [Ar.] 3 *ποιεῖν* AF || *εἰ δὲ*] *εἰς δὲ τὸ* [Ar.] || *καταφέρεται* APY [Ar.] Buss. Us.: *καταφαίνεται* cett. || *ἡ ὀδμή* [Ar.] || *διὰ ταῦτα* [Ar.] || *δὲ* AFMPTe [Ar.] Buss. Us.: om. KORSbi*v* 4 *τὰ τριφθέντα* F || *τὰ* om. Oi || *τίς*

it because the air is set in motion less in the cold, so the movement from <the body> that possesses the odour does not reach [us] in the cold because of the difficulty of movement?[398]

2.102. Why do dogs not smell the tracks when a hare has died, and if they do not find the dead one, they no longer [try to] find it in the same way? Is it because they do not perceive the dead one because there is a different smell, and its proper one is destroyed, since everything is, sound and colour and flavour? For all these things change. So when it is alive they perceive it because the scent from the creature is continuous, but when it has died this ceases being given off; for it does not leave it behind, as Empedocles [says] that "it left [traces] of its feet on the tender grass".[399] For it is not possible for the smell or the colour to be preserved, but when it dies, both those things and this are destroyed.

2.103.[400] Why are the odours both of incense and of flowers less fragrant close at hand? Is it because earthy matter travels with [the fragrance] to those close at hand, so that, being mixed in, it makes the power [of the

---

398 [Aristotle] adds here "... of the effluence and of the air in which it is".

399 Line 2 of Empedocles fr.31B101 DK, the whole being restored by Diels as "seeking out with its nostrils fragments of the limbs of animals, <as many as> they left from their feet on the soft grass <while alive>." Cf. the anonymous commentary on Plato's *Theaetetus* 70.48-71.6 = Bastiniani and Sedley 1995, 458, which Diels restored as 'Empedocles accepts effluences and says that dogs track down "fragments of the limbs of animals", but that this is impossible when the creatures are dying'. If accepted, this reconstruction would suggest that "left" in the line of Empedocles quoted in our text has the neuter plural subject "creatures", rather than the singular which would be required by Usener's conjecture (1859, xiii-xiv) of "as it traversed the tender grass" (reading *polôn* in place of *podôn* "of feet"). However Bastiniani and Sedley 1995, 556 reject the last clause in Diels' restoration, pointing out that it fails to take into account the fact that our present text, while drawing a contrast indeed between living creatures and dead ones, is *rejecting* Empedocles' account. On similar grounds Pearson, reviewing the second edition of Diels' *Vorsokratiker* at *CR* 23 (1910) 50, proposes, in the text of Empedocles, in place of "<as many as> they left from their feet on the soft grass <while alive>", "<it smells as many as> they left from their feet on the soft grass". Cf. also Plutarch, *De curiositate* 11 520e, *Quaest. nat.* 23 917e.

400 = [Ar.], *Probl.* 12.9 (the first solution here; cf. id. 12.2) and 12.4 (the second solution here); Theophrastus, *CP* 6.17.1; Hieronymus of Rhodes fr.10 White/53 Wehrli. Cf. Sharples 1985, 193 and Sharples (forthcoming).

τριφθέντα τὰ ἄνθη ἀπόλλυσι τὴν ὀσμήν. ἢ ὅτι ἡ ὀσμὴ θερμότης ⟨τίς⟩ ἐστι καὶ τὰ εὐώδη θερμά, τὸ δὲ θερμὸν κοῦφον, ὥστε διὰ τοῦτο πορρωτέρω γίνεται ἡ ὀσμή;

2.104. *Διὰ τί πάντα μᾶλλον ὄζει κινούμενα; ἢ ὅτι ἀναπίμπλησι τὸν ἀέρα, διόπερ ἀναπέμπεται θᾶττον πρὸς τὴν αἴσθησιν;*

2.105. *Διὰ τί ἐνίοτε ναυτιάσαντές πως μᾶλλον πεινῶσιν; εἰκὸς γὰρ ἦν ἐναντίως ἔχειν· πλήρωσις γὰρ ἡ ναυτίασις ἐστίν· ἢ ὥσπερ ἀφόδου γενομένης συμβαίνει γενέσθαι τοὺς τόπους κενοὺς οἷς πεινῶσι· πρὶν δὲ ναυτιάσαι τούτους ἐπλήρουν;*

2.106. *Διὰ τί ἐν μὲν τοῖς πλοίοις ἡ ναυτία χολώδης· ἦν δ' ἄλλως, φλεγματώδης; ἢ διότι ἐν μὲν τοῖς ἄλλοις ἔμπονος, ἐνταῦθα δὲ μετὰ φόβου ἢ ταραχῆς· ὁ δὲ φόβος συγκινητικὸς τῶν συμφύτων; τὸ μὲν οὖν φλέγμα*

---

add. Us. ex [Ar.]: *quaedam v* 5 *πορρωτέρω* AFMPRTe Buss. Us.: *πόρρωθεν* KOSbe[mg]iv: *πορρωτέρω διιόντων ἀμιγεστέρα* [Ar.] 6 *ἡ ὀσμὴ*] *ὀσμὴ* F: *ὀσμῆ* A. – Totum problema ita exhib. GY: *Διὰ τί αἱ ὀδμαὶ ἧττον εὐώδεις καὶ θυμιαμάτων καὶ ἀνθῶν οὐ τοῖς πόρρω, ἢ τοῖς πλησίον; ἢ ὅτι τοῖς πλησίον συναποφέρεται, τὸ γεῶδες· ὥστε καὶ τανύμενον* (*ταννύμενον* Y) *ἀσθενῆ ποιεῖ τὴν δύναμιν· ἡ* (*εἰ* Y) *δὲ πόρρω καταφαίνεται* (*καταφέρεται ἡ* Y) *ὀδμή· διὰ τοῦτο δὲ καὶ τριφθέντα τὰ ἄνθη ἀπόλλυσι τὴν ὀσμὴν* (*ὡσμήν* G)*· ἢ ὅτι ἡ ὀσμή, θερμότης ἐστὶν καὶ τὰ εὐώδη θερμά· τὸ δὲ θερμόν, κοῦφον· ὥστε διὰ τοῦτο· πορρωτέρω γίνεται ὀσμή* (*ἡ ὀσμὴ* Y)*· δεῖ δὲ καὶ τοῦτο εἰδέναι· ὅτι ἐκεῖνα ἡμᾶς χρὴ προβάλλειν* ... (sequitur [Alex.], *Probl.* 1 Ideler, praef. 76-84): ita D: *Διὰ τῶν μὲν ἀνθέων αἱ ὀδμαί, μᾶλλον μὲν τοῖς ἐγγὺς εἰσὶν εὐώδεις· τοῖς πόρρω δὲ ἧττον· τῶν δὲ θυμιαμάτων τὸ ἀνάπλεον· ὅτι ἐκεῖνα μὲν συνεπιφέρεταί τι καὶ γεῶδες· διὸ καὶ ἀσθενεῖ πόρρω ἐλθεῖν ἡ τούτων εὐωδία· διατοῦτο καὶ τριφθέντα τὰ ἄνθη, εὐθὺς ἀπόλλυσι τὴν ὀσμήν· ἢ καὶ διὰ τὸ εἶναι ψυχρά· τὰ δὲ θυμιάματα θερμά εἰσί· καὶ διατοῦτο καὶ πλέον εὐώδη· τὸ δὲ θερμὸν κοῦφον· ὥστε διατοῦτο πορρωτέρω γινόμενον τὴν οἰκείαν δείκνυσι εὐωδίαν μάλιστα.*

2.104 (AFKMOPRSTbei*p* Buss. Us.; *v* 2.42;[Ar.] 12.5 et 13.12; 2.103 Buss.)
1 *τὸν*] *πλείω* [Ar.] 2 post *ἀέρα* (1) habet haec [Ar.]: *ἢ ἡσυχάζοντα; διαπέμπεται οὖν ἡ ὀσμὴ θᾶττον* (12.5: *θᾶττον ἡ ὀσμὴ* 13.12) *οὕτως πρὸς τὴν αἴσθησιν ἡμῶν* (cf. Marenghi ad 12.5: *ἡμῶν* om. 13.12) || *διόπερ* KOSbe[mg]i*v* Us.: *διότι* AFMPRe Buss.: *ἢ ὅτι* T

2.105 (AFKMOPRSTbei Buss. Us.; paraphrasin tantum habet *p*; *v* 2.43; 2.104 Buss.)
1 *πως* FK Buss. Us.: om. M*v*: *πῶς* cett. || *ἦν* om. KOi: *ἦν δεῖ* A 2 *ἐναντίως*] *διεναντίως* FT: *δὴ ἐναντίως* P || *ἔχειν*] *ἔχῃ* A || *πλήρωσις – ἐστίν* om. T || *ναυτία* P Buss. Us. || *ἢ ὅτι* A || *ἐφόδου* P 4 *τούτου* P || *τούτους ἐπλήρουν*] *του*[.........] A: *τοὐναντίως* Buss.

2.106 (AFKMOPRSTbei*p* Buss. Us.; *v* 2.44; 2.105 Buss.)
1 *ἐν μὲν*] *ἡ μὲν* e || *ἦν*] *ἡ* K*v* 2 *φλεγματώδης*] *φθεγματώδης* O[ac]i: *φλέγματος* T || *ἔμπονως* sic A 3 *ἢ*] *καὶ* T[ac] 4 *γενόμενον* A Buss.: *γιγνομένη* M || *γιγνόμενον*

fragrance] weaker, but if[401] further away, the smell arrives? And for this reason also flowers that are ground up lose their fragrance.[402] Or is it because the odour is a type of heat and fragrant things are hot, and what is hot is light, so that for this reason the odour occurs further away?

2.104.[403] Why do all things have a stronger smell when in movement? Is it because [the odour] fills the air,[404] and for this reason is conveyed to the sense more swiftly?

[Vomiting, excretion and purgation][405]

2.105. Why do some people, when suffering from nausea, feel more hungry? For it would be reasonable for the opposite to be the case; for nausea is being full. Is it because there is as it were a departure and the places where they feel hunger become empty, while before developing nausea they were filling these?

2.106. Why is the vomit of those on ship bilious, but otherwise phlegmatic? Is it because in the other cases it is painful, but here it is accompanied by fear or anxiety, and fear moves things that are natural [in a person]? Phlegm

---

401 [Aristotle] has *eis* "to further away" rather than *ei* "if further away".

402 Cf., with Marenghi 1991, 101 n.2, [Ar.], *Probl.* 13.11 and Theophrastus, *Od.* 13.

403 This is a single statement of the explanation which, as Marenghi 1991, 134 notes, appears *twice* in [Ar.], *Probl.*, at 12.5 in connection with good odours and at 13.12 in connection with bad ones. It seems more likely that the editorial process which led to the present state of the [Aristotle] *Problems* produced the duplication there than that our text has combined two *Problems* from there.

404 [Aristotle] has "more air", and continues "than things that are at rest? So the odour is in this way transmitted more rapidly to our sense."

405 Cf., perhaps, Theophrastus, *On Secretion* (or *On Elimination, Peri ekkriseôs*), **328** no.14 FHS&G; but also Sharples 2004, 148.

ἐπιπολάζον καὶ ἀλλότριον, ἡ δὲ χολὴ σύμφυτον γιγνόμενον ἐκ συμφύτου.

2.107. Διὰ τί ἐν ταῖς φαρμακοποσίαις σημεῖον ὅταν ἀριθμεῖν δύνηται τὰς δοκούς, ὅτι κεκάθαρται; ἢ ὅτι οὐκ ἔνεστι περιττωματικὴ ὑγρότης; ἐποίει γὰρ τὸν ἴλιγγον ὥσπερ τοῖς μεθύουσι· τὸ γὰρ αὐτὸ σημεῖον τοῦτο καὶ τοῖς μεθύουσιν ἤδη ἐστὶ τοῖς μὲν γιγνόμενον, τοῖς δὲ μὴ γιγνόμενον.

2.108. Διὰ τί ὑπέρινοι μᾶλλον γίνονται κάτω ἢ ἄνω καθαρθέντες; ἢ διότι εὐροωτέρα ἡ κάτω ὁδός, ἡ δ' ἄνω παρὰ φύσιν, ὥστε καὶ ἀπ' ἐλάττονος φαρμάκου δυνάμεως ἔλαττον καὶ πλεῖον κινεῖσθαι, καὶ κατὰ φύσιν ἢ παρὰ φύσιν; ἢ ὅτι τὰ μὲν περιττώματα ⟨ἃ⟩ ἐν τῇ κάτω καθάρσει ⟨ἐκκρίνεται⟩ ἐστὶ τῶν χρησίμων καὶ ἐχόντων τῷ σώματι ἀπόλαυσιν [ἐκκρίνεται], τὰ δὲ ἄνω ἀλλότρια; ἔνθα μὲν γὰρ τῆς τροφῆς πλείονα λείψανα τὰ οὔπω ἐξικμασμένα ἀλλ' ἔτι ἔχοντα τροφήν, ἔνθα δὲ χολὴ καὶ φλέγμα ἄτροφα· ἡ μὲν οὖν κάτω κάθαρσις τὴν τροφὴν ἀφαιρουμένη εὐλόγως ἐκλύει, ἡ δ' ἄνω ἢ οὐδὲν ἢ ἧττον· δηλοῖ δὲ ὁ ἐλλέβορος· ἄνω τε γὰρ καθαίρει· καὶ εὐθὺς εὐθετεῖ τὰ σώματα, ὡς οὐδενὸς τῶν χρησίμων ἀφαιρουμένου ἐν τῇ ἄνω καθάρσει.

2.109. Διὰ τί οἱ μέλανες καὶ οἱ λεῖοι μᾶλλον ἢ οἱ λευκοὶ καὶ δασεῖς γίνονται ὑπέρινοι; ἢ ὅτι ὑγρότεροί τε οἱ λεῖοι τῶν δασέων (δηλοῖ δὲ ἡ τῶν γυναικῶν φύσις ὑγρὰ καὶ λεία οὖσα) καὶ οἱ μέλανες τῶν λευκῶν ψυχρότεροι (τὸ δὲ ψυχρὸν τοῦ ὑγροῦ ἐγγύς· τὸ γὰρ θερμὸν ξηρόν)· ὑπέρινοι δὲ γίνονται συντηκομένου τοῦ σώματος, ἀλλ' οὐ μόνον ἀφαιρουμένου τοῦ περιττώματος.

ἐκ συμφύτου om. F || *ἐκ τοῦ συμφύτου* MRe

2.107 (AFOPTi Buss. Us.; 2.105* Buss.)

1 *φαρμακοποιίαις* FOi || *ἀριθμεῖς* A || *τὰς*] *τοὺς* F 2 *ὅτι κεκάθαρται*] *ὃ κεκάθαρται* P || *ὑγνότης* T 3 *τοῦτο*] *τούτοις* FOi 3-4 *τὸ γὰρ αὐτὸ – ἤδη*] *τὸ γὰρ αὐτὸ σημεῖον ἤδη καὶ τοῖς μεθύουσι καὶ τούτοις* T: om. A Buss. 4 *ἐστὶ*] *ἡ δ' ἐστὶ* A Buss.

2.108 (AFKMOPRSTbei*p* Buss. Us.; *v* 2.45; 2.106 Buss.)

1 *γίνονται μᾶλλον* P || *διότι*] *ὅτι* KOi 2 *κάτω* MR^{pc}e*p* Us.: om. cett. || *ὁδὸς* om. M 4 *ἢ* om. AFPRTe^{ac} || *ἃ* add. Us. || *ἐκκρίνεται* add. Us. 5 *ἐστὶν*] *εἰσὶ* M || *ἔχοντος τοῦ σώματος* coni. Buss. || *ἀπολαύουσιν* F || *ἐκκρίνεται* secl. Us.: *καὶ ἐκκρίνεται* KOSbi 6 *τῆς τροφῆς* trasp. post *λείψανα* K || *τὰ* om. K 7 *ἔτι*] *ἔστιν* T 8 *εὐλόγως*] *εὐκόλως* Re^{ac}: om. M 9 *ἢ* om. AFMPe: *δὲ* R || *δὲ*] *καὶ* S: *δὲ καὶ* KORbei || *τε*] *τὸ* F: om. T || *εὐθετεῖ* M Us.: *confirmat v*, ex quo *ἰσχύει* coni. Buss.: *ἔτι* cett. 9-10 *τὰ σώματα*] *τὰ σώ* T 10 *οὐθενὸς* FPT || *τῶν χρησίμων* om. *v* || *ἀφηρημένου* FPTe Buss.

2.109 (AFKMOPRSTbei*p* Buss. Us.; *v* 2.46; 2.107 Buss.)

1 *οἱ δασεῖς* P Us. 2 *τε* om. A Buss. 3 *φύσις*] *φύσι* sic A 5 *δὲ γίνονται* om. A

is superficial and alien, bile is natural and comes from what is natural.
2.107.[406] Why, in taking drugs, is being able to count the roof-beams a sign of having been purged? Is it because there is no moist residue present? For this produces dizziness, as in those who are drunk; for this same sign also applies to those who are drunk, occurring for some but not for others.
2.108.[407] Why are people purged more violently if they are purged downwards than upwards? Is it because the downwards passage flows more naturally, while the upwards one is unnatural, so that even by a lesser power of the drug it is moved less [in the one case] and more [in the other],[408] and naturally rather than contrary to nature? Or is it because the residues that are eliminated in the downwards purging are useful and beneficial to the body, those upwards alien? For in the former place there are many remnants of food which have not yet been digested but still possess nourishment, in the latter bile and phlegm which contain no nourishment. So the downwards purging, which removes nourishment, reasonably causes weakness, the upwards one either not at all or less so. This is shown by hellebore;[409] for it purges upwards and at once puts bodies in a good condition, since nothing of what is useful is removed in upwards purging.
2.109. Why are dark and smooth-skinned people purged more violently than pale and hairy ones? Is it because smooth-skinned people are moister than hairy ones (this is shown by women, who are naturally moist and smooth-skinned) and dark ones are colder than pale ones (cold is close to moisture; for what is hot is dry), and they are purged violently when the body is liquefied, rather than just the residue being removed.

---

406 A fuller version, in different words, of Theophrastus, *Vert.* 12. Visual disturbances as a test of drunkenness are discussed in [Ar.], *Probl.* 3.9 and 3.20, where the explanation is in terms of heat, and also (without specific reference to testing) in [Alex.], *Probl.* ii.33, where the explanation is in terms of *pneuma* rising up from the wine. This problem occurs only in MSS AFOPTi; see Introduction, §2.1.13.

407 Cf. 2.117. The term *huperinos* for "violently purged", here and in the following problems, was used by Hippocrates but was apparently obsolete by the Roman period; it is explained by Galen, *Ling. Hippocr. explic.*, 19 148.15-149.1 Kühn. It is used by Aristotle at *GA* 3.1 750a29, of chickens which die after laying too many eggs (formed from residue) and by Theophrastus at *HP* 9.14.2, of the effect of the "driver" (below, n.427) purging upwards. Above, 9 and n.58.

408 So the MSS. But perhaps we should delete "less and".

409 I.e. white hellebore; see below, 2.126.

2.110. *Διὰ τί μέλαιναν μὲν χολὴν καθαιρόμενοι καὶ φλέγμα πολὺ οὐ γίνονται ὑπέρινοι, τὴν δὲ ξανθὴν χολὴν πολὺ ἐλάττω καθαρθέντες; ἢ ὅτι τὴν μέλαιναν μὲν χολὴν εἰώθασι καθαίρεσθαι ῥᾳδίως οἱ πολλοί, καὶ μὴν τὴν τρυγώδη, διὰ τὸ κεχωρίσθαι ἤδη τοῦ σώματος (διὸ καὶ μεμέλανται); οὐδεμία οὖν ἰσχὺς ἀφῄρηται τοῦ σώματος καθαρθέντος· τὸ δὲ φλέγμα περίττωμά ἐστιν, ὥστε ⟨ὅσῳ⟩ ἄν τις ἀφέλῃ μεῖζον, κουφίζει μᾶλλον ὥσπερ ἐπικειμένου βάρους· ἡ δὲ ξανθὴ ἐκ συστήματός ἐστι τῆς κακῶς ᾠκοδομημένης σαρκὸς καὶ οὐ προϋπάρχουσα οὐδὲ κεχωρισμένη πρότερον ἡ πολλή, ὥστε οὐδὲν διαφέρει πολλὴν ἀφελεῖν χολὴν ἢ σάρκας πολλὰς ἢ αἷμα· τούτου δὲ συμβαίνοντος ἀνάγκη γίνεσθαι ὑπερίνους· μέχρι μὲν γὰρ τούτου ῥεῖν ἐνδέχεται, ἕως ἂν ἔχῃ αἷμα καὶ σάρκας ὡς εἰπεῖν, θάτερα δὲ ἔχει πέρας διὰ τὸ κεχωρίσθαι.*

2.111. *Διὰ τί λυγγιῶσι καὶ ὑπερκαθαιρόμενοι καὶ ὑπερπληρούμενοι; ἢ ὅτι δι' ἀμφότερα καταψύχονται; ἥ τε γὰρ ὑπερκάθαρσις ἀφελοῦσα τὸ ὑπέκκαυμα ψύχει, καὶ ἐὰν πλείων εἰσέλθῃ τροφὴ τῆς δυνάμεως, μαραίνει τὴν οἰκείαν θερμότητα· ὁ δὲ λυγμός ἐστιν ὅταν πνεῦμα ἐναπολαμβάνηται διὰ τὸ ψυχθῆναι τὸ ὑγρὸν ἐν τῷ πνευματικῷ τόπῳ.*

2.112. *Διὰ τί τὰ παιδία καὶ οἱ γέροντες μάλιστα ἁλίσκονται διαρροίαις ἐναντίας ἔχοντες τροφάς; ἐπειδὴ τὰ μὲν παιδία εἰ καὶ μὴ διὰ τὴν τροφὴν τοῦτο πάσχουσιν, ἀλλὰ διότι ὑγρὰ τὰ σώματα αὐτῶν ἐστι καὶ πυκνά, ὥστε εἴσω τὰ περιττώματα συρρέοντα διὰ τὴν τῶν σαρκῶν πυκνότητα*

2.110 (AFKMOPRSTbei*p* Buss. Us.; *v* 2.47; 2.108 Buss.)
1 *μέλαιναν μὲν*] [...............]*να* A 2 *γένονται* sic A 3 *τὴν μέλαιναν μὲν* RSb Us.: *τὴν μέλαιναν* KOPi: *τὴν μὲν μέλαιναν* cett. || *οἱ πολλοὶ* om. Buss. 3-4 *καὶ μὴν τὴν τρυγώδη* nos: *καὶ μὴ τὴν τρυγώδη* AFMOPRSTbei Us.: *μὴ* del. Buss. (in var. lect.): *καὶ – τρυγώδη* om. K*v* 4 *διὰ τὸ* Us.: *διὰ τὸ τρυγώδη εἶναι καὶ* K*v*: om. cett. Buss. || *<τῷ> ἤδη κεχωρίσθαι* Buss. 5 *οὐδὲ μία* e 6 *ὅσῳ* add. Buss. (in var. lect.) Us., ita quoque vertit *v* || *ἄν*] *ὅτάν* KOSbi || *μείζων* AT 7 *ξανθὴ χολὴ* T; ita quoque *v* (ex suppl.?) 8 *οὐ* om. KOSbi || *πρότερον* FK*v* Buss. Us.: *πότερον* AMOPRSTbei 9 *πολλή*] *πολλοὶ* T || *ἀφελεῖν*] *ἀφαιρεῖν* KOi || *σάρκα πολλὴν* K*v* 10 *ὑπεροίνους* e 11 *ἔχῃ* om. A
2.111 (AFKMOPRSTbei*p* Buss. Us.; *v* 2.48; 2.109 Buss.)
1 *λυγγιῶσι* Us.: *λυγγῶσι* Sbe[mg] Buss.: *λυγκῶσι* KOi*v*: *λυττῶσι* AFMPRTe || *ὅτι* AFMPT Buss. Us.: *διότι* KORSbei 2 *δι'* om. KMORSbei || *ἥ τε*] *εἴτε* MRe 3 *ὑπερκκαυμα* sic A 3 *πλεῖον* A 5 *τῷ* om. OSi
2.112 (AFKMOPRSTbei*p* Buss. Us.; *v* 2.49; 2.110 Buss.)
1 *ἁλίσκονται* om. A 2 *ἔχοντες*] *ἔχοντας* Oi: *ἔχοντα* K || *τὰς τροφὰς* Se[mg] ||

2.110. Why are those who are purged of black bile and much phlegm not purged violently, while those who are purged of yellow bile [are], having been purged of much less? Is it because most people are accustomed to be purged easily of black bile, and of[410] what is thick like dregs, because it has already been separated from the body (and this is also why it has become black), so no strength is removed when the body is purged. Phlegm is a residue, so the more one removes, the more relief it gives, as if a weight were resting on [the person]. But yellow [bile] is from the constitution of badly constructed flesh, and most of it does not exist beforehand and has not been separated beforehand, so it is no different from taking away much bile or flesh or blood; and when this happens it is necessary for them to be violently purged; for it can flow up to the point where it involves blood and flesh so to speak, whereas the others have a limit because they are separate.

2.111. Why do people have hiccoughs both when they are excessively purged and when they are excessively replete? Is it because they are chilled by both? For excessive purging chills by taking away the fuel [for the inner heat], and if more nourishment enters than the power [of digestion can cope with], it weakens the [person's] own proper heat; and hiccoughs are when *pneuma* is trapped because the moisture in the respiratory region is chilled.[411]

2.112. Why are children and the elderly most afflicted by diarrhoea when they have opposite nourishment? Because if it is not because of their nourishment that children are afflicted in this way, [it is] because their bodies are moist and dense, so that the residues, flowing inwards because of

---

[410] Reading *kai mên* for *kai mê*, "and not of what is thick like dregs".

[411] See above, 2.53.

ὑγραίνει τὰς κοιλίας, οἱ δὲ γέροντες ψυχρὰ τὰ σώματα ἔχουσι καὶ τὰς κοιλίας· ὥστε τὰ μὲν δι' ὑγρότητα τῶν σωμάτων καὶ πυκνότητα, οἱ δὲ διὰ ψυχρότητα πάσχουσι τοῦτο καὶ τὰ μὲν διὰ τὸ ἀναπνεῖν ἔξω τὰ περιττώματα, οἱ δὲ διὰ τὸ μανὰ τὰ σώματα ἔχειν καὶ τὸ θερμὸν ἀποπνέον ἐξ αὐτῶν ἐξατμίζειν· τὸ αὐτὸ οὖν οὐ διὰ τὸ αὐτὸ πάσχουσι· συναίτιον δὲ τοῖς παιδίοις ἐστὶ καὶ τὸ τὴν τροφὴν ποσὴν καὶ ποιάν τινα εἶναι· πολλά τε γὰρ προσφέρονται καὶ γλυκέα καὶ παντοδαπά· τὰ μὲν γὰρ νέα παντελῶς γάλακτι τρέφεται ὅ ἐστι δύσπεπτον, τὰ δὲ μικρὸν ἐπηυξημένα ποικίλοις καὶ γλυκέσι χρῆται ἐδέσμασι τροφῶν καὶ μητέρων πρὸς χάριν ὁμιλουσῶν αὐτοῖς· τὰ δὲ τοιαῦτα δυσδιαχώρητά ἐστιν.

2.113. *Διὰ τί καὶ τὸ πολὺ ποτὸν ἐνίοτε καὶ τὸ ὀλίγον τὰς διαρροίας ἴσχει; ἢ διότι τὸ μὲν διὰ τὸ καταξηραίνειν εἰκότως ἅτε μικρὸν ὄν, τὸ δὲ διὰ τὸ πλῆθος εἰς τὴν οὐρήθραν χωροῦν συνεπισπᾶται καὶ τὸ ἐκ τῆς κοιλίας ὑγρόν; ἅμα γὰρ κατ' ἄμφω οὐκ εἴωθεν ὑποχωρεῖν, ἀλλ' ἐὰν μὲν ἡ κοιλία ὑποχωρῇ μᾶλλον ἡ οὐρήθρα ἀποξηραίνεται· ἐὰν δὲ ἡ οὔρησις πολλὴ γίνηται, ἡ κοιλία καταξηραίνεται· οἷον ἄπαρσιν γὰρ ἐποίησεν ἐκ τῆς κοιλίας εἰς τὴν οὐρήθραν τὸ πολὺ ποτόν.*

2.114. *Διὰ τί οἱ πίνοντες εὐτραφεῖς ὄντες ὑγρὰς τὰς κοιλίας ἔχουσι μᾶλλον ἢ σκληράς, διὸ καὶ διαρροίαις ἀποθνήσκουσι μᾶλλον τῶν τεθραμμένων, ἔδει δ' εὐτρόφους ὄντας ἐκλαμβάνειν μᾶλλον τὸ τρόφιμον, τούτου δὲ γινομένου ξηραίνεσθαι τὰς κοιλίας αὐτοῖς; ἢ ὅτι τὸ εὐτραφές ἐστιν [οὐκ] ἐν*

ἐπειδὴ] *ἢ ὅτι* K*ν* || *μὲν* om. MORSbei || *διὰ* om. F 5-6 *οἱ δὲ γέροντες – τὰς κοιλίας* om. FP 7 *τοῦτο καὶ* K*ν* Us.: *τοῦτο δὲ καὶ* cett. Buss. || *ἀναπνεῖν*] *πνεῖν* T 8 *μανὰ* Us.: *μόνα* codd. Buss.: *firma* *v* 9 *ἐξ αὐτῶν*] *τὰ ἀπ' αὐτῶν* A: *ἀπ' αὐτῶν* FP 10 *ποσήν τινα* OSbe[mg]i*ν*: *τοσήν τινα* K || *ποιάν*] *τοιαύτην* K 11 *προσφέρονται* AMRe Buss. Us.: *προφέρονται* cett. || *παντοδαπᾶ* FOi || *τὰ μὲν γὰρ νέα* AFMRTe[ac] Buss.: *τὰ μὲν καὶ νέα* OSbe[mg]i: *τὰ μὲν νέα* P: *τὰ μέντοι νέα* K: *τὰ μὲν γὰρ καὶ νέα* Us. 12 *ὅ*] *ὅπέρ* K || *δύσπεμπτον* M || *σμικρὸν* A Buss. || *ποικίλοις* FKOPSbe[mg]i*ν* Buss. (in var. lect.) Us.: *ποικίλως* AMRTe 13 *ἐδέσμασι*] *καὶ δέσμασι* A || *ὁμιλουσῶν*] *ὁμιλοῦντων* A: *ὁμιλοῦσιν* P

2.113 (AFKMOPRSTbei*p* Buss. Us.; *ν* 2.50; 2.111 Buss.)

1 *καὶ*[1] om. Buss. || *ἴσχει*] *ἔχει* PT[ac] 2 *ὂν* om. KMORSbei 3 *συνεπισπεύεται* e || *καὶ* om. A Buss. 4 *ἅμα*] *αἷμα* sic A || *ἐὰν μὲν ἡ* MP Buss. Us.: *εἰ μὲν* A 5 *ὑποχωρῇ*] *ὑποχωρεῖ* Buss. 5-6 *ἐὰν δὲ – οἷον ἄπαρσιν γὰρ*] *ἡ γὰρ φύσις* F: *ἐὰν δὲ – καταξηραίνεται* om. *ν* 6 *ξηραίνεται* K || *οἷον* om. K*ν* || *ἄπαρσιν* Us.: *ἄπασι* AMOPRSTbei: *ἄπωσιν* Buss.: *λάπασιν* K: *exinanitionem* v || *τὴν* om. T

2.114 (AFKMOPRSTbei*p* Buss. Us.; *ν* 2.51; 2.112 Buss.)

2 *διὸ καὶ*] *διὸ* KORSbei: *διότι* P || *τῶν τεθραμμένων*] *qui parum cibi sumpserint*

the denseness of the flesh, make the bowels moist; while old people have both cold bodies and [cold] bowels. So the former are afflicted in this way on account of the moistness and denseness of their bodies, the latter because of the coldness; and the former because the residues blow outside, the latter because they have rare bodies and the heat which blows away from them vapourises[?]. So they are afflicted in the same way for the same reason. A contributory cause for the children is also that their nourishment is of a certain quantity and quality;[412] for they consume many things which are sweet and of every sort. For the young are fed entirely on milk, which is hard to digest, and those that have grown a little make use of varied and sweet foods because their nurses and mothers interact with them with a view to gratifying them; and such things are indigestible.

2.113.[413] Why do both a large quantity of drink, sometimes, and a small quantity stop diarrhoea? Because the one reasonably does so by drying, since it is small, and the other, because of the quantity, passes into the urethra and draws along with it also the moisture from the bowel. For it is not usual to pass simultaneously by both, but if the bowel moves, the urethra dries up more, while if there is much urination, the bowel dries up; for the large quantity of drink caused as it were a departure from the bowel to the urethra.

2.114. Why do those who drink and are fat[414] have moist bowels rather than hard ones, and for this reason die of diarrhoea more than those who are [merely] nourished, when as they are fat they ought to extract more nourishment, and when this happens their bowels ought to become dry? Is it because being fat is[415] a matter of extracting much, and the residue becomes

---

[412] Children's diarrhoea due to failure to digest is mentioned, without further details, by Aretaeus, *Caus. acut. morb.* 2.7.5 (*CMG* 2 75.2-4 Hude).

[413] [Ar.], *Probl.* 3.21 argues that a large amount of wine has a drying effect because it cannot be absorbed by the stomach and passes directly to the urethra.

[414] Literally "well-nourished" (*eutrapheis*); but the term refers to the effect of food on the body rather than to the amount of food consumed.

[415] Deleting *ouk* with Bussemaker.

τῷ πολὺ λαμβάνειν, σκληρὰ δὲ ἡ περίττωσις γίνεται, ὅταν μόλις καταβαίνῃ τὰ προσενεχθέντα εἰς τὴν κοιλίαν, τοῦτο δὲ γίνεται διὰ τὸ μὴ ταχὺ ἀπολαύειν τὸ σῶμα καὶ μὴ ἐκκρίνεσθαι τὸ χρήσιμον ἐξ αὐτῶν καὶ διὰ τὸ ἐν τῇ κάτω κοιλίᾳ μένειν πολὺν χρόνον ἅτε μὴ ἀθρόα καταβαίνοντα (συγκαίεται γὰρ μᾶλλον)· οὐχ αἱ μάλιστα δὲ σκληρὰ ἀποδιδοῦσαι κοιλίαι εὔτροφοί εἰσιν, ἀλλ' αἱ ταχὺ λαμβάνουσαι τὸ χρήσιμον ἐκ τῶν προσενεχθέντων· σημεῖον δὲ τὸ τοὺς ξηροὺς τὰς κοιλίας ἰσχνοτέρους εἶναι ἢ εὐσαρκοτέρους.

2.115. Διὰ τί οἱ γλυκεῖς καὶ οἱ ἁλμυροὶ τῶν χυμῶν διαχωρητικώτατοί εἰσιν; ἢ ὅτι ταρακτικώτατοι, τοιαῦτα δέ ἐστι τὰ ἄπεπτα καὶ βαρέα; διὰ γὰρ τὸ μήτε κρατεῖσθαι ὑπὸ τῆς ἐν ἡμῖν θερμότητος κάτω τε φέρεσθαι τὸ μέλλον ἐκταράξειν· ἡ γὰρ ἐκτάραξις τοῦτ' ἐστίν. ἢ ὅτι οἱ μὲν ἄλλοι χυμοὶ ἔξω τοῦ ἐλαίου εὐδιαφόρητοι καὶ εὔσηπτοί εἰσι, τὸ δ' ἔλαιον ἄπεπτον μέν (ἄσηπτον γάρ), ἀλλὰ διὰ κουφότητα ἄνω ἐπιπολάζει· διὸ οὐ διαχωρητικόν ἐστιν· τὸ δ' ἁλμυρὸν βαρὺ καὶ παχύ, τὰ δὲ τοιαῦτα μάλιστα φέρεται κάτω καὶ ἄσηπτά ἐστι;

2.116. Διὰ τί ἡ μὲν ἅλμη καὶ ὁ ὀρρὸς τὴν κοιλίαν λύουσιν, ὁ δὲ οἶνος πολὺς πινόμενος οὔ; ἢ ὅτι τῶν μὲν ἡ ὑπόστασις εἰς κοιλίαν ἔρχεται; οὐ γὰρ

---

v 3 δὲ[2]] ἐστὶ A 4 αὐτοῖς] αὐτῆς A || εὐτρεφὲς T || οὐκ del. Buss. 5 πολὺ] πολλὰ F 6 γίνεται om. T 7 ταχὺ] ταχὺν A || ἀπολαύειν] ἀπολάβειν A 7-8 διὰ τὸ] διὰ τοῦτο K 8 ἀθρόως K*v* 9 συγκαίεται MOPRSbei Us: συγκάεται FKT Buss.: συγκάθεται A || σκληρὰ] σκληραὶ M: ξηρὰ AFPT Buss. Us. || ἀποδιδοῦσαι] ἀποδοῦσαι MRe 9-11 συγκαίεται – προσενεχθέντων] *magis enim simul exuruntur alvi que certe nutrite non sunt reddunt non magnopere dura que exceperunt; sed raptim ex detrusis et retentis utile excipiunt v* 10 εὔτροφοι M Buss. Us.: ἄτροφοι cett. || ἀλλ' αἱ FKMe Buss. Us.: ἀλλὰ cett. 11 τὸ τοὺς] τόπους A || ξηρὰς M || τὰς τῆς κοιλίας F || ἰσχνοτέρους] ἰσχνοτέρας M[ac]Re: ἰσχυροτέρους K*v* || ἢ] καὶ F

2.115 (AFKMOPRSTbei*p* Buss. Us.; *v* 2.52; 2.113 Buss.)

1 οἱ[2] om. KOPi 2 ταρακτικώτατοι PT Buss. Us.: ταρακτῶτατοι sic A: πρακτικώτατοι FMORSbei: ἀπεπτότατοι K*v* || τοιαῦτα Buss. Us.: αὐτὰ AFPTe: τὰ αὐτὰ K*v*: εἰ αὐτὰ MORSbi || δέ ἐστι] δ' εἰσὶ Buss.: δ' ἔησι sic A: ἐστὶ M 3 μήτε] μὴ K || τε APT Buss. Us.: om. FKMORSbei || φέρεται K*v* 4 μέλον A || ἐκτάραξεις A || ἢ om. A || οἱ] ἡ Re 5 εὐδιαφόρητοι Buss. (in var. lect.) Us.: εὐδιάφοροι codd.: *facile labiles v* 6 ἄσηπτον] εὔσηπτον K*v*: εὔπεμπτον M || διαχωρητικὸν AFKMOPRTei*v* Buss. Us.: διαχωρητὸν Sbe[mg] 7 κάτω φέρεται A Buss.

2.116 (AFKMOPRSTbei*p* Buss. Us.; *v* 2.53; 2.114 Buss.)

1 ὁ ὀρρὸς] ὁ ὀρὸς Re Buss.: ὁ ὁρὸς A: ἡ ὀρὸς M 1-2 πινόμενος πολὺς KOi 2 οὔ[1]] οὐ λύει F || τῶν] τὸ Re || ἡ om. MRe 3 ὁ δὲ οἶνος] ὃς A Buss. || πεμφθεὶς A 4

hard when what is consumed descends into the bowel with difficulty, and this happens on account of the body not deriving the benefit speedily and the useful [part] not being separated out of [the food], and because it remains in the lower bowel for a long time because it does not descend all at once (for it is burned up more). It is not the bowels that produce the hardest [excrement] that produce good nourishment, but those that swiftly take what is useful from what is consumed. An indication [of this] is that those with dry bowels are thinner rather than fleshier.

2.115. Why are sweet and salty flavours most laxative?[416] Is it because they cause the most disturbance, and like this are those that are unconcocted and heavy? Because they are not overcome by the heat in us, and [because] what is going to cause disturbance travels downwards; for this is what disturbance is. Or is it because the other juices,[417] apart from olive oil, are easily digested and easily decay, while olive oil is unconcocted (for it does not decay), but because of its lightness it floats on the top, and for this reason is not laxative; while what is salty is heavy and dense, and such things travel downwards most and do not decay?

2.116. Why do salt and whey[418] loosen the bowels, while drinking a lot of wine does not? Is it because the substance of the former goes into the

---

416 Salt foods are regarded as laxative by Hippocrates, *Aff.* 40 (6 250.6 Littré, where Littré translates Hippocrates as saying that, to be laxative, soup should be "sweeter rather than saltier, or warm", taking the first *ê* as "than" and the second as "or", but P. Potter in the Loeb ed., vol. 5, 1988, 62, surely rightly, has "more sweet, salty or hot") and 55 (266.1 L.), *Loc. hom.* 45 (84.3 Craik), *De diaeta* 2.45.1 (*CMG* 1.2.4 166.23 Joly); Rufus of Ephesus ap. Oribasius, *Coll. med.* 2.61.6 (*CMG* 6.1 59.30-31 Raeder). Hippocrates, *Aër.* 7.15 (*CMG* 1.1.2 40.2-6 Diller) says that people in general think that salt water is laxative, but the opposite is in fact the case (cf. Craik 215). *Sweet* food or wine is said to be laxative at Hippocrates, *Acut.* 14 (2 332.7 L.), cited by Galen, ad loc. (3.2, *CMG* 5.9.1 220.7-8 and 223.27 Helmreich) and at *PHP* 9.6.42 (*CMG* 5.4.1.2 580.29 De Lacy); Hippocrates, *Aff.* 40 (above); Oribasius, *Coll. med.* 2.58.61 (*CMG* 6.1.1 51.8-10 Raeder), 2.61.6 (59.29-30), 5.6.16 (123.10-13).

417 The "other juices" are apparently those that are sweet, so that what follows gives the explanation of why sweet flavours are laxative.

418 For the laxative effect of whey cf. Hippocrates, *De diaeta* 2.42.3 (*CMG* 1.2.4 166.4-5 Joly).

εὔπεπτά ἐστιν ὥσπερ ὁ οἶνος θερμὸς ὤν, ὁ δὲ οἶνος ταχὺ πεφθεὶς διουρεῖται· τοῦ μὲν οὖν ἡ ὑπερβολὴ τὴν κύστιν ταὐτὸ ποιεῖ ὅπερ ἐκεῖνα τὴν κοιλίαν, ἀλλ' οὐκ ἄγει ὑπόστασιν εἰς κύστιν, οὐ γὰρ ἔχει· τὰ δὲ μετὰ περιττωμάτων ἐξέρχεται ἑπομένων τῇ ὑποστάσει αὐτῶν οὔσῃ παχείᾳ καὶ ποιεῖ τὴν διάρροιαν.

2.117. *Διὰ τί ἐν τῇ κάτω καθάρσει ἢ ἐν τῇ ἄνω ὑπέρινοι μᾶλλον γίνονται; ἢ διότι εἰς μὲν τὴν κάτω κοιλίαν ἐξ ἅπαντος τοῦ σώματος πέφυκε συρρεῖν τὰ ὑποστήματα (τοῦτο δὲ φανερὸν τῇ τῶν ἐμβρύων τροφῇ· ἐν ἐκείνῳ γὰρ ὄντα τρέφεται τῷ τόπῳ· ἡ γὰρ ὑστέρα κατ' αὐτὴν ἔχει τὴν χώραν)· εἰς δὲ τὴν ἄνω ἀπὸ τῶν πλησίον χωρῶν συνέρχεται, ὥστε ἀπ' ἐλάττονος δυνάμεως εὐκινητότερα πάντα εἰς τὸ κάτω διὰ τὸ προϋπάρχειν εἰσφερόμενα; ἡ μὲν οὖν ἄνω κάθαρσις τὸ ἐπιπολάζον καὶ τὸ ἀλλότριον ἁρπάζει μόνον, διὸ καὶ ἑλλεβορισθέντες εὐθὺς ἰσχύουσιν· ἡ δὲ κάτω ἐκ πολλοῦ συνάγει τόπου καὶ ἐνίοτε συνεπισπᾶται καὶ τῶν οἰκείων πολλὰ καὶ τῶν περιττωμάτων οὐ ταμιεύεται, ὥστε συμφέρει πολλάκις μὴ πάντα ἐκκενῶσαι· γίνεται γὰρ τὰ ὑπολειφθέντα τροφὴ καὶ ἔχει ἰσχύν· ἐὰν δὲ μηδὲν ὑπολειφθῇ, ὁτὲ μὲν κένωσιν ποιεῖ ἰσχυράν· ὁτὲ δὲ ἀναιροῦνται καθαρθέντες καλῶς, ἐὰν βραδυτέρα γένηται ἡ προσφορὰ τῆς συντήξεως ἐπὶ τὸν ἐπικαιρότατον τόπον.*

2.118. *Διὰ τί τοῖς ὑγιεινοῖς καὶ γέρουσιν αἱ κάτω καθάρσεις οὐ συμφέρουσι (τάχιστα γὰρ ἀπόλλυνται καὶ ὑπέρινοι γίνονται); ἢ ὅτι τοῖς μὲν νοσεροῖς περίττωμά ἐστι τὰ ὑπιόντα, τοῖς δὲ ὑγιεινοῖς αὐτῶν τῶν εὖ ᾠκοδομημένων*

ταὐτὸ] *ταὐτὸν* P Us.: *ταυτῷ* sic A 5 *εἰς*] *εἰ* O || *μετὰ* om. Re || *περιττωμάτων*] *περιττώματα* M 6 *ἑπόμενον* T || *παχεία*] *ταχεία* K*v* || *καὶ ποιεῖ* FKOSbe^mg^i*v* Us.: <*ὃ*> *ποιεῖ* Buss.: *ποιεῖ* AMPRTe || *διάρροιαν*] *διάνοιαν* T

2.117 (AFKMOPRSTbei*p* Buss. Us.; *v* 2.54; 2.115 Buss.)

1 *ἢ*] *καὶ* A 2 *ἢ* om. A || *συρεῖν* T 3 *ἐν τῇ τῶν* F 4 *τρέφεται*] *τρέφονται* Buss.: *τρέφωνται* A || *ὑστέρα*] *ὑστεραία* Oi || *κατ' αὐτὴν*] *κατὰ ταύτην* Buss. 5 *πλησίον*] *πλησίων* AFMR || *συνεισέρχεται* P 6 *εἰς* om. P || *προϋπάρχειν*] *ὑπάρχειν* Me^ac^ 6-7 *διὰ – εἰσφερόμενα*] *quod percellant ad sese deferendum v* 7 *ἡ μὲν*] *εἰ μὲν* OTi || *ἄνω* om. Oi || *τ' ἀλλότριον* P 8 *ἁρπάζειν* T 8-9 *ἐκ τοῦ πολλοῦ* A 9 *καὶ τὰ τῶν* Buss. (in var. lect.) 10 *οὐ ταμιεύεται*] *εὐταμίευτα* Us.: *εὐταμιεύεται* A || *ὥστε* A Buss. Us.: *quapropter v*: om. cett. || *συμφέρει δὲ* KOSbe^mg^i || *πάντα*] *πάντας* A 11 *τροφὴ*] *τρυφῇ* F: *τροφὴν* T 12 *μηδὲν*] *μηθὲν* AFPT Buss. || *ἀναιροῦνται*] *ἀναιροῦντες* F 13 *ἐὰν δὲ* FP || *γένηται* AFPT Buss. Us.: *γίνηται* KMORSbe: *γίνεται* i || *προφορὰ* S || *συντήξεως* Buss. Us.: *συντάξεως* codd.

2.118 (AFKMOPRSTbei*p* Buss. Us.; *v* 2.55; 2.116 Buss.)

2 *ὅτι* om. Buss. 3 *περιττώματα* FK || *ὑπιόντα*] *ὑπιέντα p*: *ἀπιόντα* Us. || *αὐτῶν*]

bowel? For they are not easily concocted as wine is, being hot, while wine is swiftly concocted and passed as urine. Excess of [wine] has the same effect on the bladder as they do on the bowel, but it does not take substance into the bladder, for it does not possess any; but they pass out along with the residues which follow their substance, which is dense, and produce diarrhoea.

2.117.[419] Why are people violently purged downwards more than upwards? Is it because the sediment from all the body naturally flows together into the lower bowel (this is clear from the nourishment of embryos; they are nourished in that region; for the womb is in that place), but into the upper from the neighbouring region [only], so that everything is more easily moved downwards by a lesser power because it is already present having been conveyed there? So the upwards purging seizes only on what is on the surface and alien, and for this reason those who have taken hellebore[420] at once regain their strength; the downwards [purging] draws and sometimes drags [matter] along with itself from a large area, and many proper [ingredients] and residues are not stored, so that it is often beneficial that not everything is emptied out; for what is left behind is nourishment and has [some] strength. If nothing is left behind, sometimes it produces a forcible emptying; at other times people die when they have been purged well, if the colliquescence is applied too slowly to the most crucial region.

2.118. Why do downwards purgings not benefit the healthy and the old (for they perish very quickly and are purged violently)? Is it because in the case of the sick what passes out[421] is residues, but in the case of the healthy there

---

419 Cf. 2.108. Whereas there the fact that downwards purging is natural and the fact that it removes nourishment are presented as altermative explanations, here they form part of the same explanation.

420 I.e. white hellebore; see below, 2.126.

421 *hupienai* is a technical term for the passing of excrement (cf. LSJ s.v. *huperkhesthai*, VII); Usener ought not to have changed *hupionta* of the MSS to *apionta*.

γίνεται ἡ ἀφαίρεσις, ὥστε σύντηξις, οὐ περίττωσις ἡ ἀπὸ νόσου; ὅταν οὖν ὁρμήσῃ, μέχρι θεμελίων κινεῖται.

2.119. Διὰ τί τοῦ θέρους μᾶλλον αἱ διάρροιαι γίνονται ἢ τοῦ χειμῶνος, ἔστι δὲ τοῦτο τὸ πάθος ὑγρασίας ὑπερβολή, καὶ ὁ χειμὼν ὑγρότερος, ὥστ' ἔδει μᾶλλον ἐν τῇ ὁμοίᾳ ὥρᾳ συμβαίνειν τὸ πάθος ἢ ἐν τῇ ξηρᾷ; ἢ ὅτι ἐν τῷ θέρει ἀντιπεριίσταται τὸ ὑγρόν, ἐν δὲ τῷ χειμῶνι, τὸ θερμόν; οἷον γὰρ ἂν τι ᾖ τὸ ἔξω, ἐναντίως ἔχει τὸ ἐντός· διὸ καὶ τὰ φρέατα τοῦ μὲν θέρους ψυχρὰ ἐν θερμῇ τῇ ὥρᾳ, τοῦ δὲ χειμῶνος θερμὰ ἐν ψυχρᾷ. τὸ αὐτὸ δὴ καὶ ἡ κοιλία πάσχει, ὥστε ἐν μὲν τῷ χειμῶνι θερμοτέρα οὖσα πέττει ἱκανῶς, ἐν δὲ τῷ θέρει ψυχροτέρα οὖσα οὐ πέττει ἱκανῶς, ἡ δὲ διάρροια ἀπεψία ἐστίν· συμβάλλεται δέ τι καὶ ἡ τροφὴ τοιαύτη οὖσα ἐν τῷ θέρει, οἷα ταρακτική, καὶ αἱ νύκτες ἐλάττους.

2.120. Διὰ τί ἐν τῇ νυκτὶ ἀπ' ἐλάττονος καθαίρει τὰ φάρμακα ἢ ἐν τῇ ἡμέρᾳ; ἡ μὲν γὰρ σκαμμωνία καθαίρει ἀπὸ ὀγδοημορίου, ἐλατήριον δὲ ἀπὸ

ἀπὸ AF: αὐτὸ MRTe^ac || εὖ] εἶναι e^ac 4 περίττωμα P

2.119 (AFKMOPRSTbei*p* Buss. Us.: *ν* 2.56; 2.117 Buss.)

1 τὸ θέρος P || τὸν χειμῶνα P 2 χειμὼν] χυμὸν T || ἔδει F*ν* Us.: ἀεὶ AKMOPRSbei Buss.: ἂν T 3 ἐν τῇ[1]] ἐν τε P || συμβαίνει PRe || post συμβαίνειν add. ἔδει Buss. 5 ἐναντίας FT || τὸ μέν θέρος P 6 ἐν θερμῇ τῇ ὥρᾳ om. *ν* || ἐν ψυχρᾷ om. *ν* || δὴ Us.: δὲ codd. Buss. 7 πάσχοι S || πέττει] οὐ πέττει FM: οὐ πέπτει e || ἱκανῶς] *apposita* *ν* (vel suppl. et ἱκανῶς om.?) 7-8 ἐν δὲ τῷ θέρει ψυχροτέρα οὖσα (οὖσα om. F) οὐ πέττει ἱκανῶς AFT*p*(ἱκανῶς om. *p*) Buss. Us.: om. cett. 8 δὲ[2] om. A 9 τι om. F*ν* || οἵα e

2.120 (AFKMOPRSTbei*p* Buss. Us.; *ν* 2.57; 2.118 Buss.)

2 γὰρ om. O || ἀπὸ] ἀπ' Buss. Us. || ὀγδοημερίου Oi || τὸ δὲ ἐλατήριον K 3

is a removal of the [parts] that have been well constructed themselves, so that [what happens] is colliquescence, not [removal] of residue from disease? So when it sets out, it causes movement as far as the foundations.

2.119. Why do attacks of diarrhoea occur more in summer than in winter,[422] when this affliction is an excess of moisture, and the winter is more moist, so that the affliction should occur more in the season that is similar than in the dry one? Is it because in the summer the moisture is concentrated inwards, in the winter the heat? For whatever anything is like on the outside, it is the opposite on the inside; and for this reason wells are cold in summer, in a hot season, but hot in winter, in a cold one.[423] So the same thing happens to the bowel, so that in winter, being hotter, it concocts adequately,[424] but in the summer, being colder, it does not; and diarrhoea is failure of concoction. The nourishment also contributes something, being in summer of such a sort as to cause disturbance, and the nights are shorter.[425]

2.120. Why do drugs produce purging from a smaller dose at night than in the day? For scammony[426] purges with an eighth part, and the Driver[427] with

---

422 Summer diarrhoea is mentioned by Hippocrates, *Aër.* 7.4 (*CMG* 1.1.2 36.8-10 Diller), *Aph.* 3.21 (4 496.1 Littré), *Epid.* 1.2.4 (2 616.13 L.); Aretaeus, *Caus. acut. morb.* 2.7.5 (*CMG* 2 75.4 Hude); Oribasius, *Coll. med.*, lib. incert. 42.12 (*CMG* 6.2.2 149.22 Raeder). The cause is given as yellow bile by Galen, *In Hippocr. Aph.* 3.21 (17.2 619.10-14 Kühn).

423 Cf. (of springs) Theophrastus, *Ign.* 16 and fr. 173 FHS&G, and Alexander, *In Meteor.* 202.33-203.3, all with the same explanation as here, that heat is compressed inwards by cold: Sharples 1998b, 122. For the concentration of heat by cold see also below on 3.42.

424 Philoponus, *In An. Post.* 419.30-420.6 gives as examples of problems with the same explanation, concentration of heat, as here, (a) why we digest better in winter, (b) why we breathe rapidly when asleep (see above, 2.40), and (c) why wells are hot in winter (see the previous note); the same three examples at Anon., *In An. Post.*, *CAG* 13.3 590.13-16 and Eustratius, *In An Post*. 231.13-19). Philoponus explains the mechanism in (a) by the compression of pores in the body compressing the heat within. (b) appears, together with the question why we digest better when *asleep*, and the explanation of both in terms of internal heat, at Galen, *In Hippocr. Epid. 6* 4.15 (*CMG* 5.10.2.2 220.11-15 Wenkebach and Pfaff).

425 Allowing less time for digestion during sleep.

426 *Convolvulus scammonia* L. Cf. Amigues 1988-, vol.2 233.

427 A purgative prepared from squirting cucumber, *Ecballium elaterium* (L.) A. Richard. Cf. Amigues loc. cit.

τεταρτημορίου ὅταν δοθῇ καταπότια. ἢ διὰ τρία ταῦτα, δι' ὑγρότητα, διὰ θερμότητα, διὰ τὸ συμμένειν πολὺν χρόνον; ὅτι τε γὰρ οἱ ὑγρότεροι μᾶλλον καθαίρονται· ὑγρὰ γὰρ ἡ κάθαρσις· τὸ γὰρ περίττωμα * * * καὶ ἐν τῷ πλείονι χρόνῳ μᾶλλον καθαίρει· ἐν δὲ ταῖς νυξὶ συμβαίνει τοῦτο μάλιστα, διότι πέττει ὁ ὕπνος, ὥστε θᾶττον καὶ μᾶλλον αἰσθάνεσθαι τὸ σῶμα τοῦ φαρμάκου.

2.121. Διὰ τί τὰ φάρμακα τὰ καθαρτικὰ φλέγμα μὲν ἄγει καὶ χολὴν ἐκ τῶν φλεβῶν εἰς τὴν κοιλίαν, αἷμα δ' οὔ, λεπτὸν ὄν; ἢ ὅτι ἄγει τὸ ἀλλότριον ὂν διὰ τὸ μὴ προσπεφυκέναι, τὸ δὲ περίττωμα ἀλλότριον; ἢ ὅτι παχύτερον τὸ αἷμά ἐστι, τὰ δὲ ἐξελθόντα παχύνεται, τὸ φλέγμα καὶ ἡ χολή; δῆλον δ' ὅτι τὸ αἷμα παχύτερον, ὅταν ψυχθῇ καὶ παγῇ.

2.122. Διὰ τί τοῖς ἐκ τῶν γυμνασίων καὶ φαρμακοποσιῶν οὐκ εὐθὺς προσφέρομεν τροφήν; ἢ ὅτι καθαίρεται τὸ σῶμα ἔτι καὶ οὐκ ἀναπέπαυται πονοῦν καὶ ἀποκέκριται τὰ περιττώματα; ἔτι καὶ ἡ μεταβολὴ μεγάλη οὖσα ἀσύμφορον, γίγνοιτο δὲ μετὰ πολλὴν κένωσιν πλήρωσις ἀθρόως.

2.123. Διὰ τί λέγουσι μὲν πάντες, σεμνύνονται δὲ οἱ πλούσιοι ἐπὶ τῷ φαγεῖν ἢ πιεῖν ἡδέως, ἐπὶ δὲ ταῖς κενώσεσιν οὐκέτι· καίτοι οὐδὲν ἧττον ἡδύ, ὅταν ἐπιθυμῇ τις, καὶ προσέτι ὠφέλιμον τὸ δεόμενον τυχεῖν καὶ λυπηρὸν τὸ μὴ τυγχάνειν; πότερον ὅτι κάθαρσις, ὥστ' ἐνδιατρίβειν τῷ λόγῳ ἀηδὲς περὶ αὐτῶν; ἢ ὅτι οὐκ ἐφ' ἡμῖν, ἐκεῖνα δὲ ἐφ' ἡμῖν;

τεταρμορίου sic O || δοθῇ] δοκῇ e 4 συμμένειν K Buss. (in var. lect.) Us.: σημαίνειν FMOPRSTbei: σημένειν sic A: om. *v* 5 "post κάθαρσις multa desunt" Buss. || lacunam post περίττωμα statuit Us. 6 νυξὶ] γυναιξὶ M

2.121 (AFKMOPRSTbei*p* Buss. Us.; *v* 2.58; 2.119 Buss.)
4-5 τὰ δὲ ἐξελθόντα – χολή om. *v* || τὸ φλέγμα καὶ ἡ χολὴ] ἡ χολὴ καὶ τὸ φλέγμα F: τὸ αἷμα καὶ ἡ χολὴ K 5 τὸ αἷμα om. T

2.122 (AFKMOPRSTbei*p* Buss. Us.; *v* 2.59: 1-3 = [Ar.] 5.28; 2.120 Buss.)
1 τοῖς om. K || γυμνασιῶν Us. || καὶ φαρμακοποσιῶν] μὴ φαρμακοποιῶν F 2 προσφέρομεν] προφέρομεν K: προσφέρουσι [Ar.] || καθαίρεται] καθαίρε T || οὐκ ἀναπέπαυται [Ar.] Us.: οὐχ ἅμα περὶ αὐτὰ (αὐτὸ P) codd.: οὐχ ἅμα περὶ ταῦτα Buss. (in var. lect.) 3 πονοῦν] πονεῖ K*v*: πονοῖεν ἂν Buss. (in var. lect.) || καὶ ἀποκέκριται] ἀλλ' ἀποκρίνεται K*v* 4 γίγνοιτο] γίνεται K || πολλὴν] πολὺν OSi || πλήρωσις] πλήρωμα Buss.

2.123 (AFKMOPRSTbei Buss. Us.; 2.121 Buss.)
1 τῷ] τὸ T 2 πιεῖν] ποιεῖν A 3 προσέτι ὠφέλιμον Us.: προσῇ τὸ ὠφέλιμον K Buss. (in var. lect.): πρόσεστι τὸ ὠφέλιμον cett. || τὸ] καὶ T|| τοῦ δεομένου Buss. (in var. lect.) 4 ὥστ'] ὡς T 5 περὶ αὐτῶν ἀηδὲς F || αὐτῶν] αὐτὸν P

a fourth part if they are given as drinks. Is it on account of these three things, moisture, heat, and remaining for a long time? For those who are more moist are purged more; for purgation is moist. For the residue ... and it purges more in a longer time; this happens most at night, because sleep produces concoction, so that the body is aware of the drug more speedily and to a greater extent.

2.121. Why do purgative drugs draw phlegm and bile from the veins into the bowel, but not blood, although it is thin? Is it because they draw [out] what is alien because it is not naturally connected, and residue is alien? Or is it that blood is [in fact] thicker, but the things that pass out *become* thick, the phlegm and the bile? It is clear that blood is thicker, when it is chilled and congeals.

2.122.[428] Why do we not give food immediately to those who have engaged in gymnastics or consumed drugs? Is it because the body is still being purged and has not stopped labouring, and the residues have been separated off? Moreover change which is great is not beneficial, and after much emptying there would be repletion all at once.

2.123. Why does everyone speak about eating or drinking with pleasure, and the rich pride themselves on it, but not also in the case of evacuations?[429] And yet it is no less pleasant, when someone desires it, and in addition it is beneficial to get what one needs and painful not to get it? Is it because they are purging, so it is not pleasant to spend time on them in conversation? Or is it that they are not up to us, and the former things are?

---

428 The first two sentences = [Ar.], *Probl.* 5.28.

429 Petronius' Trimalchio (*Satyricon* 47) is perhaps the comic exception that proves the rule.

2.124. Διὰ τί τοῦ θέρους μᾶλλον γίνεται χολὴ τῶν περιττωμάτων ἐλαττόνων ὄντων; διότι τὸ μὲν θερμὸν σηπτικόν, ἡ δὲ χολὴ ἐκ σήψεως.

*2.125. Διὰ τί οἱ χυλοὶ πάντων οὕτω διαχωροῦσι μᾶλλον ἢ μετὰ τῶν φύλλων καὶ ῥιζῶν καὶ καρπῶν; ἢ * * * τῶν φύλλων εἰσίν; ἐξῆκται δ' αὐτῶν τὸ θερμὸν εἰς τὸ ὑγρόν, τῶν μὲν ἑφθῶν ὑπὸ τοῦ πυρός, τῶν δὲ ὠμῶν αὐτοὶ οἱ χυλοὶ ἔχουσι· διὸ οἱ μὲν ἰσχυροὶ καθαίρουσι τὰς κοιλίας, οἱ δ' ἀσθενεῖς λύουσι· καὶ γὰρ οἱ τῶν φαρμάκων ὀποὶ μᾶλλον καθαίρουσι τῶν ῥιζῶν καὶ καρπῶν ἐξ ὧν εἰσι, καὶ οἱ τῶν γαλάκτων ὀρροὶ ὡσαύτως μᾶλλον ἢ τὸ γάλα· τὸ γὰρ σαρκῶδες τοῦ γάλακτος ὅπερ γῆς ἐστι μέρος παχύ ἐστι· σημεῖον δέ· οὐ γὰρ ⟨ἂν⟩ ηὐξάνετο τὰ ζῷα τῷ γάλακτι, εἰ μὴ σαρκῶδες ἦν· νῦν δὲ οἱ μὲν ὀρροὶ οὐροῦνται μᾶλλον καὶ διαχωροῦσιν ὑπὸ λεπτότητος, αἱ δὲ σάρκες ὑπολειπόμεναι αὔξουσι καὶ πηγνυμένων τῶν γαλάκτων τυροὶ γίνονται, ἑψομένων δὲ πυρὶ καὶ ὀπτωμένων προσειομένων τε βούτυρος.*

2.126. Διὰ τί ἐναντίως ὁ λευκὸς καὶ ὁ μέλας ἐλλέβορος καθαίρει; ἢ ὅτι θερμὰ τὰ καθαίροντα; θερμῷ γὰρ ἅπαντα λύεται, ψυχρῷ δὲ πήγνυται μᾶλλον· εἰ οὖν τὰ μὲν κάτω, τὰ δ' ἄνω καθαίρεται, ἀφομοιοῖ ἑκάτερον αὐτῷ τῶν φαρμάκων τὴν κάθαρσιν· βαρὺς μὲν γὰρ καὶ πυκνός ἐστιν ὁ μέλας, ὁ δὲ λευκὸς ἀραιὸς καὶ κοῦφος· τὰ γοῦν ὑφ' ἑκατέρου τακέντα ὑγρὰ

2.124 (AFKMOPRSTbei*p* Buss. Us.; *v* 2.60; 2.122 Buss.)
1 *τὸ θέρος* P || *μᾶλλον* om. A || *περιπτωμάτων* i 2 *διότι*] *ἢ ὅτι* A Buss.: *ἢ δι' ὅτι* K*v*

2.125 (AFKMOPRSTbei*p* Buss. Us.; *v* 2.61; 2.123 Buss.)
2 lacunam statuit Us. || *ἢ – εἰσίν*] *aut cum foliis sunt v* || *εἰσίν* del. Buss. || *δ'*] *δι'* *p* 3 *ὅτι τῶν μὲν* A Buss. || post *πυρός* lacunam statuit Buss. in var. lect. 6-11 *ἐξ ὧν – βούτυρος* om. *p* 6 *ὀροὶ* sic T Buss.: *οροὶ* sic A 8 *γὰρ* om. A || *ἂν* add. Us. || *ηὔξανε* A 9 *ὀροὶ* sic T Buss.: *οροὶ* sic A || *διαχωροῦσιν*] *δυσχωροῦσιν* F 10 *πηγυμένων* F 11 *ἑψημένων* F: *ἑψωμένων* T: *ἑψομένων* Oi || *πυρὶ*] *περὶ* i || *προσειομένων* Us.: om. M Buss.: *coagulato* (sicut 10 *πηγνυμένων*) *v*: *προσιεμένων* cett. || *τε*] *δὲ* AFPRTe: om. M Buss. || *βότυρος* Oi

2.126 (AFKMOPRSTbei*p* Buss. Us.; *v* 2.62; 2.124 Buss.)
1 *ἐναντίων* KO 3 *εἰ*] *ἢ* A(= *εἰ*, ut saepe in hoc cod.): *ἢ* FPRT 4 *αὐτῷ* Buss. Us.: *αὐτῶν* AF: *αὐτῶν* cett. (post *φαρμάκων* P) || *ἔστιν* post *γὰρ* T 5 *γοῦν*

2.124. Why is there more bile in summer when the residues are less?[430] Because heat produces decay, and bile results from decay.

2.125. Why do all juices pass through more easily on their own than with the leaves and roots and fruits? Is it that they are <hotter than> the leaves?[431] The heat of [the leaves] is drawn out into the moisture, by the fire in the case of those that are boiled, while in the case of those that are raw the juices themselves possess this; and for this reason those that are strong purge the bowels, those that are weak loosen them. For the juices of drugs purge more than do the roots and fruits from which they come, and in the same way the whey of milk more than the milk.[432] For the meaty part of milk,[433] which is a portion of earth, is dense. An indication: living creatures would not grow on milk, if it were not meaty. But as it is the whey is passed more as urine and excrement because of its thinness, while the meat, being left behind, produces growth. And when milk is congealed cheese is produced, when it is boiled on a fire and roasted and churned, butter.

2.126.[434] Why do white[435] and black[436] hellebore purge in opposite ways? Is it because things that purge are hot? For everything is relaxed by heat, but congealed more by cold. So if some things are purged downwards, others upwards, each of the drugs makes the purgation similar to itself. For the black [hellebore] is heavy and dense, the white rare and light; so when the

---

430 Prantl 1851, 366 n.71 observes that the statement that the residues are less in summer contradicts [Ar.], *Probl.* 2.33. But the argument there is that residues are less in winter because nourishment is concocted better then, whereas in summer it putrefies; and the latter point is in agreement with our text. For the connection of (yellow) bile with summer cf. Hippocrates, *Nat. Hom.* 7 (6 50.12 Littré), and above n.422.

431 Lacuna postulated by Usener; we have supplied "hotter than" *exempli gratia.*

432 Cf. above, n.418.

433 This may sound less odd if we think of protein.

434 [Ar.], *Probl.* 1.41 considers the same question for drugs more generally, and not just hellebore, but answers it in terms of differences in degree of heat between the different drugs, rather than differences in density.

435 *Helleborus cyclophyllus* Boiss. Amigues, loc. cit.

436 *Veratrum album* L. 2.108 and 117 above seem to say of hellebore in general that it purges upwards. According to Riddle 1985, 120 no.148 white hellebore is in fact an emetic and also lowers blood pressure, while black hellebore (ibid. 124, no.162) is a purgative (downwards).

ὅταν αὐτοῖς ἀφομοιωθῇ, τὰ μὲν ὑπὸ τοῦ μέλανος βαρέα καὶ πυκνὰ τὴν κάτω φορὰν ἔλαβεν, ἡ δ' ἀπὸ τοῦ λευκοῦ ἀραιὰ γενομένη καὶ κούφη εὐχερῶς ὑπὸ τῆς προϋπαρχούσης θερμασίας ἀνηνέχθη.

127. Διὰ τί τῶν τετραπόδων οὐδὲν ὑπὸ ὑστέρας πνίγεται ἔχοντά τε καὶ ζῳοτοκοῦντα; ἢ διότι μεγάλων ὄντων τῶν ἄνω πρὸς τοῖς αἰδοίοις ἡ ὑστέρα ἐστὶ πᾶσι· διὰ δὲ τὸ πολὺ ἀπέχειν οὐκ ἀναβαίνουσιν ἄνω, διὰ δὲ τὸ μὴ ἀναβαίνειν οὐκ ἀπολαμβάνουσι τὸ πνεῦμα, πνίγονται δὲ τούτου γινομένου;

128. Διὰ τί τὰ πιαινόμενα τῇ ὠμῇ λύσει τῶν αἰρῶν μιγνυμένων, τάχιστα πιαίνεται; διὰ τὸ τὴν μῖξιν ἔχειν τινὰ λόγον· ἐὰν γὰρ λίαν πολὺ μιχθῇ, ἀποκτείνει· ἢ ὅτι αἱ αἶραι βαρὺ ἔδεσμα καὶ καρηβαρεῖν ποιεῖ, τῷ οὖν καρώματι ὕπνον συνεχῆ ποιεῖ ὥσπερ ὁ μανδραγόρας, ὁ δ' ὕπνος πιαίνει μάλιστα συνέχων τὰ σώματα;

129. Διὰ τί τῆς κόπρου ἡ τῶν βοῶν ἥκιστα δυσώδης τῶν τετραπόδων καὶ διπόδων; πότερον ὅτι πέττεται μάλιστα (σημεῖον δ' ἡ λειότης), καὶ ὅτι ἁπλῆ ἡ τροφὴ (ἄχυρα γάρ); ἡ δὲ ποικίλη δυσώδης, διὸ καὶ ἡ τῶν ἀνθρώπων δυσωδεστάτη. ἢ ὅτι σημεῖον ὅτι μάλιστα μηρυκάζει; τὸ γὰρ μηρυκάζειν ἐστίν, ὅταν μεταβάλλῃ ἐκ τῆς πρώτης εἰς τὴν δευτέραν, διὸ καὶ λεῖον γίνεται τὸ περίττωμα· διὸ καὶ πλεῖστον χρόνον ἐν τῇ κοιλίᾳ μείναντα ἥκιστα ὄζει, τὰ δὲ προσφατώτατα μάλιστα.

KOPSi Us.: *οὖν* AFMRTbe Buss. || *τακέντα* AFMPT Buss. Us.: *κέντρου* KOSbe^{pc}i*v*: *τὰ κέντρα* Re^{ac} 6 *τὰ*] *τὸ* AMRTe || *ὑπὸ*] *ἀπὸ* F 7 *ἔλαβεν*] *ἔλαχεν* P 8 *προϋπαρχούσης*] *excellente v*(?)

2.127 (AFKMOPRSTbei*p* Buss. Us.; 2.125 Buss.)

in margine haec exhibet K: *τοῦ αὐτοῦ περὶ τετραπόδων ἀπορίαι καὶ λύσεις διάφοροι· τόμος ε* 1 *ὑπὸ*] *ἀπὸ* FKORSbei*p* || *ἔχοντά*] *ὠοτοκοῦντά* K || *καὶ* om. AP 3 *ἐστὶ*] *ἐπὶ* K 4 *γενομένου* KOi Us.

2.128 (AFKMOPRSTbei*p* Buss. Us.; *v* 3(5).1; 2.126 Buss.)

1 *ὠμῇ λύσει*] *λύσει* tantum M: *ὁμιλήσει* AKSb*v*: *ὁμιλήσει* sic Oi || *αἰρῶν* Buss. Us.: *ἀέρων* codd. *pv* 2 *πεπαίνεται* MR^{pc}e^{ac} || *διὰ τὸ τὴν*] *διὰ δὲ τὸ τὴν* AFMRe Buss.: *διά δὲ τὴν* T || *ἔχει* T || *πολὺ λίαν* P 3 *ἀποκτένει* sic AT: *ἀποκτενεῖ* KMORSbei*v* || *αἱ αἶραι p* Us.: *αἶραι* FOSbei Buss.: *αἶρε* sic A: *αἴρει τὸ* K*v*: *ἔραι* PT: om. MR

2.129 (AFKMOPRSTbei*p* Buss. Us.; *v* 3(5).2; 2.127 Buss.)

1 *τῶν δὲ τετραπόδων* T 3 *ἀπλῶ* sic A || *δυσωδία* P 4 *δυσωδεστέρα* F || *ὅτι*^{1} om. F || *ὅτι*^{2}] *ἢ* A 5 *μηρυκάζειν*] *μηρυκάζει* F || *μεταβάλη* A Buss. 6 *διὸ καὶ πλεῖστον χρόνον* PT Us.: *καὶ πλεῖστον· διὸ χρόνον* A Buss.: *καὶ πλεῖστον χρόνον* FKMORSbei*v* 7 *μείναντα*] *μεῖναν* KOi: *μήναντα* A: *μένοντα* S

moisture produced by each is made like to it, that produced by the black, being heavy and dense, takes the downward route, while that from the white, becoming rare and light, is easily moved upwards by the pre-existing heat.

[Four-footed creatures][437]

2.127. Why is no four-footed animal choked by its womb,[438] though they possess them and produce live young? Is it because their upper parts are large and they all have the womb near the genital organs, and because it is a long way away it does not travel upwards, and because it does not travel upwards it does not cut off the breath, and [women] are choked when this happens?
2.128. Why do those that are fattened on raw corn-meal grow fat most quickly when darnel is mixed with it? Because the mixture has a certain proportion; for if too much is mixed in, it kills them. Or is it because darnel is a heavy food and makes them drowsy,[439] and by the drowsiness it produces continuous sleep, like mandragora, and sleep fattens because it composes bodies most?
2.129. Why is the dung of cattle the least ill-smelling of four-footed and two-footed animals? Is it because it is most concocted (its smoothness is an indication) and because their food is simple (for it is bran); varied [food produces excrement that is] ill-smelling, and for this reason that of human beings is most ill-smelling. Or is it an indication that they chew the cud most? For chewing the cud is when it changes from the first to the second [stomach? type of excrement?], and for this reason also the residue becomes smooth. And for this reason, having remained in the bowel for a long time, it smells least, while what is freshest [smells] most.[440]

---

437 A new book starts here in cod. K with the title "Various problems and solutions by the same author concerning four-footed animals: book 5"; Valla, who used K as the source for his translation in this part of the collection, has a new book headed "Doubts about four-footed animals".

438 Referring to the Hippocratic belief in the "wandering womb" in women.

439 Cf. Aristotle, *Somn. vig.* 3 456b31: Preus 1988, 83.

440 That solid excrement smells worse the more quickly it has passed through the body (as is in fact the case), while urine smells worse the longer it has been in the body, is the theme of [Ar.], *Probl.* 13.1.

2.130. Διὰ τί τὰ Λιβυκὰ θηρία τοῦ μὲν θέρους οὐ πίνει ὅτε μάλιστα δίψα γίνεται, χειμῶνος δὲ πίνουσιν ὅτε ἧττον διψῶσιν; ἢ διὰ τὸ μὴ γίνεσθαι ὕδατα τοῦ θέρους ἐν τῇ Λιβύῃ; συνήθεις γὰρ μὴ πίνειν· τοῦ δὲ χειμῶνος διὰ τὸ γενέσθαι πίνει· ὃ οὖν ἐκεῖ εἴθισται, καὶ ἐν ἄλλοις τόποις ἐλθόντα τὸ ἔθος σῴζει.

2.131. Διὰ τί τὰ πλεῖστα τῶν τετραπόδων ῥᾷον τοῦ ὑγροῦ τὸ ξηρὸν περίττωμα προΐεται (ἀφοδεύει μὲν γὰρ καὶ βαδίζοντα, οὐρεῖ δὲ ἐφιστάμενα τὰ πολλὰ αὐτῶν); ἢ ὅτι τοῖς μὲν θήλεσιν ἡ μήτρα τῆς κύστεως ἔμπροσθέν ἐστι, τοῖς δ' ἄρρεσιν οὔ; τῇ γὰρ ξηρᾷ περιττώσει κατ' εὐθυωρίαν ἐστὶν ἐκ τῆς κοιλίας εἰς τὴν ἕδραν ἡ φορά, διὸ ῥᾳδίως ἐκρεῖ· τὰ ⟨δ'⟩ οὖρα καὶ καμπὴν ποιεῖ τοῦ γοῦν αἰδοίου εἰς τοὔμπροσθεν ἀνενηγμένου· ὁρμήσαντα οὖν ἐκ τῆς κύστεως τὰ ὑγρὰ καὶ προσπίπτοντα τοῖς καταντικρὺ τόποις ἀναχωρεῖ πολλάκις, ἂν μὴ ἐπιστάντα τὰ ζῷα τῷ πνεύματι συναπώσῃ· διὸ καὶ οἱ ἱπποκόμοι ἐφιστᾶσιν τὰ ὑποζύγια ἐν τῷ οὐρεῖν· σημεῖον δέ· τὰ μὲν θήλεα τῶν ἀρρένων ὑποζυγίων ἐστὶν οὐρητικώτερα διὰ τὸ τῇ ἕδρᾳ τὸ αἰδοῖον ἐν ταὐτῷ τόπῳ κεῖσθαι· τῶν δὲ τετραπόδων τὰ μονώνυχά ἐστι δυσουρητικώτερα διὰ τό τε τοῦ οὔρου πάχος καὶ διὰ τὸ χόρτον καὶ καρπὸν ἀργὸν ἐσθίοντα λάβρως ἐσθίειν· οὐ γὰρ λεαίνοντα τοῖς ὀδοῦσι τὴν τροφὴν φυσᾶται τὴν κοιλίαν οἰδησάντων τῶν εἰσιόντων, ἧς διατεινούσης καὶ πνευματωθείσης οὐκ ἐμπίπτει δι' αὐτῆς τὰ ὑγρὰ εἰς τοὺς τῆς κύστεως πόρους· σημεῖον δὲ τό τε στραγγουρίᾳ καὶ στρόφῳ ἔχεσθαι καὶ τὸ τοὺς ἰωμένους τὴν χεῖρα εἰς τὴν ἕδραν καθιέντας ἐκφέρειν κόπρον· τοῦ γὰρ πνεύματος μετὰ ταῦτα ἐξελθόντος ἀνοίγονται οἱ τῶν ὑγρῶν πόροι·

2.130 (AFGKMOPRSTbei*p* Buss. Us.; *v* 3(5).3; 2.128 Buss.)
1 Λιβυκὰ] λευκὰ S: λικυκὰ T || τὸ μὲν θέρος P 2 χειμῶνος δὲ] τὸν χειμῶνα P || πίνου T πίνωσιν A || ἥττω ORbi 3 ὕδατα] ὕδατος M || τὸ θέρος GP || λικύῃ T || σύνηθες K || γὰρ om.K || τὸν δὲ χειμῶνα P 4 πίνειν T || ἐκεῖ om. A || ἤθισται Sb || ἔθος] ἔθνος F
2.131 (AFOPTi Buss. Us.; 2.128bis Buss.)
4 τῇ γὰρ ξηρᾷ περιττώσει Us.: ἡ μὲν ξηρὰ περίττωσις cett. (ἡ γὰρ ξηρὰ περίττωσις P: δὲ pro μὲν Buss. in var. lect.) 5 ⟨οὐ⟩ ῥᾳδίως Buss. in var. lect. || δ' add. Us. 6 γοῦν P Us.: γοῦν vel γε Buss. (in var. lect.): οὖν cett. || ἀνενηγμένου Us.: ἀνηνεγμένου cett. 7 οὖν Us.: γοῦν codd. Buss. || καὶ om. Buss. 8 συναπώσῃ PT Buss.: συναπτῶσῃ A: συναναςπάσῃ FOi: ἀναπώσῃ Us. 9 ἐν τῷ οὐρεῖν τὰ ὑποζύγια P Us. 11 μονώνυχα Fi Buss. Us.: μονόνυχα cett. 12 διά τε το FOi || καρπῶν A 13 λεαίνονται PT 14 ἧς] εἷς sic A 15 δι' αὐτῶν FOi 16 στρόφωσι P 16-17 τοὺς ἰωμένους] τοῦς ειωμένους T 17 ἐκφέρειν] ἀφαιρεῖν Buss. 18 ταῦτα] ταύτης AT Buss. || ἐξελθόντας A || τε οἱ A: θ' οἱ Buss. 19 λούονται A || συχνὰ ἐλαύνοντες Us.: συχνὰ λεαίνοντες FOi:

2.130.[441] Why do wild animals from[442] Libya not drink in summer when there is the greatest thirst, but drink in winter when they are less thirsty? Is it because there is no water in Libya in the summer? For they are used to not drinking; but in winter the drink because there is [water]. So when they go to other places they preserve the habit which they learned there.

2.131. Why do most four-footed animals pass dry residue more easily than moist? For they excrete even while walking, but most of them pass urine while standing still. Is it because in the females the womb is in front of the bladder, but not in the males? For the dry residue has a straight path from the bowel to the rump, and so it passes out easily; but the urine makes a bend as the genital organ is in a forward position. So the moisture, setting out from the bladder and striking the opposite parts, returns frequently, if the creatures do not stand still and expel it along with the *pneuma*. And it is for this reason that grooms make yoke-animals stop when they urinate. An indication: female yoke-animals pass more urine than males because the genital organ is in the same place in the rump. Four-footed animals that have solid hoofs pass urine with more difficulty because of the thickness of the urine and because they eat straw and unprepared fruit greedily: for they do not grind their food with their teeth, and so they get wind in their bowel when what enters it swells up, and when [the bowel] is distended and inflated the moisture does not pass through it into the passages of the bladder. An indication is that they are subject to strangury and colic and that those who cure them insert their hand into the rump and extract dung; for when the wind passes out after this the passages of the moisture are opened.

---

441 So too, apart from the last sentence here, Ar., *HA* 8.28 606b24; Richter 1885, 30.

442 The solution shows that the reference is to animals imported from Libya to other regions, not just to the behaviour of animals in Libya itself.

βοηθοῦσι δὲ τῷ πάθει καὶ λούοντες καὶ συχνὰ ἐλαύνοντες· πάντα γὰρ ταῦτα ἀπορροὴν πνευμάτων ποιεῖ ἀνοίγοντά τε τοὺς κατὰ σάρκα πόρους καὶ λεπτύνοντα τὰ ἐν αὐτοῖς ὑγρά· ἁλίσκονται δὲ τοῖς εἰρημένοις πάθεσι μᾶλλον τῶν ἄλλων τὰ μονώνυχα καὶ διὰ τὸ δύσριγα εἶναι.

2.132. Διὰ τί τὰ ἄγρια θηρία μονόχρα; ἢ ὅτι μᾶλλον ἁπλῆ ἡ δίαιτα, αἱ δὲ ταραχώδεις καὶ ποικίλαι διαθέσεις χρωματίζουσι τὰ σώματα μᾶλλον; σημεῖον τούτου ἀλεκτρυόνες ὅτι πολύχροοι· παμφαγώτατοι γὰρ τῶν ὀρνίθων.

2.133. Διὰ τί οἱ πυροὶ τοῖς τετράποσι δυσδιαχωρητότεροι τῶν κριθῶν; ἢ διὰ τοὺς ἐχίνους; ταχὺ γὰρ λεαινόμενον τὸ ἄχυρον αὐτῶν λεπτυνόμενον, ἔχεται ἐν τοῖς ἐχίνοις, διὰ τὴν τραχύτητα τῶν ἐχίνων· τῶν δὲ κριθῶν ἅμα συμπέττεται διὰ τὸ τῇ πέψει γίνεσθαι χρόνον πλείω.

2.134. Διὰ τί τὰ μονώνυχα γηράσκοντα μείζους ἴσχει τοὺς ὀδόντας ὥσπερ τὰ ἄλλα ἐλάττους; ἢ ὅτι τε φύσει εὐαυξεῖς διὰ τὸ μὴ ἔχειν κέρατα, καὶ ἡ χρῆσις ὀλίγη ὑφ᾽ ἧς τρέφονται, καὶ ἐπιπαραλλάσσουσι τὰς σιαγόνας, οὐχ ὥσπερ ὁ ἄνθρωπος ἀρτίστομοι.

2.135. Διὰ τί οὔτε πώγωνα οὔτε δασεῖαν βλεφαρίδα ἔχει οὐθὲν τῶν τετραπόδων ὅσα μονώνυχά ἐστιν; ἢ ὅτι χαίτην ἔχει ταῦτα καὶ εἰς ταύτην καταναλίσκεται αὐτοῖς ἡ ὑγρότης ἐξ ἧς αἱ τρίχες γίνονται; ἀδύνατον γάρ ἐστι τὴν περίττωσιν ταύτην κινεῖσθαι ἅμα εἰς τἀναντία ἥ ἐστι τροφὴ

συνλεαίνονταις sic A: *συνελεαίνοντες* P: *συναλεαίνοντες* T: *συλλεαίνοντες* Buss. 22 *μονώνυχα* FOi Buss. Us.: *μονόνυχα* cett.

2.132 (AFKMOPRSTbei*p* Buss.Us.; *v* 3(5).4; 2.129 Buss.)

1 *μονόχρα* M*p* Us.: *μονόχρωα* A: *μονόχροα* cett. || *ἢ ὅτι μᾶλλον* Us.: *ἢ ὅτι* KM*v* Buss.: *ἢ ἄλλον* sic A: *μᾶλλον* cett. || *διαίτη* Buss. 2 *χρηματίζουσι* S 2-4 *μᾶλλον – ὀρνίθων* om. *p* 3 *τοῦτο* AFMPRTe Buss.: *τούτου οἱ* K: *τὸ τῶν* coni. Buss. || *ἀλεκτρυόνων* AFMPRTe[ac] Buss. || *πολύχροα* AFMPRTe[ac] Buss. || *παμφαγώτατα* F: *παμπαγώτατοι* Oi: *παμφαγώτατος* e[ac] || *γὰρ* om. P

2.133 (AFKMOPRSTbei*p* Buss. Us.; *v* 3(5).5; 2.130 Buss.)

1 *διαχωρητότεροι* e 2 *τὸ τοὺς ἐχίνους* Re || *λεπτυνόμενων* A 3 *τοῖς ἐχίνοις*] *τῇ ἐχίνῃ* P 4 *τὴν πέψει* e || *πλείω χρόνον* MR[pc]e Buss.: *πλείω χρόνω* R[ac]: *χρόνων πλείω* sic A: *χρόνου πλείω* Oi: *χρόνον πλείονα* P

2.134 (AFKMOPRSTbei*p* Buss. Us.; 2.131 Buss.)

1 *μονόνυχα* APST[pc]*p* 2 *ὅτι φύσει τὲ* K 3 *τὰς*] *τὰ* A 4 *ἀρτίστομοι* Us.: *ἀρτιόστομος* P: *ἀρτιστόματι* sic A: *ἀρτιστόματοι* Buss.: *ἀρτιόστομον* cett.

2.135 (AFKMOPRSTbei*p* Buss. Us.; 2.132 Buss.)

1 *οὐδὲν* FKT 2 *μονόνυχα* APST: *μώνυχα* M || *ταῦτα*] *τοιαύτην* P || *ταύτην*] *τὴν* O 3 *αὐτοῖς*] *αὐτῆς* Oi 3-5 *ἐξ ἧς – ὄπισθεν* om. *p* 4 *τροφὴ*] *τροπὴ* M 5 *ἀλλὰ*]

They also help [them in the] affliction by washing them and driving them frequently: for all these things make the wind pass out by both opening the passages in the flesh and making the moisture in them thinner. Animals with solid hoofs are subject to the aforementioned afflictions more than the others also because they are sensitive to cold.

2.132. Why are wild animals of a single colour? Is it because their diet is more simple, and disturbed and complex conditions colour bodies more?[443] An indication of this is that cockerels are many-coloured: for they are the most omnivorous of birds.

2.133. Why does wheat pass through four-footed animals less easily than barley? Is it because of the husks? For the bran of the [wheat] is swiftly ground up and made fine, but is held in the husks because of the roughness of the husks; but that of barley is digested at the same time because the digesting occupies more time.[444]

2.134. Why do solid-hoofed animals when they grow older have bigger teeth, just as other [animals] have smaller ones? Is it because they naturally grow easily because they do not have horns, and [the teeth] are not much used on their nourishment, and their jawbones pass alongside each other, not closing their jaws like humans?[445]

2.135.[446] Why do none of the solid-hoofed four-footed animals have either a beard or thick eyelashes? Is it because these have a mane and the moisture from which the hairs come is used up on this? For it is impossible for the residue that nourishes these to be moved simultaneously in opposite

---

443 As Aristotle argues at *GA* 5.6 786a35ff.: Richter 1885, 31.

444 That is, in the case of barley the digestion of the bran and the husks is completed simultaneously, while in the case of wheat the roughness of the husks prevents this.

445 See below, 2.158.

446 The converse in [Ar.], *Probl.* 10.25; the reason human beings do not have manes is that they have beards.

τούτοις, ἀλλὰ ἢ ἄνω ἢ κάτω καὶ ἢ πρόσθεν ἢ ὄπισθεν.

2.136. Διὰ τί τὰ τοιαῦτα μονώνυχα οἷον ἵπποι ἢ ὄνοι καὶ ὀρεῖς ἡνίκα ἂν γένηται, μένει καὶ οὐθὲν αὐξάνεται τὸ ἀπὸ τῆς ὁπλῆς ἄνωθεν (μὴ συμβαλλομένης δὲ τῆς ὁπλῆς μένει ἕως τοῦ γόνατος καὶ τοῦ ὀπισθίου καὶ τοῦ ἐμπροσθίου σκέλους), οἱ δὲ ἄνθρωποι τοὐναντίον· ἐπὶ τὰ κάτω γὰρ αὐξάνονται μᾶλλον; ἢ ὅτι τοὺς μὲν δεῖ ὀρθοὺς πορεύεσθαι· τὰ κάτω οὖν αὐξηθῆναι ἀνάγκη, ἵνα ῥᾷον φέρῃ τὰ ἄνω, ἢ οὐκ ἀναστήσονται· κατὰ δὲ τοὐναντίον τὰ τετράποδα διατελεῖν, ὥστε τὰ ἄνω αὐξάνεσθαι ἀνάγκη, τὰ δὲ κάτω μένειν [ᾧ ἀναστήσονται ὀρθά]· ὥστε διὰ τὸ μὴ ὀρθὸν ἀνάγκη ταῦτα γίνεσθαι; διὸ καὶ πωλία μὲν ὄντα ῥᾳδίως τὸ οὖς ξύονται τῷ ὀπισθίῳ ποδί, πρεσβύτερα δὲ γενόμενα οὔ· οἱ δ' ἄνθρωποι ἀντεστραμμένως· οὐ γὰρ δύνανται μέχρις ἡλικίας τινὸς τὰ παιδία κύκλῳ θιγεῖν τοῦ ὠτός.

2.137. Διὰ τί τὰ μὲν μονώνυχα τὸ ἀπορρέον ὕδωρ πίνει, τὰ δὲ δίχηλα, οἷον βόες τὰ συρρέοντα μᾶλλον; ἢ ὅτι οἱ βόες χαίρουσι μᾶλλον τοῖς ἁλμυρωτέροις, διὸ καὶ εἰς τὴν τροφὴν παρεμβάλλονται αὐτοῖς ἅλες καὶ συμμίγνυται ὁ τοιοῦτος χυμός· τὰ δὲ συρρέοντα ἔχει τὸ τοιοῦτον μᾶλλον (ὑπόστασιν γὰρ ἔχει, τὸ δ' ἁλμυρὸν ὑφίσταται διὰ τὸ βάρος)· τὰ δὲ μονώνυχα τὰ γλυκέα μᾶλλον καὶ πίνει καὶ ἄλλως προσφερομένων χαίρει.

ἀλλ' Buss. Us.: ἅμα F || ἢ ἄνω] καὶ ἢ ἄνω KOi: ἄνω A || καὶ om. FKOi
2.136 (AFKMOPRSTbei*p* Buss. Us.; *ν* 3(5).6; 2.133 Buss.)
1 διὰ τί οὖν A Buss. || τὰ τοιαῦτα] τὰ A Buss.: τοσαῦτα P: ταῦτα FMRTe^{ac} || μονόνυχα APST*p* || ἢ ὄνοι] ὄνοι KOSbi: οἱ ὄνοι A: καὶ ὄνοι T: om. *ν* || καὶ ὀρεῖς Buss. Us.: καὶ ὄρνις AFSb: καὶ ὄρνεις MRTe: ἢ καὶ ὄρνεις P: καὶ ὄρνιθες KOi*ν* 2 τὸ om. KOi*ν* 2-3 μὴ – μένει om. *ν* 3 συμβαλλομένης] συλλαμβανομένης FPT Buss.: συλλαβανομένης A || μένει ⟨δὲ⟩ ἕως Buss. 5 αὐξάνονται] αὐξάνεται S || δεῖ ὀρθοὺς Buss. Us.: ὀρθοὺς δεῖ K: δι' ὀρθοῦ AMORSbei*p*: διορθοῦ F: διορθοῦν T: διορθοῦται P || πορεύεσθαι K Buss. Us.: ὡς πορεύεσθαι cett. 6 ἵνα] εἶναι AMP^{ac}RT^{ac}e: post ῥᾷον habet P || φέρῃ Buss. Us.: φέρει A: φέρηται FOPSTbi: φέρωνται K*ν*: φαίνεται MRe || τὰ om. FMOPRSbei || κατὰ δὲ Us.: τὰ δὲ codd. Buss. 7 τὰ^{1} om. K Buss. || τὰ^{2} om. KORSbei*ν*: ὄντα *p* || αὔξεσθαι F 7-11 τὰ δὲ – ὠτός om. *p* 8 ᾧ – ὀρθά del. Us., om. *ν* || ᾧ] οἷς K || ὀρθά AM Buss. Us.: ὀρθῶς K: ὀρθῷ cett. || διὰ K*ν* Us.: om. cett. || τὸ] τῷ Buss. (var. lect.) || μὴ Buss. (var. lect.) Us.: om. K*ν*: μὲν cett. || ὀρθῷ F^{pc} Buss. (in var. lect.) 9 ταῦτα] τοιαῦτα Us.: om. P || πωλία Buss. Us.: *iuvenilia* *ν*: πολιᾶ AMRTe: πολιὰ FPSb: πολλὰ KOi || post πολλὰ add. νέα in marg. K || μὲν ὄντα K*ν* Buss. Us.: μένοντα cett. || ξύοντα P 11 κύκλῳ θιγεῖν om. F: κύκλῳ om. T
2.137 (AFKMOPRSTbei*p* Buss. Us.; *ν* 3(5).7; 2.134 Buss.)
1 μονόνυχα APST*p* || δὲ om. A Buss. || δίχηλα AT*p* Buss. Us.: διχηλῶ F: διχηλὰ KMRbe: διχηλᾶ cett. 3 ἑαυτοῖς P || ἅλες M Buss. Us.: add. sup. lin. R: ἅλας e: om. AFKOPSTbi*ν* 4 συμμίγνυνται M 4-5 τὸ τοιοῦτον – ἔχει om. F 6 μονόνυχα AOPSi

directions, but either upwards or downwards and either forwards or backwards.

2.136.[447] Why when solid-hoofed animals such as horses or asses and mules are born does the part upwards from the hoof stay [the same] and not grow (for with the hoof not contributing it stays [the same] as far up as the knee both of the back and the front leg), but with human beings the opposite; for they grow more in the lower part? Is it because the latter have to proceed upright; so it is necessary for the lower parts to grow, so that they may more easily carry the upper parts, or they will not stand up; but four-footed animals continue in the opposite [condition], so that it is necessary for their upper [parts] to grow, their lower to stay [the same] [[by which they stand upright]];[448] so that it is necessary for these things to happen because [they are] not upright. And for this reason while they are young foals they easily scrape their ears with their back feet, but not when they grow older. But human beings are the reverse; for children cannot touch their ear in a circle until they reach a certain age.

2.137. Why do solid-hoofed animals drink flowing water, cloven-hoofed ones like cattle rather water that has collected? Is it because cattle take more pleasure in what is more salty, for which reason salt is also added to their food and this sort of flavour is mixed in, and water that has collected has this quality more, for it has sediment, and what is salty settles out because of its weight. while solid-hoofed animals both drink sweet things more and take pleasure in them when administered in other ways?

---

[447] Cf. Ar., *PA* 4.10 686b7ff.

[448] Deleted by Usener. In Aristotle's view standing humans are upright, standing quadrupeds are not: *PA* 2.10 656a13.

2.138. Διὰ τί αἱ ὕες πιαινόμεναι δυσριγότεραι γίνονται; ἢ διότι καὶ οἱ παχεῖς ἄνθρωποι ὅσῳ ἂν μᾶλλον παχύνωνται, πορρωτέρω γίνονται τοῦ οἰκείου θερμοῦ, ὅσῳ δ' ἂν μᾶλλον τοῦτο πάσχωσιν, εὐψυκτότερα τὰ ἔσχατα ἀνάγκη εἶναι, τὰ δὲ εὔψυκτα ταχὺ ἀνάγκη αἰσθάνεσθαι τοῦ ψύχους;

2.139. Διὰ τί ἄνθρωπος μέν οὐ χαλαζιᾷ, οὐδὲ τῶν ἄλλων ζῴων οὐδέν, αἱ δὲ ὕες τοῦτο πάσχουσι; ληπτέον δὴ τὸ πάθος τί ἐστιν ἡ χαλάζησις· ἔστι δ' οἷον τοῖς ἀνθρώποις οἱ ἴονθοι, τοῦτο ταῖς ὑσὶν ἡ χαλάζησις. ᾧδε μετεστραμμένον· οἱ δ' ἴονθοί εἰσι μικροὶ δοθιῆνες διότι πολλοὶ καὶ διεσπαρμένοι· μεγάλα γὰρ πολλὰ χαλεπώτερον ποιῆσαι καὶ πλείονος περιττώματος ἔργον, ὥστε πρὶν μερισθῆναι οὕτως ἕτερον ἂν νόσημα προσηγέρθη· γίνονται δὲ οἱ ἴονθοι οὔτε ὡς ἀρρώστημα οὔτε ὡς τῶν κατὰ φύσιν, ἀλλὰ τοῖς εὐτραφέσι μέν, τροφῆς δὲ φαύλης ὑπερβολῇ· διὸ γίνονται οὐ τῇ τυχούσῃ ἡλικίᾳ· οὔτε γὰρ παιδίοις οὔτε γέρουσι γίνονται ἴονθοι οὔτε τοῖς λεπτοῖς· τοῖς μὲν οὖν ἀνθρώποις ὥσπερ εἴρηται, οἱ ἴονθοι γίνονται ἔξω διὰ τὴν λεπτότητα τοῦ δέρματος· τοῖς δὲ ὑσὶ γίνονται μὲν δι' εὐτροφίαν διὰ πάχος δὲ τὸ τοιοῦτον ἔσω τοῦ δέρματος τρέπεται· εἰ δέ εἰσι δι' εὐτροφίαν, ὑπερβολῇ περιττώματος· ὥσπερ καὶ οἱ ἴονθοι φυσικαῖς ἡλικίαις γίγνονται, οὕτω τοῖς τοιούτοις τῶν ζῴων ἐγγίνονται αἱ χάλαζαι, ὅταν γένηται τὸ ἔτος ἔγκαιρον· τότε γὰρ ἥκιστα δύναται πέττειν ἡ ὗς τὴν τροφήν, τὸν δὲ ἄλλον χρόνον ῥᾷστα πέττει τῶν ζῴων· διὸ καὶ πιότατόν ἐστι καὶ τὴν προχώρησιν

|| *καὶ πίνει*] *καταπίνει* T || *προσφερομένων* Us.: *προσφερόμενα* codd. || *χαίρειν* O[ac]

2.138 (AFKMOPRSTbei*p* Buss. Us.; *v* 3(5).8; 2.135 Buss.)

1 *πιαινόμεναι*] *αι* add. sup. lin. O: *πιαινόμενοι p* || *δυσριγώτεραι* F: *δυσριγότεροι p* 1-2 *καὶ οἱ παχεῖς ἄνθρωποι*] *καὶ* APT: *οἱ παχεῖς ἄνθρωποι* Oi: *καὶ οἱ ἄνθρωποι οἱ παχεῖς* F Buss. || *οἱ παχεῖς – πορρωτέρω*] *οἱ ἄνθρωποι* * * * *πόρρω* tantum *p* (lacunam statuit Duffy) 2 *πορρώτερον* K 3 *τοῦτο πάσχουσιν* AT[ac]: *πάσχωσι τοῦτο* K: *πάσχουσι τοῦτο* Oi 4 *αἰσθάνεται* Oi

2.139 (AFKMOPRSTbei Buss. Us.; *v* 3(5).9; 2.136 Buss.; paraphrasin tantum habet *p*)

1 *αἱ*] *οἱ* Oi 2 *ληπτέον*] *ρηπτέον* sic A || *δὴ*] *τί* T || *χαλάζωσις* A Buss. 3 *ἀνθρώποις* om. T || *ἴινθοι* Se[mg] || *τοῦτο ταῖς ὑσὶν*] *τοῦταις ἡσὶν* sic A 4 *μετεστραμένων*, *μ* tert. add. sup. lin., A || *οἴονθοι* MR: *ἴινθοι* Se[mg] || *δοθιῆνες*] *δοθῆναι* O[ac]: om. F 5 *διεσπαρμένοι*] *διεστραμμένοι* KOi*v* || *πολλῶ* K*v* 7 *οἱ* om. F(K add. sup. lin.) || *ἴινθοι* e[mg] 8 *εὐστραφέσι* MR: *ἐστραφέσι* e || *δὲ* om. KOSbi || *ὑπερβολὴ* KMOei 10 *ἀνθρώποις* om. O || *οἱ ἴονθοι* KOSbi Buss. Us.: *οἴονθοι* A: *ἴονθοι* cett. 11 *δέρματος*] *αἵματος p* || *εὐτραφίαν* KOPSbi 11-12 *διὰ πάχος – δι' εὐτροφίαν* om. A 12 *πάχος*] *τάχος* K || *εὐτραφίαν* KOSbi 13 *ὑπερβολὴ* KOSbei || *καὶ*] *δὴ* M Us.: *δὲ καὶ* K || *οἱ* om. FKMORSbei 14 *αἱ* om. AF Buss. 15 *ἡ ὗς* Us.: *ὗς* FKObi*v*: *μῦς* MPRSTe: om.

<Pigs>

2.138. Why are pigs when they are fattened more sensitive to cold? Is it because fat people too, the fatter they get, are further away from their own proper heat, and it is necessary that, the more this happens to them, the more easily cooled the outermost [parts] are, and what is easily cooled necessarily is more swiftly aware of the cold?

2.139. Why does neither man nor any other of the animals suffer from tubercles in the flesh, but pigs do? We must grasp what the affliction is, what having tubercles in the flesh is. As eruptions on the face are for humans, so is having tubercles in the flesh for pigs, but changed round in this way. Eruptions are little boils, and for this reason they are many and scattered; for it is more difficult and is the work of more residue to create many big ones, so that before they were distributed up in this way another disease would be stirred up. Eruptions occur neither as a weakness nor as [a characteristic] of those in a natural condition; rather in those who are fat,[449] but through an excess of poor nourishment. And for this reason they do not occur at just any time of life; for neither children nor old people suffer eruptions nor thin people. So in humans, as was said, eruptions occur externally because of the rare texture of their skin; in pigs they occur because they are fat, but because of their fat, this sort of thing turns inside the skin. If they are because [the pigs] are fat, they are from an excess of residue. As eruptions too occur at a natural age, so tubercles in the flesh occur in these animals when the year is seasonable; for at that time the pig is least able to concoct its nourishment, while for the rest of the time it concocts it most easily of any animal; and for this reason it is fattest and

[449] Literally "well-nourished". The contrast is between animals that put on weight because they have and absorb just the right amount of good nourishment, and those that do so because of an excessive amount of poor nourishment; both will put on weight, and so can be described as "well-nourished", but in the latter the excess of poor nourishment gives rise to larger amounts of residue. See below, n.451.

ἐξηλλαγμένην ποιεῖται πλείστην· σημεῖον δ' ὅτι χαλαζώδεις εἰσὶν αἱ ὑγρόσαρκοι τά τε περὶ τὰ σκέλη καὶ τὸν τράχηλον καὶ τοὺς ὤμους ἐν οἷς μέρεσι καὶ πλεῖσται γίνονται χάλαζαι· κἂν μὲν ὀλίγας ἔχῃ, γλυκυτέρα ἡ σάρξ γίνεται διὰ τὸ εὐτροφίας εἶναι σημεῖον, ὥσπερ εἴρηται· ἐὰν δὲ πολλὰς ὑγρὰ λίαν καὶ ἄχυμος· ἄπεπτος γὰρ ἡ τροφὴ ἡ λίαν πολλὴ καὶ οὐ κρατεῖται ὑπὸ τῆς φύσεως· ἐκβάλλουσι γὰρ τὰς χαλάζας μάλιστα ταῖς τιφαῖς· ἔστι δὲ τοῦτο καὶ πρὸς τὴν τροφὴν χρήσιμον· ὥσπερ δὲ οὐδὲ τοῖς παιδίοις γίνονται οἱ ἴονθοι οὐδὲ γέρουσιν, οὕτως οὐδὲ τοῖς χοίροις τοῖς γαλαθηνοῖς χάλαζα οὐδὲ τοῖς νέοις εἰ μὴ σπάνιον καὶ τότε ἐὰν ἡ μήτηρ πολλὰς ἔχῃ· οὐ γίνονται δὲ ταῖς λίαν γραίαις ὑσίν, ἀλλὰ ταῖς παλαιοτέραις.

2.140. *Διὰ τί οἱ ὕες μάλιστα τῶν ζῴων ἀφρὸν ἀφιᾶσι καὶ ὀχεύοντες καὶ μαχόμενοι καὶ πορευόμενοι, καὶ οἱ ἐνόρχεις μᾶλλον ἢ οἱ ἐκτομίαι; ἢ ὅτι στενὸν ἔχει τὸν βρόγχον, ὡς δηλοῖ ἡ φωνή, καὶ φύσει πίων ἐστίν, ὥστε πυκνὸν τὸ πνεῦμα ὂν καὶ πολὺ δι' ὀλίγου ἔρχεται; διὰ μὲν οὖν τὸ ὑγρὸν εἶναι σιαλοχοεῖ θερμαινόμενον, διὰ δὲ τὸ στενόπνουν ἀσθματικόν· ἀφρὸς δ' ἐστὶ πνεύματος πυκνοῦ ἄνοδος ἐν ὑγρῷ, διὸ καὶ οἱ ἐπιληπτικοὶ ἀφιᾶσιν ἀφρόν· κατεχόμενον γὰρ τὸ πνεῦμα ἐν τῷ πνεύμονι πνιγόμενον ἀνάγει βίᾳ τὸ ὑγρόν· μαχομένοις μὲν οὖν ὁ θυμός, θέουσι δ' ἡ κίνησις, ὀχεύουσι δ' ἡ προθυμία ἐκθερμαίνει τὸ πνεῦμα καὶ ποιεῖ πυκνόν, τὸ δὲ πάχος διὰ τὴν στενότητα ποιεῖ πνιγμόν, ὁ δὲ πνιγμὸς ἀφρόν.*

A Buss. 16 *τὴν προχώρησιν*] *προχώρησιν* KOi: *πρὸς τὴν χῶρισην* sic A 17 *πλείστην* om. F, del. Us. || *αἱ*] *οἱ* F 18 *ὑγόσαρκοι* typorum errore Us. || *τὸν* om. A 18-19 *τά τε περὶ τὰ σκέλη – μέρεσι* om. F 19 *ὀλίγοις* M 20 *εὐτροφία* AT: *εὐτραφίας* KOSbi || *δὲ* om. KORSi || *πολλὰς* Buss. Us.: *πολλαὶ* T: *πολλῇ* sic A: *πολλὰ* cett. 21 *καὶ ἄχυμος*] *ἢ καὶ ἄχυμα* K || *ἄπεττος* Oi 22-23 *ταῖς τιφαῖς* T Us.: *ταῖς τίφαις* Buss.: *τὰ στίφα* A: *ταῖς στιφαῖς* FP: *ταῖς τροφαῖς* KOSi*v*: *ταῖς τρυφαῖς* MRe: *ταῖς φαῖς* b 23 *ἔστι*] *ἔτι* Oi || *δὲ*[2]] *δὴ* F || *οὐδὲ*] *οὐ* KOi*v* 24 *οἱ ἴονθοι*] *ἴονθοι* A Buss.: *ἴινθοι* S || *χοίροις*] *γέρουσιν* S[ac] 25 *γαλαθηνίοις* F: *γαλαθινίοις* T: *γαθηνοῖς* Oi || *χάλαζα* om. A: *χάλαζαι* FK*v* 26 *πολλὰς ἔχει* A: *ἔχῃ πολλὰς* F || *δὲ οὐδὲ* K || *ὑσὶν*] *εἰσὶν* Oi: om. K

2.140 (AFKMOPRSTbei*p* Buss. Us.; *v* 3(5).10; 2.137 Buss.)

1 *τῶν ἄλλων ζώων* F || *ὀχένοντες* Oi 2 *μαχόμενοι*] *βαχόμενοι* MRe[ac] || *οἱ ἐνόρχεις*] *οἱ ἐνόρχοις* A: *οἱ ἐν ὄρχοις* Oi: *ἐνόρχοις* K || *οἱ*[2] om. F 3 *στενὸν* om. F || *ἔχει*] *ἔστιν* sic A || *τὸ βρόχον* A || *βρόγχον ὁ ὗς* FP: *βρόγχον ἡ ὗς* AT || *ὡς* MOSb[pc]e[pc]i Us.: om. cett. || *δηλοῖ ἡ φωνὴ* om. K*v* || *φύσιν* MRe || *πίων*] *ποῖον* T 4 *δι' ὀλιγου* per *sensim* male vertit *v* || *διέρχεται* F || *οὖν* om. AP || *ὑγρὸν*] *πνεῦμα* F 5 *σιαλοχοεῖ*] *σίαλον χέεται* A || *δὲ*[1] om. A || *στενόπνουν*] *στενόπουν* FKOSbe: *στενόνπνουν* i: *στενόπορον p* 7 *πνιγόμενων* A per errorem ut saepe 8 *δ' ἡ*] *δὲ* (bis) A Buss. 9 *καὶ ποιεῖ*] *ποιεῖ* PT: om. A || *πυκνὸν*] *θερμὸν* T[ac] 10 *στενότητα*] *θερμότητα* M[ac]

changes most of what it consumes. An indication is that it is the [pigs] with moist flesh[450] that have tubercles, and the parts around the legs and neck and shoulders, in which parts the greatest number of tubercles occur. If there are fewer, the flesh is sweeter, because they are a sign of being fat, as has been said; if there are many, it is excessively moist and tasteless, for the excessive nourishment is unconcocted and not mastered by nature.[451] They get rid of the tubercles mostly by [feeding] einkorn (*Triticum monococcum* L.); this is also useful for nourishment.[452] As neither children nor old people get eruptions, so neither do sucking-piglets get tubercles, nor the young ones except rarely and if the mother has many; nor do they occur in excessively old pigs, but in those that are rather old.

2.140.[453] Why do pigs most of [all] animals emit foam when mating and fighting and travelling, and entire ones more than castrated ones? Is it because [the pig] has a narrow windpipe, as its voice shows, and is naturally fat, so that its breath is frequent and passes in large quantity through a narrow space? So because it is moist it salivates when it grows hot, and because it has a narrow windpipe it pants; and foam is the rising up of much breath in moisture, which is why epileptics too emit foam,[454] for the breath is trapped in the lung and, being choked, brings up the moisture forcibly. So when [the pigs] fight it is their anger, when they run the movement, and when they mate the desire that heats up the breath and makes it frequent, and the fatness causes choking because of the narrowness, and the choking [causes] the foam.

---

[450] Reading *hugrosarkoi* with Ar., *HA* 8.21 603b17. This sentence and the next (as far as "tasteless") = Ar., *HA* 8.21 603b16-20; cf. Richter, 1885, 31.

[451] 2.152 below, saying that pigs with tubercles are not fat, is only superficially in contradiction with this, for the term there is not *eutrapheis* but *pakheiai*, implying much dense or sturdy flesh. A pig can grow large while having watery and tasteless flesh; in that case it is *eutraphes* (above, n.449) but not *pakhus*.

[452] Both these statements about einkorn from Ar., op. cit. 603b25-26.

[453] See below, 2.143.

[454] Hippocrates, *Morb. sacr.* 10 explains foaming at the mouth in epileptics rather by the breath *not* being able to enter the lung, which therefore "foams and boils as if dying".

2.141. Διὰ τί ὑσὶ καπρώσαις ἐάν τις προσχέῃ ὄξος εἰς τὰ αἰδοῖα, λήγουσιν ἐπὶ χρόνον τινὰ τῆς κάπρης; πότερον ὅτι διὰ τὸν πόνον ἐκκόπτει τὸ ἀλγῆσαν τὴν ἐπιθυμίαν; δάκνει γὰρ προσχεόμενον. ἢ ὅτι θερμότης αἰτία τοῦ πάθους καὶ ἡ φλεγμονή, εἰ ὑπὸ θερμοῦ γίνονται ⟨τὰ⟩ τῆς ἐπιθυμίας, τὸ δ' ὄξος σβέννυσι τὴν θερμότητα, τουτέστι τὴν αἰτίαν τοῦ πάθους;

2.142. Διὰ τί τὰ μὲν ἄλλα τετράποδα οἷον ἵπποι καὶ τὰ ὑποζύγια ἐκβάλλουσι τοὺς ὀδόντας καὶ κύων περὶ ἑπτάμηνον καὶ αἶγες καὶ πρόβατα καὶ ἄνθρωπος, ὗς δὲ οὐκ ἐκβάλλει; ἢ ὅτι πλατὺ ῥύγχος ἔχει καὶ ἰσχυρόν; ἢ ὅτι τοῖς ὀδοῦσι τοῖς ἐμπροσθίοις, οὕσπερ βάλλουσι μόνους, οὐ πονεῖ μόνον, τῷ γὰρ ῥύγχει ἐργάζεται.

2.143. Διὰ τί ἡ ὗς ὀχευθεῖσα ἀφρὸν ἀφίησιν ἐκ τοῦ στόματος; ἢ ὅτι στενὸν τὸν φάρυγγα καὶ τὰς σύριγγας ἔχει, διὸ καὶ ὥσπερ πνιγομένης αὐτῆς ἡ φωνή ἐστιν, ὅταν οὖν ἐν τῇ ὀχείᾳ διὰ στενῶν ἐπανίῃ τῶν πόρων ⟨τὸ πνεῦμα⟩· καὶ μεθ' ὑγρότητος, ἀφρὸς γίγνεται· μάλιστα γὰρ ἐν τοῖς τοιούτοις ἀφρίζει.

2.144. Διὰ τί αἱ ἄγριαι ὕες ἔλαττον τίκτουσι τῶν ἡμέρων; ἑπτὰ γὰρ τὸ πλεῖστον τίκτουσιν, αἱ δὲ τρισκαίδεκα. ἢ ὅτι πλεονάκις τίκτουσιν αἱ ἥμεροι διὰ τὴν εὐπορίαν τῆς τροφῆς καὶ τὴν ἀλέαν τῶν τόπων, δι' αὐτὸ δὲ τοῦτο πλείω τίκτουσιν;

2.141 (AFKMOPRSTbei*p* Buss. Us.; *v* 3(5).11; 2.138 Buss.)
1 *προχέῃ* FP: *προχέει* A*p* || *ὄξος προχέῃ* F || *λήγουσιν*] *λέγουσιν* Oi 2 *τῆς κάπρης*] *τῆς κάπρου* FM: *τῆς καπρίας* K: *τοῖς κάπρης* Oi: *a coitu v* || *ὅτι* om. AFP 3 *προσχεόμενων* A 4 *καὶ ἡ φλεγμονὴ* Us.: *καὶ φλέγμα* F Buss.: *καὶ η φλέγμα* T: *ἢ φλέγμα* A: *καὶ ἢ φλέγμα* AMOPRSbei: om. K*v* || *εἰ ὑπὸ θερμοῦ* Buss. (in var. lect.): *οἱ δ' ὑπὸ θερμοῦ* FS: *οἵ δ' ὑπὸ θερμοῦ* ORbi: *οἱ δ' ὑποθερμοῦ* Te: *οἱ δ' ὑπόθερμοι* M: *ὑπὸ θερμότητος γὰρ* K*v*: *οἱ δ' ὑπὲρ θερμοῦ* AP: *δ' * * ὑπέρθερμοι* Us. || *τὰ* add. Buss. (in var. lect.)
2.142 (AFKMOPRSTbei*p* Buss. Us.; *v* 3(5).12; 2.139 Buss.)
1 *ἱπποζύγια* Oi 2 *ἐκβάλλουσι*] *dentiunt* perperam *p* 3 *ἄνθρωποι* AF Buss. Us. || *οὐκ ἐκβάλλει*] *edentulus est* perperam *p* || *ἢ*[2] om. AFMPRTe[ac] 4 *μόνους*] *μόνον* F || *οὐ πονεῖ μόνον*] *οὐ ποιεῖ μόνον* KOi: *οὐ πομονεῖ μέσον* sic e
2.143 (AFKMOPRSTbei*p* Buss. Us.; *v* 3(5).13; 2.140 Buss.)
1 *ἀφίησιν*] *ανίεισιν* sic A: *ἀνίησιν* PT || *ἢ* A Buss. Us.; om. cett. || *στενῶν* A 2 *ἔχει* KOSbi*p* Us.: add. sup. lin. P: post *φάρυγγα* exhibent MRe Buss.: om. AFT 3 *ὅταν μὲν οὖν* T || *διαστενῶν* OTei || *τὸ πνεῦμα* add. Us.
2.144 (AFKMOPRSTbei*p* Buss. Us.; *v* 3(5).14; 2.141 Buss.)
1 *ἔλαττω* Buss. (in var. lect.) || *ἡμερῶν* O: *ἡμετέρων* F 1-2 *τὸ πλεῖστον τίκτουσιν*] *πλεῖστον τίκτουσιν* Oi: *τίκτουσι τὸ πλεῖστον* MRe 2 *ἢ* om. A 3 *διὰ τὴν ἀλέαν* F || *τοῦτο* om. A 4 *πλεῖον* A

2.141. Why if one applies vinegar to the genitals of sows that are on heat do they cease to be so for a time? Is it that, because of the pain, the discomfort removes the desire? For it bites when it is poured on. Or is it that heat is the cause of the condition and the inflammation, if it is by heat that <what is characteristic> of desire comes about, and the vinegar quenches the heat, that is, the cause of the condition?

2.142. Why do other four-footed animals, like horses and yoke-animals, shed their teeth, and dogs at about seven months and goats and sheep and humans, but the pig does not?[455] Is it because it has a flat and strong snout? Or because it does not work only with its front teeth, the only ones which they shed? For it uses its snout.

2.143.[456] Why does the sow when mounted emit foam from its mouth? Is it because it has a narrow throat and bronchial passages, and for this reason its voice is as if it were being stifled? So when, in the course of being mounted, it emits breath through narrow passages accompanied by moisture, foam is produced. It is in such circumstances that it foams most.

2.144. Why do wild pigs bear fewer young than tame ones? For they bear seven at most, but [tame ones] thirteen. Is it because tame ones bear young more often because of the abundance of food and the heat of the places [where they are kept], and for this same reason they bear more young?

---

[455] Cf. Ar., *HA* 2.1 501b4-5 (Richter 1885, 31) and *GA* 5.8 788b15; but the claim is false (Peck 1942, 554 n. (a)).

[456] See above, 2.140.

2.145. Διὰ τί ἄρρενες μὲν ὕες ἄγριοι πολλοί εἰσι, κάπροι δ' ὀλίγοι; ἢ ὅτι κνησμὸς νέοις οὖσιν ἐμπίπτει εἰς τοὺς ὄρχεις διὰ μέγεθος; ξυόμενοι οὖν πρὸς τὰ δένδρα ἐκθλίβουσι καὶ γίνονται οὓς καλοῦσι χλούνας, ὅθεν καὶ Ὅμηρος πεποίηκεν "ὦρσεν ἔπι χλούνην σῦν ἄγριον".

2.146. Διὰ τί ὕες πιαινόμεναι λειποτριχοῦσιν; ἢ ὅτι πιαινόμεναι σαρκοῦνται μᾶλλον, τοιαῦται δὲ πέττουσι ⟨μᾶλλον⟩, μᾶλλον δὲ πέττουσαι ἔλαττον τὸ περίττωμα λείπουσιν, αἱ δὲ τρίχες ἐκ τῶν περιττωμάτων γίνονται;

2.147. Διὰ τί οἱ ἄγριοι ὕες ὡς ἐπὶ τὸ πλεῖστον πάντες μέλανές εἰσιν, οἱ δὲ ἥμεροι οὔ; ἢ διότι οἱ μὲν ἀεὶ ἐν τοῖς ἕλεσιν ὄντες ὑπὸ τοῦ βορβόρου μέλανες γίνονται (σημεῖον δ' ὅτι βάπτει ὁ βόρβορος· καὶ γὰρ αἱ βακτηρίαι λευκόταται κατορυττόμεναι εἰς αὐτὸν μέλαιναι γίνονται)· τῶν δ' ἡμέρων ὀλίγοι οὕτω διάγουσιν.

2.148. Διὰ τί αἱ ὕες ἐὰν μὴ φθέγγωνται θηλαζόμεναι, οὐ προΐενται γάλα; διὸ καὶ οἱ βουλόμενοι βδάλλειν αὐτὰς κεντοῦσιν· ἢ ὅτι ἀποβιάζονται τῷ πνεύματι πάντα, καὶ φθέγγεται πάντα κατέχοντα τὸ πνεῦμα· τὸ δὲ τῆς ὑὸς γάλα οὐ ῥᾳδίως ἐκκρίνεται (ὀλίγον γὰρ ὂν εἰς πολλοὺς διέσπασται μαστούς), ὥστε κατὰ τὴν ἔξοδον δεῖται βίας.

2.145 (AFKMOPRSTbei*p* Buss. Us.; 2.142 Buss.)
1 ἄγριοι post εἰσι F 2 κνισμὸς T || ἐμπίπτοι T || ξυόμενοι] ξυόμενει e[ac]: διωκόμενοι F || οὖν] γοῦν KOi*p* 3 ἐκθλίβουσι Us.: ἐκφεύγουσι codd. *p* Buss. 4 πεποίηκεν om. P || ὦρσεν Homerus Buss. Us.: ὦσεν KMOPRSbei: ὡς FT: ὃς A || χλούνην σῦν] χλούνι σιν sic A: χλούνη σῦν FP: χλοῦνι σῦν T
2.146 (AFKMOPRSTbei*p* Buss. Us.: *v* 3(5).15; 2.143 Buss.)
1 αἱ ὗες T || πιαινόμεναι[1]] πιαινόμενοι P Buss. || λιποτριχοῦσιν Buss. 2 τοιαῦτα Oi || μᾶλλον[2] add. Us. || πέττουσι P
2.147 (AFKMOPRSTbei*p* Buss. Us.; *v* 3(5).16; 2.144 Buss.)
1 οἱ om. F || ἄγροι O[ac] || ὕες om. O || πάντες post ὕες AFT Buss. 2 οἱ μὲν FPT Buss. Us.: ἡ μὲν A: αἱ μὲν KMORSbei || ἀεὶ ἐν KOSbi*v* Us.: ἀεὶ διὰ τὸ ἐν AFMPRe Buss.: οἱ ἄλλοι διὰ τὸ ἐν T || ὄντες Us.: εἶναι AFMPRTe(*p*) Buss.: οὖσαι KOSbi || μέλανες FP Buss. Us.: μέλαιναι cett. 3 βάπτει] βλάπτει KMOi (sed non *v*) || γὰρ om. MRe Buss. 3-4 σημεῖον – γίνονται om. A 4 κατορυττόμεναι] ὀρυπτόμεναι O || δ' εἰς FKMORSbei Buss. 5 ὀλίγαι FT
2.148 (AFKMOPRSTbei*p* Buss. Us.; *v* 3(5) 17; 2.145 Buss.)
1 ὗς T || φθέγγονται Ae*p* || θηλαζόμεναι post ὕες KOi || προΐονται T*p* 2 βδάλλειν] βγάλλειν e[ac]: βδάλειν AS || κεντοῦσιν Buss. Us.: κινοῦσιν codd. *v* 2-3 ἀποβιάζονται πάντα τῷ πνεύματι S 3 κατέχονται T 4 ἐκρίνεται T[ac] || τε γὰρ AFPT || διέσπασθαι S: διέσπαται e 5 μασθοὺς MPRSbe || διέξοδον KOi

2.145. Why are there many wild male pigs, but few [entire] wild boars? Is it because when they are young their testicles itch because of their size? So rubbing them against the trees they squeeze them out and become what they call castrated boars, as Homer too wrote in his poem "she sent against [him] a savage castrated boar".[457]

2.146. Why do pigs that are fattened lose their hair?[458] Is it because when they are fattened they grow more flesh, and growing more flesh they concoct [their food] more, and concocting it more they leave less residue, and hairs come from residues?

2.147. Why generally are all wild pigs black, but tame ones not? Is it because, always being in the marshes, they become black because of the mud (an indication is that mud dyes; for if the whitest sticks are buried in it they become black), but few tame ones live thus?

2.148. Why do sows not give milk if they do not make a sound when they are sucked? (And for this reason those who want to milk them prick them.) Is it because all [creatures] use breath when employing force, and all [creatures] make a sound when they restrain their breath? The sows' milk is not given out easily (for there is not much of it and it is drawn into many separate teats), so it needs force to be given out.

---

457 *Iliad* 9.539, of the Calydonian boar. For this alleged habit of boars cf. (with Usener 1859, x-xi; Rose 1863, 219) Ar., *HA* 6.28 578a32-b5 (where the itching is said to be due to disease), Plutarch, *Quaest. Nat.* 21 917d, [Aristoph. Byz.] *HA Epit.* 2.570 143.9-12 Lambros. Bussemaker xviii argued that the passage in Plutarch, which mentions Aristotle by name, is a reference to the present *Problem*, but Flashar 1962b, 360 that the reference in Plutarch is to the *HA* itself. Neither text actually contains the statement made by Plutarch that Aristotle says that the Homeric term applies to boars with only *one* testicle; cf. Pearson and Sandbach 1965, 197 nn. b,c, and also Flashar 1962b, 307; Sandbach 1982, 225, on the possibility that Plutarch's source is the original Aristotelian *Problems* from which the Bekker *Problems* derive. Roselli 1979, 11 argues that the citation of only this line in the present problem supports the contention of Luppe 1970, 31-34 that the line that follows it in the text of *HA* and is based on *Odyssey* 9.191 is a later addition there.

458 Prantl 1851, 366 n.71 notes that this conflicts with the statement in [Ar.], *Probl.* 10.21 that one reason why pigs do not shed their winter hair may be that the amount of fat they have prevents them shedding their hair for any reason.

2.149. Διὰ τί τῆς ὑὸς οἱ πρῶτοι μαστοὶ οἱ πρὸς τὸ στῆθος εὔκρατοι πρὸς τὸ βδάλλειν καὶ οἱ ἐφεξῆς κατὰ λόγον, διὸ καὶ ὁ γεννώμενος εὐθὺς τῶν χοίρων ἐπὶ τὸν πρῶτον ὁρμᾷ καὶ ὁ ἀεὶ ἐχόμενος κατὰ λόγον, καὶ θηλάζουσιν οὕτω γε, ἐὰν μὴ ὁ κρείττων βιάσηται; πότερον ὅτι θερμότατος ὁ τόπος περὶ τὰ στήθη καὶ αἴτιος καὶ τοῖς ἄλλοις τῆς θερμότητος (ἀρχὴ γὰρ οὗτος); ἐπεὶ δ' ἕλκεται πλείστην τροφὴν ὁ θερμότατος καὶ πέττει μάλιστα, οὗτος δὲ μάλιστα τοιοῦτος, ἀνάγκη πλεῖστον γάλα καὶ γλυκύτατον ἐνταῦθα εἶναι. ἢ οὐ μόνον τοῦτο αἴτιον, ἀλλ' ὅτι τὸ μὲν γάλα ἐστὶν αἷμα πεπεμμένον, ὁ δὲ τόπος ὁ περὶ τὰ στήθη αἷμα ἔχει πεπεμμένον καὶ καθαρώτατον (ἐνταῦθα γὰρ αὐτοῦ ἡ ἀρχή), ὥστε ἐν ᾧ τὸ αἷμα καθαρώτατον, ἐν τούτῳ τὸ γάλα ἄριστον εἶναι.

2.150. Διὰ τί ἡ ὗς χαίρει τοῖς δυσώδεσιν; ἢ διότι τὰ πλατύρυγχα οὐκ αἰσθητικὰ τὴν ὄσφρησιν, τὰ δὲ μὴ αἰσθητικὰ τοῖς δριμυτέροις χαίρει; τούτων γὰρ μάλιστα μόνων αἰσθάνεται· ταὐτὸ δὲ πάσχουσι καὶ οἱ διεφθαρμένοι τὴν ὄσφρησιν ὑπὸ δυσωδίας οἷον σκυτοδεψοὶ καὶ κοπρολόγοι.

2.151. Διὰ τί αἱ μὲν θήλειαι τῶν ἀρρένων * * * ὥσπερ καὶ ἐν τοῖς ἄλλοις γένεσι τῶν ζῴων δειλότερον τὸ θῆλυ τοῦ ἄρρενος πλὴν ἄρκτου καὶ παρδάλεως· ὥστε διὰ τὸν φόβον κατ' ἐνίους θέουσι μᾶλλον ὥσπερ οἱ

2.149 (AFKMOPRSTbei*p* Buss. Us.; *v* 3(5).18; 2.146 Buss.)
1 *υἱὸς* A || *μασθοὶ* MPRSbe || *εὔκρατει* R 2 *βδάλλειν*] *βγάλλειν* e^{ac}: *βδάλειν* A || *κατὰ λόγον*] *καταλόγον* T: *κατ' ὄλιγον* A || *γεννώμενος*] *γεννηθεὶς* *p*: *γενόμενος* AP Buss. Us.: *qui se ad lactendum erigit v* 3 *ἐπὶ*] *πρὸς* P || *ὁ ἀεὶ* M Us.: *ἀεὶ* ⟨*ὁ*⟩ Buss.: *ἀεὶ* cett. || *καὶ*^{2} om. P 4 *γε* om. F || *κρείσσων* M || *θερμότατος*] *θερμότητος* Fe^{ac}: *θερμότερος* KOSbe^{pc}i || *ὁ περὶ* K*v* Us. 5 *τῆς ἄλλης* A || *τῆς*] *τοῖς* R || *ἐπειδὴ* Oi 6 *θερμότητος* e^{ac} || *πέττεται* P 7 *καὶ γλυκύτατον ἐνταῦθα εἶναι*] *εἶναι ἐνταῦθα καὶ γλυκύτατον* KOi 8 *ἢ ὅτι οὐ* P || *μόνον* om. *v* || *πεπεμένον* F^{ac}T^{ac}: *πεπαμμένον* P 9 *αἷμα ἔχει* APT Buss. Us.: *ἆσθμα ἔχει* F: *ἔχει αἷμα* KORSbei || *πεπεμένον* F^{ac}: *πεπαμμένον* P 10 *αὐτοῦ ἡ ἀρχὴ* AFR(*αὐτοῦ* scr. sup. lin. R)*p*: *ἡ αὐτοῦ ἀρχὴ* KOSbi: *αὐτῇ ἡ ἀρχὴ* e 11 *εἶναι*] *ἔστι* F

2.150 (AFKMOPRSTbei*p* Buss. Us.; *v* 3(5).19; 2.147 Buss.)
1 *ἡ* om. A Buss. || *τοῖς πλατυρύγχοις* A 2 *αἰσθητικὰ*^{1}] *αἰσθάνονται* A || *τὴν ὄσφρησιν* Buss. Us.: *τῆς ὀσφρήσεως* AFMPRTbe*p*: *τῆς αἰσθήσεως* KOi: *sensus odoris v* || *τὴν ὄσφρησιν, τὰ δὲ μὴ αἰσθητικὰ* om. S 3 *μάλιστα μόνων*] *μάλιστα μόνον* RT: *μόνων μάλιστα* KOi: *μάλιστα μόνος* e^{ac}: *tantum vel precipue v* || *ταὐτὸ*] *τοῦτο* A Buss.: *ταὐτὰ* P 4 *σκυτοδέψαι* e^{ac}: *σκυτοδεψεῖν* T

2.151 (AFKMOPRSTbei*p* Buss. Us.; 2.148 Buss.)
1 *μὲν* om. FK || lacunam statuit Us. 2 *δηλότερον* AObi || *τὸ θῆλυ δειλότερον* P || *τῆς ἄρκτου* F: *ἄρκου* T 3-6 *ὥστε – ἐστίν* om. *p* 3 *ὥστε*] *ὅτι* coni. Buss. in var.

2.149. Why are the teats of the sow adjacent to the chest well-tempered for milking and the subsequent ones in order, for which reason the piglet that has just been born[459] makes for the first one and the subsequent ones in order, and this is how they suckle, unless a stronger one forces its way in? Is it because the place around the chest is hottest and the cause of heat for the others (this being the principle), and since the hottest [place] attracts most nourishment and concocts it most, and this is most like this, it is necessary that there should be most, and the sweetest, milk here. Or is not only this the cause, but that milk is concocted blood, and the region around the chest has blood which is concocted and most pure (for this is where the principle [of blood] is), so that where the blood is most pure, there the milk is best?

2.150.[460] Why does the pig delight in things that smell bad? Is it because [creatures] with broad snouts do not perceive smells well, and [creatures] that do not perceive [smells] well take pleasure in what is pungent? For they more or less perceive only these. The same thing happens to those whose sense of smell has been destroyed by bad smells, like tanners and dung-collectors.

2.151. Why do the female [pigs] <make more sound> than the males? <Is it that>[461] as in other kinds of animal too the female is less courageous than the male, apart from the bear and the leopard?[462] So it is more on account of fear, according to some, that they run more, like those who swing their arms

---

459 *gennômenos* of the majority of MSS is supported by *p* and was probably the reading of α. *genomenos* of AP Buss. Us. is more strictly correct (it is being born that is at issue rather than being engendered), but the loose usage may go back to the original text.

460 Cf. below 2.156, and the second answer, attributed to Aristotle, in [Alex.], *Probl.* i.141 (the first, attributed to "the doctors", being that they naturally like dung because they have large livers); Rose 1863, 219. Bussemaker 1857, xviii and Usener 1859, x argues that in citing "Aristotle" [Alex.], *Probl.* i.141 is actually citing the present text and 2.156.

461 Our supplements *exempli gratia*. Ar., *HA* 6.28 578a31 comments that female wild swine grunt more than domestic pigs, wild boars less; and see below 2.168.

462 The same exceptions noted at Ar., *HA* 9.1 608a34: Richter 1885, 31.

παρασείοντες τὰς χεῖρας καὶ οἱ ἀποβάλλοντες τὰ ὅπλα κἂν τι ἄλλο βάρος ἔχωσιν, οἱ δ' ἄρρενες ὡς ποιήσοντές τι · ὅμοιον γὰρ τῷ στρέφειν τὴν οὐρὰν τὸ φρίττειν ἐστίν.

2.152. Διὰ τί χαλαζωθεῖσαι ὕες οὐ γίνονται παχεῖαι; ἢ ὅτι τῆς τροφῆς ἧττον κρατεῖ τὸ σῶμα αὐτῶν; σημεῖον δ' ὅτι περίττωμα πλεῖον γίνεται αὐταῖς· αἱ γὰρ χάλαζαι ἐκ περιττωμάτων εἰσί, πιαίνεται δὲ τὸ ἀπολαῦον τῆς τροφῆς μᾶλλον.

2.153. Διὰ τί αἱ ὕες ὕπτιαι φερόμεναι σιγῶσι μᾶλλον; ἢ ὅτι οὕτω φερόμεναι ἰσχύουσιν ἧττον φθέγγεσθαι μετὰ ἰσχύος· αἱ δὲ φύσει ἄφωνοι οὖσαι πλείονος ἰσχύος δέονται; ὅτι δὲ φύσει ἄφωνοι, δῆλον· εἰσὶ γὰρ ἔνιαι ἄφωνοι ὅλως ὕες.

2.154. Διὰ τί αἱ ὕες ἀφοδεύουσι πρότερον ἢ οὐροῦσιν; ἢ ὅτι καὶ τῶν ἀνθρώπων ὅσων εὔπεπτοι αἱ κοιλίαι; τῆς δὲ ὑὸς πεπτικὴ ἡ κοιλία μάλιστα.

2.155. Διὰ τί αἱ μὲν ἄγριαι ὕες ἅπαξ τίκτουσιν, αἱ δὲ ἥμεροι δίς; ἢ ὅτι ἐκείναις μὲν ἁπλῆ τροφὴ γίνεται, καὶ ἅπαξ βοτάνη, ταύταις δὲ ποκίλη καὶ πολλή⟨, διὸ δὶς τίκτουσι τοῦ ἐνιαυτοῦ⟩· ὥσπερ πλεονάκις δίς καὶ τὰ

---

lect. || *φόβων* T || *κατ' ἐνίους θέουσι* FP Buss. Us.: *κατενίους θέουσι* T: *κατ' ἐνιουσθένουσι* sic A: *κατ' ἐνίους* KMORSbei || *οἱ* om. KORSbei 4 *οἱ* om. A || *τι*] *καὶ* F 5 *ἔχουσιν* OT[ac]bi || *ποιήσονται* Oi || *τι* Us.: *τὸν* Oi: *τὸ* cett. || *τὸ γὰρ* K || *στρέφειν*] *τρέφειν* T[ac] 6 *φρίττον* OSbi

2.152 (AFKMOPRSTbei*p* Buss. Us.; *v* 3(5).20; 2.149 Buss.)

1 *χαλαζωθεῖσαι*] *χαλαζῶσαι* A[ac]: *χαλαζωθεῖσιν* F || *οὐ*] *οὐχὶ* A || *παχεῖαι οὐ γίνονται* F 2 *περιττώματα* F || *πλείω* AFPT 3 *αὐταῖς* add. in marg. O: *αὐτοῖς* FS: om. A || *αἱ γὰρ χάλαζαι ἐκ περιττωμάτων εἰσὶ*] *αἱ δὲ περιττωμάτων εἰσὶ σημεῖον, ἤγουν* (sic) *αἱ χάλαζαι* A: *αἱ γὰρ χάλαζαι περιττωμάτων εἰσὶ σημεῖον* Buss. || *περιττώματος* P 3-4 *μᾶλλον* post *δὲ* exhibent S*p*, et illuc transtulit e[pc]

2.153 (AFKMOPRSTbei*p* Buss. Us.; *v* 3(5).21; 2.150 Buss.)

1 *αἱ* om. F || *σιτῶσι* Oi 2 *ἰσχύουσιν* om. K*v* || *ἧττον* K Us.: *ἥττον* T: *ἥττονος* Buss.: *ἥττονα* cett. || *φθέγγεσθαι*] *φθέγγονται* K*v*: *φέρεσθαι* A || *φύσει*] *φείσει* O 3-4 *πλείονος ἰσχύος – ἄφωνοι* om. A 4 *ὅλως* Us.: *prorsus* *v*: *ὡς* codd. (scr. sup. lin. M): om. K Buss.

2.154 (AFMOPRSTbei*p* Buss. Us.; 2.151 Buss.)

1 *αἱ* om. A 2 *ὅσων*] *νόσων* A || *τῆς*] *τοῖς* A || *δὲ* om. A || *υἱὸς* S

2.155 (AFKMOPRSTbei*p* Buss. Us.; *v* 3(5).22; 2.152 Buss.)

1 *μὲν* om. A || *ἥμεραι* KMORSbei*p* 2 *ἐκείναις μὲν* F Us.: *ἐκείνοις μὲν* AKMORSbei Buss.: *μὲν ἐκείνοις* T: *ἐκείν*[...] P || *ἁπλῆ ἡ τροφὴ* Us. || *ταύταις* MP Us.: *τούτοις* cett. Buss. 3 *διὸ δὶς τίκτουσι τοῦ ἐνιαυτοῦ* huc transtulit Us, post *δίαιταν* (4) habent codd. *v* Buss. || *ὥσπερ* Us.: *ὥστε* codd. (*ὅστε* A ut saepe) *v* Buss. || *δὶς καὶ* Us.: *καὶ δὶς* A: *ἢ διὰ* Buss. (in var. lect.): *καὶ διὰ* FMORSbei: *καὶ*

beside them[463] and those who throw away their weapons and any other weight they are carrying, but the males [run] in order to do something; for shuddering is like turning one's tail.

2.152.[464] Why do pigs with tubercles in their flesh not grow fat? Is it because their body masters their nourishment less? An indication is that they have more residue; for the tubercles come from residue, and it is those that gain more benefit from their food that grow fat.

2.153.[465] Why are pigs more silent if they are carried upside-down? Is it because when they are carried in this way they have less strength to make a sound with strength, and being naturally without sound they need more strength. That they are naturally soundless is clear, for some pigs are completely without sound.

2.154. Why do pigs excrete solid waste before passing urine? Is it because humans too who have bowels that digest well [do this]? The bowel of the pig digests very well.

2.155. Why do wild sows bear once, tame ones twice? Is it because the former have simple nourishment, and just pasture, while the latter have much varied [nourishment], and for this reason bear twice in the year, as sheep too often [bear] twice because of a good diet?

---

463 Connected with running quickly by Ar., *IA* 3 705a17, *EN* 4.3 1123b31 (the latter in the context of fear), and [Ar.], *Probl.* 5.8 881b6.

464 See above, 2.139, and n.451.

465 Cf. [Alex.], *Probl.* i.140, where the answers are either that they are normally more turned towards the earth than other creatures because of their delight in ill-smelling things, or that their windpipes collapse when they are turned upside down.

πρόβατα διὰ τὴν εὖ δίαιταν [διὸ δὶς τίκτουσι τοῦ ἐνιαυτοῦ];

2.156. Διὰ τί ὗς τοῖς δυσώδεσι χαίρει; ἢ διὰ τὸ πλατὺν ἔχειν τὸν μυκτῆρα καὶ οὐκ ὀξύν, οὐκ αἰσθητικ(ός) ἐστι, διὰ (δ)ὲ τὸ (μὴ αἰσθητ)ικὸν (εἶν)α(ι διάκειται ⟨πράως⟩) π(ρ)ὸς τὰς ὀσμάς; ὥσπερ οἱ Φρύγες τοῖς φαύλοις οἴνοις χαίρουσι, καὶ αὐτὴ ταῖς διεφθαρμέναις ὀσμαῖς καὶ σαπραῖς· αἱ τοιαῦται γὰρ ἅπτονται δριμεῖαι οὖσαι τῶν μὲν Φρυγῶν ὡς μεμολυσμένων διὰ τὸ ἀεὶ πίνειν, τῶν δὲ ὑῶν διὰ τὴν παχύτητα τῶν πόρων.

2.157. Διὰ τί οἱ ἄτμητοι τῶν τομιῶν μείζους ἔχουσι τοὺς χαυλιόδοντας; ἢ ὅτι αἱ θήλειαι ἐλάττους τῶν ἀρρένων ἔχουσιν, οἱ δὲ ἐκτετμημένοι μεταβάλλουσιν εἰς τὸ θῆλυ τὴν φύσιν [ὥσπερ ἂν θήλεις ἐξ ἀρρένων γινόμενοι];

2.158. Διὰ τὶ τοῖς ὑσὶν οἱ χαυλιόδοντες γίνονται; πότερον ὥσπερ τινὲς λέγουσι, διὰ τὸ μὴ συμπίπτειν οὐ τριβόμενοι αὐξάνονται; ὥσπερ καὶ τοῖς ἀνθρώποις οἳ ἂν οὕτω φύωσιν ὥστε παραλλάττειν τοὺς ὀδόντας, πολὺ

διὸ PT: δύνανται τίκτειν ὥσπερ καὶ K 4 τὴν αὐ⟨τὴν⟩ δίαιταν Buss.: τὴν καλὴν δίαιταν K || τίκτει K

2.156 (AFMOPRSTbei 𝔓 Buss. Us.; 2.153 Buss.)

1 ἡ ὗς RSbe et 𝔓 ut vid. (cf. Luppe): ὗς cett. || πλατὴν AO || ἔχει T 2 καὶ om. A Buss. 2-3 οὐκ αἰσθητικὸν – ὀσμάς ex 𝔓 restituit Luppe: οὐκ αἰσθητικὸν εἶναι διάκειται πρὸς τὰς ὀσμάς tantum codd. || αἰσθητικὸς] αἰσθητὸς sensu activo auctorem ipsum scripsisse censuit Luppe || αἰσθητικὸν] αἰσθητὸν M Luppe (cf. supr.): αἰσθητικῶς Buss. (in var. lect.) || εἶναι om. b, del. Buss. in var. lect. 3 διάκειται] συμβαίνει MP Us.: διακεῖσθαι FOi || τὰς om. AFOi Buss. || οὖν post ὥσπερ add. Us. || φλυγες 𝔓 || τοῖς] τῆς A: om. O 4 αὐτὴ] αὐτοὶ AFOTi Buss. 5-6 δριμεῖαι οὖσαι – τῶν πόρων] ita codd.: plura olim habebat 𝔓 quae tamen legi non iam possunt 5 δριμεῖαι οὖσαι Us.: διτταὶ οὖσαι A: δίαι οὖσαι P: διαιοῦσαι T: λιταὶ οὖσαι Buss. (in var. lect.): om. cett. || τῶν μὲν φρυγῶν Se[pc] Us.: τῶν μὲν Tb Buss.: τῶν μέντοι AMPRTe[ac]: om. FOi || ὡς Sb[2] Us.: τοῖς AFOPi Buss.: τῆς T 𝔓: om. MRe || μεμολυσμένων Sbe[pc] Us.: μεμολυσμένης MRTe[ac] 𝔓: μεμολυσμένοις AFOPi Buss. 6 ὑον A

2.157 (AFKMOPRSTbei*p* Buss. Us.; *v* 3(5).23; 2.154 Buss.)

1 ἄτμητοι] ἄτρητοι K: om. F ||τομοιῶν Oi || *maiales* (i.e. *castrati*) *verribus maiores v* || ἔχουσι] εἰσὶν ἔχοντες F || χαυλιώδοντας MOei: καυλιόδοντας T 2 ἐλάττους] *maiores v* || ἔχουσαι Oi || ἐκτετρημένοι K 3 εἰς τὴν φύσιν τὸ θῆλυ F || ἂν θήλεις Us.: θήλειαι K*v*: ἀνήβεις Buss.: ἀνίβεις sic A: ἀήθης M: ἀνήθεις cett. 3-4 ὥσπερ – γινόμενοι del. Us. 4 γενόμενοι K*v*: γιγνόμενος e[ac]

2.158 (AFKMOPRSTbei*p* Buss. Us.; *v* 3(5).22bis(24); 2.155 Buss.)

1 ταῖς ὑσὶν OSbi || καυλιόδοντες T || γίνονται μεγάλοι *p* || πότερον] ἢ K: om. Oi 2 οὐ τριβόμενοι del. Us., exhib. *p* || ὥσπερ] ὡς καίπερ sic A 3 οἳ] οἱ F: οἷον MRe[ac] || ἂν οὕτω] ἐν τούτω F || φύωσιν P Buss. Us.: φύουσιν AT: φυῶσιν M: φιῶσιν sic

2.156.[466] Why does the pig take pleasure in things with a bad smell? Is it that, because it has a nostril that is wide and not keen, it is not able to perceive smells well; and because it is not able to perceive well it is mildly disposed towards smells?[467] So as the Phrygians take pleasure in inferior wines, so [the pig does] in corrupt and rotten smells; for these, being pungent, affect the Phrygians since they are polluted because they are always drinking, and the pigs because of the thickness of their [sensory] passages.[468]

2.157. Why do uncastrated [boars] have larger tusks than castrated ones? Is it because the females have smaller ones than the males, and those that have been castrated change into the female by nature, [as if becoming females instead of males]?[469]

2.158. Why do boars have tusks? Is it as some say that they do not meet, and grow because they are not worn down?[470] As with humans those who grow in such a way that their teeth slide past each other grow much bigger ones,

---

466 Cf. above, 2.150. This text appears, attributed to Aristotle's *Aporemata*, in an Oxyrhynchus papyrus (2744) of the 2nd century AD which Luppe 1970 argues contains a commentary by Didymus in the first century B.C.: cf. Roselli 1979 and Luppe 1979.

467 So the papyrus as restored by Luppe. The MSS give "... and not keen, it is not disposed to be able to perceive smells well."

468 In this last sentence the papyrus has a fuller, but mostly undecipherable text.

469 The last clause is bracketed by Usener and seems like a redundant gloss. "females" is in any case Usener's emendation. Bussemaker proposed "as if becoming immature instead of males", but the word he conjectured for "immature", *anhêbeis*, seems not to exist.

470 The last phrase is bracketed by Usener. See above, 2.134.

μείζους γίνονται, οἷον καὶ τῷ γαλακτοπότῃ Σωστράτῃ μεγάλοι ἐγένοντο διὰ τὸ μὴ τρίβεσθαι. ἢ ὥσπερ καὶ τὰ κέρατα φύσις ἀπέδωκεν ἀλκῆς χάριν καὶ βοηθείας, ⟨διὰ⟩ τοῦτο τοῖς ὑσὶ τοὺς χαυλιόδοντας; πάντα γὰρ ἔχει βοήθειαν ἀπὸ τῆς φύσεως, ὡσπεροῦν ἔνια μὲν κέρασιν, ἄλλα δ' ὄνυξι, τὰ δ' ὀδοῦσι, καὶ τούτων τὰ μὲν οὕτως ὥσπερ τὰ καρχαρόδοντα, τὰ δ' ἀντὶ τούτου τοὺς χαυλιόδοντας ἔχει· ἅμα δὲ ταῦτα ὑπάρχει οὐθενί· καὶ ἔστι μεταξὺ ταῦτα τῶν κερασφόρων καὶ καρχαροδόντων, ὥσπερ καὶ αἱ βάσεις πρὸς τὰς βάσεις. καὶ ἡ τροφὴ ἐκείνοις πρὸς τὰ κέρατα, τούτοις πρὸς τοὺς ἔξω ὀδόντας τρέπεται· σημεῖον δ' ὅτι χαυλιόδοντες ἀντὶ κεράτων εἰσὶ καὶ ⟨οὐχ⟩ ὅτι ἔτυχον οὐ τριβόμενοι διὰ τοῦτο μεγάλοι, ἀλλ' ἕνεκά τινος· οἱ γὰρ τῶν ἐλεφάντων λεγόμενοι ὀδόντες οὐκ εἰσὶν ὀδόντες ἀλλὰ κέρατα· ἀλλὰ [διὰ τὸ ἐγκεκλεῖσθαι] διὰ τὸ σχῆμα καὶ τὴν ὁμοιότητα καλοῦνται ὀδόντες, καὶ οἵ γ' ἐν Καρχηδόνι ὡς κέρατα οὕτως ὀνομάζουσιν· ὅτι δ' οὐκ ὀδόντες

Re: *φιλοῦσιν* Osbi: *ἔχωσιν* K*ν*: om. F 3-4 *πολὺ μείζους* Us.: *μείζους* K*ν*: *ἢ οὐ μείζους* cett. 4 *τῷ*] *τὸ* K ‖ *τῷ γαλακτιπότῳ* vel *τῇ γαλακριπότιδι* Buss. (in var. lect.) ‖ *Σωστράτῃ*] *σωκράτῃ* Ke[ac]: *Σωστράτῳ* vel *Σωστράτῃ* Buss. (in var. lect.): om. *ν* 5 *ἢ* K*ν* Buss. Us.: om. cett. ‖ *ἡ φύσις* K*ν* Buss.: *φυσίν* A 6-17 *καὶ βοηθείας – πάσχει* om. F 6 *διὰ* add. Us. ‖ *τοῦτο*] *οὕτω* K*ν* Buss. (in var. lect.) 7 *βοήθειαν ἔχει* A Buss. ‖ *ἀπὸ* A Buss. Us.: *ὑπὸ* cett. ‖ *ὡσπεροῦν* Us.: *οὕτως* K: *ὥσπερ* (*ὅσπερ* A) *οὖν* cett. ‖ *μὲν* om. A 8 *τὰ δ'*] *ἡ* A: *ἢ* Buss. ‖ *ὀδοῦσι ὥπλισεν* K*ν* ‖ *τούτων* om. A ‖ *ὥσπερ*] *ὡς* A Buss. Us. 9 *τούτων* A Buss. ‖ *ὑπάρχει οὐθενὶ* Us.: *οὐδαμόθεν* vel *ἀπαντᾷ οὐδαμόθεν* Buss. (in var. lect.): *πανταχόθεν* codd. *ν* 9-11 *καὶ ἔστι μεταξὺ – πρὸς τὰς βάσεις* del. Us. 10 *ταύτας* A ‖ *τῶν* om. A ‖ *κερατοφόρων* T 11 *ἐκείνοις* Buss. Us.: *κενοῖς* APT 11-12 *καὶ ἡ τροφὴ – τρέπεται* APT Buss. Us.: om. cett. *ν* ‖ *τούτοις πρὸς τοὺς ἔξω* A Buss.: *τούτους εἰς τοὺς ἔξω* T: *τοῦ εἰς τὸ ἔξω τοὺς* P, et ita quoque Us., qui lacunam ante *τοῦ* statuit 12 *τρέπεσθαι* Us. 13 *οὐχ* add. *ν* Buss. (in var. lect.) Us. ‖ *ὅτι* add. sup. lin. M ‖ *οὐ* M Buss. (in var. lect.) Us.: *μὴ* K: *οἱ* cett. ‖ *τριβόμενοι*] *renatos* v 14-17 *ἀλλὰ – πάσχει* APT Buss. Us.: om. cett. *ν* 15 *διὰ τὸ ἐγκεκλεῖσθαι* del. Us. ‖ *διὰ*[2] om. A Buss. 16 *χαρκηδόνι* AT 17 *ὃ οὐχ* Buss. Us.: *δ' οὐχ* APT

as Sostrates who lived on milk[471] had large teeth because they were not worn. Or is it that as nature has given horns for the sake of defence and assistance, so it has given tusks to boars? For all have assistance from nature, as some with horns, others with claws, and others with teeth, and of these some in the manner of saw-like teeth and others tusks instead of this; but none has these together.[472] And these [with tusks] are intermediate between those that bear horns and saw-teeth, as positions to positions;[473] and in those the nourishment is directed to the horns, in these to the external teeth.[474] An indication that tusks are instead of horns and are <not>[475] large just because they happen not to be worn down, but for a purpose, is that the so-called teeth of elephants are not teeth[476] but horns, but [[because they are enclosed]][477] on account of their shape and the similarity they call them teeth; the people in Carthage call them horns. That they are not teeth is clear

---

471 Called Sostratus in Lucian, *Demonax* 1, and Sosastrus in Plutarch, *Quaest. conv.* 4.1 660e. Cf. Kindstrand 1979-80, 51-52: Sharples 1995, 18-20.

472 Usener's emendation, following Ar., *PA* 3.1 661b23. The MSS have: "these are present together in every way".

473 "And these ... positions" is bracketed by Usener.

474 So MS A, followed by Bussemaker. Usener follows the reading of PT and postulates a lacuna: "... to the horns; * * * [the cause?] of the teeth turning towards the outside" [sc. in elephants], and this may well represent the original reading if A reflects an attempt to correct an already corrupted text.

475 Added by Usener.

476 As they *are* according to Ar., *HA* 2.5 501b32, and also in actual fact. But the point was much disputed: they are referred to as teeth by [Aristophanes of Byzantium], *Epit.* 1.7, 1.9, 2.132; Athenaeus 5.23 195A; Dionysius Periegeta 1116 (60 Bernhardy); Dioscorides, *Mat. Med.* 2.57 (1 138.17 Wellmann); schol. in Homeri *Odysseam* 19.563 (2 685.27-8 Dindorf), but as horns by Aelian, *NA* 8.10 (= [Aristophanes of Byzantium], *Epit.* 2.120), 10.12, 14.5 (= Ar. Byz. 2.129). Many authors are aware of both views: Aelian, *NA* 4.31, 11.37 (= Ar., fr.281 Rose^3; but there is no explicit attribution to Aristotle here, and the reference to horns is likely to be Aelian's own); Aretaeus, *Caus. acut. morb.* 2.13.3 (*CMG* 2 86.5-6 Hude); [Aristophanes of Byzantium], *Epit.* 2.112; Eustathius, *In Dionys. Perieget.* 1107 (307.3-5 Bernhardy); Eutecnius, *in Opp. Cyn.* 27.33-4 Tüselmann; Pausanias 5.12.1-3; schol. in Pindar, *Ol.* 3 52a (1 120.18-19 Drachmann). Of these Aelian at *NA* 11.37, Eutecnius and Pausanias regard "horns" as the correct description; the others refer to both but with no explicit preference.

477 Bracketed by Usener.

εἰσί, δηλοῖ τὸ κάμπτεσθαι καὶ γλύφεσθαι, ὃ οὐχ ἕτερον τούτων πάσχει.

2.159. Διὰ τί ποτε αἱ ὕες φρίττουσιν, ὅταν κυνηγῶνται; ἢ ὅτι τῶν ἀγωνιώντων τὸ σῶμα συστέλλεται διὰ τὸ συνάγεσθαι ὑπὸ τῆς προθυμίας καὶ βοηθεῖν αὐτῷ; καὶ αἱ τρίχες οὖν μᾶλλον ἑστήκασιν, εἰ μὴ καταπεπτώκασι διὰ τὸ ἐκλελύσθαι.

2.160. Διὰ τί τῶν μὲν προβάτων τὰ λεπτότερα λεπτοτριχώτερα, τῶν δὲ ὑῶν αἱ λεπτότεραι παχυτριχώτεραι; ἢ διότι αἱ μὲν διὰ τὸ φύσει εἶναι παχεῖαι ἐμπίπλανται τοὺς πόρους ὅταν παχυνθῶσι, δι' ὧν ἐκπίπτουσιν αἱ τρίχες, ὥστε ἧττον διαδύεσθαι δύνανται· τὰ δὲ πρόβατα διὰ τὸ πλείω τροφὴν ἔχειν ὅταν ᾖ παχέα ἢ ὅταν ᾖ λεπτά, πλείονα τρίχα ἔχουσι καὶ παχυτέραν· ὅταν δὲ λεπτά, λεπτὴν ἀπὸ μικρᾶς τῆς τροφῆς; ἡ γὰρ θρὶξ περίττωμά ἐστιν.

2.161. Διὰ τί ὕες καὶ κύνες τυπτόμενοι τῶν βοῶν μᾶλλον βοῶσι; πότερον ὅτι καὶ ἄλλως μάλιστα λυπούμενα ταῦτα; ἢ ὅτι τῶν ἄλλων μᾶλλον θυμώδη; δηλοῖ δὲ ἥ τε τῶν ἀγρίων ὁρμὴ καὶ ἡ τῶν κυνῶν ὑλάκτησις.

2.162. Διὰ τί ἵπποι δυσουριῶσιν; ὅτι ὅσα τὴν ξηρὰν ὑπόστασιν εὐχερῶς προΐεται, τὴν ὑγρὰν χαλεπῶς, οἷον πρόβατα καὶ ὕες, οἱ δ' ἵπποι οὐδὲν ⟨ἧττον⟩ τούτων· ὅσοι δὲ τὴν ὑγρὰν ῥᾳδίως, τὴν ξηρὰν ἐργωδῶς· ὁ δὲ κύων ἄμφω χαλεπῶς.

2.159 (AKMOPRSTbei*p* Buss. Us.; *v* 3(5).23(25); 2.156 Buss.)
1 *αἱ ὕες*] *αἱ οἶες* sic P: *ὕες* A || *φρύττουσιν* Oi 2 *τῆς* om. A Buss. 3 post *προθυμίας* add. *τὸ θερμὸν* Us. || *αὐτῷ* Buss.: *sibi v*: *αὐτὸ* A: *αὗται p*: *αὐτῷ* cett. Us. || *εἰ μὴ* KOSbi*v* Us.: *εἰ δὲ μὴ* AMPRTe Buss.
2.160 (AKMOPRSTbei*p* Buss. Us; *v* 3(5).24(26); 2.157 Buss.)
1 *μὲν* om. Buss. || *λεπτότερα*] *χαλεπότερα* M: om. O || *λεπτοτριχότερα* AT Buss. 2 *παχυτριχότεραι* A Buss. || *ἢ* om. A || *παχύαι* sic A: *παχὺ p* 4 *διὰ τὸ δύεσθαι* e || *πλείονα* A*p* Buss. 5 *παχεῖα* AT || *ἢ ὅταν ᾖ λεπτά*] *aut cum gracilia non fuerint* male vertit *v* 6 *λεπτὴν*] *λεπτῶν* AT: om. MRe^{ac}: *λεπτάς p*: *λεπτὸν* Buss. || *γὰρ*] *δὲ* A Buss.
2.161 (AKMOPRSTbei*p* Buss. Us.; *v* 3(5).25(27); 2.158 Buss.)
2 *λυπούμενοι* ARTe^{ac} Buss.: *λυποῦντα p* 2-3 *ἄλλων μᾶλλον θυμώδη* Us.: *ἄλλων θυμώδη μᾶλλον* K*v*: *ἄλλων θερμότερον θυμώδη* A Buss.: *ἄλλων θυμώδη* cett. 3 *ἀγρίων ὑῶν* KOSb^{2}e^{pc}i*v* Us. || *ὁρμὴ*] *ὀργὴ* APT Buss. || *ὑλάκτησις*] *ὑλακὴ* P Us.
2.162 (AKMOPRSTbei*p* Buss. Us.; *v* 3(5).26(28); 2.159 Buss.)
1 *οἱ ἵπποι* Us.: *αἱ ἵπποι p* || *δυσουριοῦσιν* M || *ἢ ὅτι* Ke*v* || *ὅσα* om. K^{ac}OPi 2 *προΐενται* KP 3 *ἧττον* add. Us. 4 *ἄμφω*] *ἄφω* K^{ac} || *χαλεπῶς* AK*pv* Buss. Us.: *χαλεπὸς* MPRSbe: *χαλεπτὸς* T: *χαλεπὸν* Oi

from the fact that [ivory] can be bent and carved, which other [teeth] cannot.

2.159. Why do boars shiver when they are hunted with hounds? Is it that the body of those who are in distress contracts because <the heat>[478] is drawn together by their eagerness [to escape] and assists itself? And their hair stands more on end, if it does not fall out because it is loosened.

2.160. Why do thinner sheep have thinner hairs, but thinner pigs thicker hairs? Is it because the latter, because they are naturally fat, when they are fattened fill up the pores through which the hairs emerge, so that they are less able to pass through; but sheep, because they have more nourishment when they are fat than when they are thin, have more and thicker hair; when they are thin, [they have] thin [hair] from their small amount of nourishment? For hair is a residue.

2.161. Why do pigs and dogs make more noise when beaten than cattle do? Is it because in other respects too these feel most pain? Or is it because they are more spirited than other [animals]? This is shown by the eagerness of wild [boars] and the barking of dogs.

<Horses, asses and mules>

2.162. Why do horses pass urine with difficulty? Because as many creatures as excrete the dry substance easily [excrete] the moist with difficulty, for example sheep and pigs, and horses no <less>[479] than these; as many as [excrete] the moist easily, [excrete] the dry with difficulty; but the dog [excretes] both with difficulty.

---

[478] Added by Usener.
[479] Added by Usener.

2.163. Διὰ τί τὰς ἵππους ἐν τοῖς ἱπποφορβίοις ἀποκείρουσι τὴν χαίτην πρὸς τὰς ὀχείας; ἵνα ταπεινωθεῖσαι ὑπομείνωσι τὴν ὀχείαν τῶν ὄνων· κομῶσαι γὰρ οὐ θέλουσιν ὑπομένειν, τότε δὲ ὡς ὅμοιαι οὐκ ἀτιμάζουσιν.

2.164. Διὰ τί οἱ ἵπποι ὅταν μὲν ἐκφεύγωσιν, οὐ κινοῦσι τὰς οὐράς· ὅταν δὲ γυμνάζωνται, πολύ; ἢ ὅτι κίνησις τῆς οὐρᾶς διὰ πόνον τοῦ σώματος ἢ διὰ κακίαν τῆς ψυχῆς; διὸ καὶ μαλακῶς ἔχοντες ἢ οἱ μαλακοὶ τὰς ψυχὰς κινοῦσιν ἄκοντες (καὶ γὰρ βίᾳ πρὸς τὴν ἱππασίαν ἄγονται)· οἱ δὲ εὐγενεῖς οὔ, ἀλλὰ ἀκινήτους καὶ ὀρθὰς ἔχουσιν· ἐν δὲ τῷ ἐκφεύγειν πάντες, ἅτε ἑκουσίως καὶ τῇ ψυχῇ ὁρμῶντες εἰς τὸ πόρρω καὶ οὐδὲ ὑποδιστάζοντες, ὥστε εἰκότως ἀκίνητος ἡ οὐρὰ ὀρθή.

2.165. Διὰ τί οἱ μακρὰν ἔχοντες τὴν οὐρὰν ἀσθενῆ καὶ μὴ δασεῖαν ἐκ τῆς φύσεως κακοὶ τὰς ψυχὰς ὡς ἐπὶ τὸ πολύ; ἢ ὅτι σημαντικὸν ἡ οὐρὰ τοιάδε ἢ τοιάδε λεγομένη, λέγω δ' οἷον ὀρθὴ καὶ ἀνάσιμος ἢ κινουμένη ἢ ἑστῶσα; τὰ μὲν γὰρ ἀρετῆς, τὰ δὲ κακίας σημεῖα· ἐπεὶ οὖν φύσει τοιοῦτοι οἱ κακοὶ τὰς ψυχάς, εὐθὺς ἡ ἔκφυσις αὐτῶν ἀσθενὴς καὶ βραχεῖα, καὶ ἐν τῇ ἀρχῇ ἐνέλιπεν ἡ φύσις· ὡς γὰρ μὴ δέον χρήσασθαι τούτῳ, ἢ οὐ προστίθησιν ἢ οὐχ ὅμοιον ὃ καὶ τοῖς χρησομένοις, οἷον κέρας καὶ ὀδόντας [καὶ] ταῖς θηλείαις.

2.163 (AKMOPRSTbei*p* Buss. Us.; *v* 3(5).27(29); 2.160 Buss.)
1 *τὰς ἵππους* AKPT*v* Buss. Us.: *τοὺς ἵππους* MORSbei || *ἱπποφορβίοις* Buss. Us.: *ἱπποισφορβείοις* A: *ἱπποφορβείοις* cett. || *τὴν χαίτην*] *τὴν χήτην* e^{ac}: *τὴν χέτην* AT: *τὰς χαίτας* P: *πρὸς τὴν χαίτην p* 2 *τὰς ὀχείας*] *τὰς ὀχείας τῶν ὄνων* K*pv* || *ἢ ἵνα* K*v* || *ταπεινωθεῖναι* Re^{ac} 3 *οὐ θέλουσιν*] *οὐ θέλωσιν* A: *οὐκ ἐθέλουν* M || *δὲ*] *γὰρ* T^{ac} || *ὅμοια* P

2.164 (AKMOPRSTbei*p* Buss. Us.; *v* 3(5).28(30); 2.161 Buss.)
1 *ἐκφεύγωσιν* A*p* Buss. Us.: *ἐκφύγωσιν* cett. 2 *γυμνάζονται* AT*p* || *ἡ κίνησις* P || *τῆς* om. P || *τὰς οὐρὰς* A || *οὐρᾶς ἐστι* K || *τοῦ* om. P 3 *οἱ μαλακῶς* K 4 *ἄκοντες*] *ἀκούοντες* e^{ac} 5 *ἀκινήτους v* Buss. Us.: *εὐκινήτους* codd. *p* || *ἔχοντας* A || *πάντες θέουσιν* K 5-7 *ἐν δὲ – ὀρθή* om. *p* 6 *τῆς ψυχῆς* P || *οὐδὲν* K (sed non *v* ut vid.) 7 *ὀρθὴ* del. Us.: *ὁρμὴ* sic e: *καὶ ὀρθὴ* K*v* Buss.

2.165 (AKMOPRSTbei*p* Buss. Us.; *v* 3(5).29(31); 2.162 Buss.)
1 *ἀσθενῆ* del. Us.: *καὶ ἀσθενῆ* K*v* Buss. || *καὶ μὴ* P*p* Buss. Us.: *εἰ μὴ* KMORSbei*v*: *ἢ μὴ* AT || *δασεῖαν* KOPSbi*pv* Buss. Us.: *δασεῖαι* AMRTe || *τῆς* om. T 2 *σημαντικὴ* Pe^{ac} 3 *λεγομένη*] *γινομένη* K*v* || *λέγω δ' οἷον*] *λέγω δ' ἡ* Buss.: *λέγο δ' ἥ* sic A: *λέγω δὴ* P || *ἀνάσημος* KOSbi: *ἀνέσιμος p*: *nihilque significans v* || *ἢ*^{1}] *ἡ* T Buss. || *ἢ*^{2}] *ἡ* T^{ac} Buss. 4-2.166.4 *ἐπεὶ οὖν – ἑαυτόν* transilit *p*, duobus problematibus in unum coniunctis 4 *τοιοῦτοι* P Us.: *τοιοῦτον* cett. || *οἱ κακοὶ* Us.: *δὲ ὃ κακοῖ* M: *ὃ κακοῖ* cett. 5 *ἡ* add. sup. lin. O || *ἀσθενεῖς* AO^{ac}i 6 *ὡς*] *ᾧ* Buss. || *μὴ δέον* P Us.: *μηδὲ* cett. *v* Buss. || *μέλλει* post *χρήσασθαι* addendum coni. Buss. || *τούτῳ* KOPSbi*v* Us.: *τοῦτο* AMRTe Buss. 7 *ὃ*] add. sup. lin. O: *ἢ* P || *χρησομένοις*] *χρησίμοις* M: *χρησόμοις* sic e^{ac} || *κέρας* Buss Us.: *κερὸς* sic A: *καιρός* cett. *v* || *ὀδόντες* K*v* || *καὶ* del. Us.

2.163. Why when mares are at stud[480] do they cut off their manes before they are mounted? So that they may be humiliated[481] and put up with mounting by the asses; for when they have long hair they are not willing to put up with it, but then, as being similar, they do not despise [them].

2.164. Why do horses not move their tails when they are in flight, but [do so] a lot when they are exercising? Is it because the movement of the tail is on account of labour of the body or evil [temper] in the soul? For this reason those that are in a soft condition or are soft rouse their souls unwillingly (for they are led to exercise by force); the noble ones do not, but hold [their tails] unmoved and straight. But when fleeing all of them [hold their tails unmoved and straight], because they are impelled to go on willingly and by their soul and are not in doubt, so that reasonably their tails are unmoved [and] straight.[482]

2.165. Why are those [horses] which have a tail which is long [and] weak,[483] and not bushy, evil-tempered in nature for the most part? Is it because the tail gives an indication when described in this way or that, I mean for example straight and turned up[484] or moving or still? For some are signs of goodness, some of badness. So since those that are evil-tempered in their souls are like this by nature, immediately the growth of their [tails] is weak and short,[485] and their nature is defective at the beginning; for since it is not necessary to use this, it either does not add it or not in the same way as in those that will use it, as with horns and teeth for females.

---

480 For the breeding of *mules,* as the sequel makes clear. The same account at Aelian, *NA* 2.10, 12.16.

481 Ar., *HA* 6.18 572b8 says that when mares' manes are shorn they are subdued and less eager for intercourse; cf. Aelian, *NA* 11.18.

482 The last word deleted by Usener.

483 Usener deletes "weak; but see the next note.

484 Psellus has "relaxed".

485 For consistency with what precedes it seems that the reference here is not to the shortness of the tail itself but to that of the hair growing from it, so that it is not bushy.

2.166. Διὰ τί ὅλως σημαντικόν; ἢ κοινότερον τοῦτο πλειόνων; καὶ γὰρ ἐπὶ κυνῶν καὶ ὑῶν καὶ λεόντων καὶ ὅλως τῶν μαχίμων καὶ θυμωδῶν, καὶ τῶν ἀρρένων μᾶλλον ἢ θηλειῶν· τὰ μὲν γὰρ ὀρθὴν ἔχει, τὰ δὲ ἀπεστραμμένην, ὥσπερ οἱ ὗες οἱ ἄρρενες· ὁ δὲ λέων καὶ τύπτει ἑαυτόν· αἴτιον δὲ πάντων ὅτι συνέχειά τις ἀπὸ τῶν κυρίων τόπων τοῦ ζῆν, ἐν οἷς ὁ θυμός, εἰς ταύτην συνάπτει· συντεινομένων μὲν οὖν καὶ θερμαινομένων ἐκείνων διικνεῖται πρὸς τὸ ἔσχατον, διὸ καὶ ἀποσημαίνει τὴν διάθεσιν· θαρρούντων μὲν γὰρ ὀρθή, τῶν δὲ ἀπεστραμμένων ἢ φοβουμένων ὑπειλημμένη καὶ ὥσπερ ἅμα τῷ θυμῷ πεσοῦσα. τὸ δ' ὅλον καὶ πρὸς εὐτολμίαν καὶ πρὸς ὕβριν καὶ φωνὴ ἔχει τινὰ δύναμιν καὶ τόνον διὰ τὴν αὐτὴν αἰτίαν.

2.167. Διὰ τί ἡ ἵππος ὀλιγοχρόνιος οὖσα τοῦ ἀνθρώπου πλείω χρόνον φέρει ἐν γαστρί; συμβαίνει τοῦτο οὖν διὰ μίαν αἰτίαν· ἡ γὰρ ἵππος πολὺ λείπεται τῶν μακροβίων ζῴων καὶ ὅτι ἀγονώτερον· μονότοκον γάρ.

2.168. Διὰ τί τῶν μὲν ἵππων οἱ ἄρρενες χρεμετίζουσι μᾶλλον τῶν θηλειῶν, τῶν δὲ προβάτων καὶ αἰγῶν τὰ θήλεα μᾶλλον βληχᾶται· αὕτη δ' ἐν ἑκάστῳ φωνή ἐστιν; ἢ ὅτι τῶν φωνῶν αἱ μέν εἰσι θυμικαί, αἱ δὲ ἄθυμοι; τὰς μὲν οὖν θυμικὰς οἱ ἄρρενες μᾶλλον ἀφιᾶσιν (ἀνδρειότεροι γάρ), τὰς δ' ἀθυμοτέρας τὰ θήλεα· οἷον κοκκύζουσι μὲν οἱ ἀλεκτρυόνες, κράζουσι δὲ αἱ ἀλεκτορίδες· ὁ μὲν οὖν χρεμετισμός ἐστι φωνὴ θυμοῦ, ἡ δὲ βλήχησις ἀσθενείας, ὥστ' εὐλόγως τὴν μὲν ἄρρενες τῶν ἵππων ἀφιᾶσι, τὴν δὲ τὰ

2.166 (AFKMOPRSTbei Buss. Us.; *v* 3(5).30(32); 2.163 Buss.; de *p* cf. ad 2.165.4)
1 *Διὰ τί – πλειόνων* om. F qui 2.158 et 2.166 in unum coniungit || *τοῦτο*] *τὸ ὑπὸ* K*v* 2 *ὅλως* M Buss. Us.: *πάντων* K: *ὅλων* cett. 3 *ἀρρένων*] *ἄρρων* O^{ac} || *ἢ τῶν θηλειῶν* K 4 *ὥσπερ οὖν* P || *οἱ*^{1} om. F || *οἱ*^{2}] *καὶ οἱ* Sb^{ac}e^{pc} || *λέων*] *καὶ λέων* F^{ac} || *πάντων* om. F 6 *μὲν* om. K || *μένων μὲν οὖν καὶ θερμαινο* add. in marg. O 8 *ὑπειλημμένη καὶ*] *ὑπειλημμένων καὶ* A: om. F 9 *πεσοῦσα* Us.: *πέττουσα* codd.: *πίπτουσα* coni. Buss. (in var. lect.): *se ... accommodans v* || *εὐτολμίαν* Us.: *ἐντονίαν* e^{pc}: *εὐτονίαν* cett. *v* || *φωνὴ* M Buss. Us.: *φωνὴν* cett.
2.167 (AFKMOPRSTbei*p* Buss. Us.; *v* 3(5).31(33); 2.163bis Buss.)
1 *ὀλιγοχρονιωτέρα* K || *χρόνω* A*p* 2 *ἢ συμβαίνει* K || *οὖν τοῦτο* FP: *οὖν* om. K 3 *καὶ* om. Buss. || *ἀγονώτερον* Us.: *ἀτονώτερον* FOPRSTbei Buss.: *ἀτονότερον* M: *ἀτονωτέρα* K: *minus efficax v*: *εὐτονώτερον* A: *ἀτονώτατον p* || *μονοτόκον* T: *μονώτοκον* O: *μονοτόκος* K
2.168 (AFGKMOPRSTbei*p* Buss. Us.; *v* 3(5).32(34); 2.163ter Buss.)
1 *χρεμετίζουσι μᾶλλον*] *μᾶλλον χρεμετίζουσι* F*p*: *χρεμετίζουσι* AGMPRb^{1}e^{ac} Buss. 2 *αὕτη* GK Buss. Us.: *αὐτὴ* cett.: *sua v* 3 *τῶν φωνῶν*] *αἱ φωναὶ* P 4 *ἀφιᾶσιν* add. sup. lin. O || *ἀνδρ(ε)ιότεραι* AFT Buss.: *ἀνδρειότατοι* G || *τὰς*^{2}] *τῆς* PT: *ταὶς* R 5 *κράζουσι*] *κακάζουσι* K*v* 6 *ἀλεκτορίδαις* A || *θυμοῦ*] *θυμικοῦ* T 7

2.166. Why in general is [the tail] an indication? Or is this shared with more [creatures]? For it applies to dogs and pigs and lions and in general to [creatures] that fight and are spirited, and to males more than females; some have it straight, others twisted aside, like male pigs; and the lion strikes itself [with its tail].[486] The reason in every case is that there is a certain continuity from the regions crucial to life, where the spirit is, connected to [the tail]; so when those [regions] are taut and heated this extends to the extremity, and for this reason it indicates [the animal's] disposition. For when they are confident it is straight, but when they are deterred or afraid it droops and as it were falls along with their spirit. In general their voice too has a certain power and tension in relation to boldness[487] and violence for the same reason.

2.167. Why is the mare, which is short-lived, pregnant for a longer time than a human?[488] This happens for a single reason:[489] the mare falls far short of long-lived animals also in that it is less productive of young; for it produces only one foal at a time.

2.168. Why do stallions neigh more than mares, but in sheep and goats the females bleat more, [when] both [sexes] have the same sound? Is it that some sounds come from spirit, others from lack of spirit? So the spirited [sounds] are produced more by the males, for they are more courageous, but those from lack of spirit by the females; for example cockerels crow but hens cackle. So neighing is a spirited sound, but bleating [a result of] weakness, so it is reasonable that the former is produced by stallions, the

---

486 Cf. Homer, *Iliad* 20.170; Hieronymus of Rhodes, fr.51 White/46 Wehrli; Aelian, *NA* 6.1. Cf. Sharples (forthcoming).

487 Usener's emendation for "good tension" of the MSS.

488 Cf. [Ar.], *Probl.* 10.9.

489 Unless something is wrong with the text, the argument would seem to be that productiveness and length of life go together. It would be easy to read *ou* for *oun*: "this happens not just for a single reason: the mare falls far short of long-lived animals, and also because it is less productive of young; for it produces only one foal at a time." But while this gives more point to the remark about a single reason, it gives a less clear answer to the problem. [Ar.] gives the explanation that the mare's uterus is hard.

θήλεα τῶν προβάτων καὶ τῶν αἰγῶν.

2.169. *Διὰ τί τῶν ἵππων ἐάν τις μίαν ἢ δύο ἢ τρεῖς σπάσῃ τρίχας, πονοῦσιν; ἢ διότι ἐκείνοις μὲν τιλμὸς γίνεται (κατὰ μίαν γὰρ ἑλκομένης ἢ δύο τὸ δέρμα οὐκ ἀκολουθοῦν τῇ ἀντισπάσει τὴν ἀλγηδόνα ποιεῖ)· ὅταν δὲ ἀθρόας ἑλκύσῃ τις, οὐ γίνεται ὁ σπασμός; τὸ γὰρ δέρμα συνακολουθεῖ.*

2.170. *Διὰ τί αἱ ἵπποι ὅταν σκυζῶσιν ἐκτοπίζουσαι οὐ θέουσι πρὸς ἕω ἢ πρὸς ἑσπέραν; ἀλλὰ πρὸς ἄρκτον ἢ πρὸς μεσημβρίαν; ἢ ὅτι λέγονται ἐν τῇ τοιαύτῃ διαθέσει ἐξανεμοῦσθαι; θέουσι γὰρ πρὸς ἄνεμον· τούτου γὰρ αἴτιον ὅτι θερμότης ἐστὶ τὸ πάθος καὶ δέονται καταψύξεως, ὥστε εὐλόγως πρὸς ἄρκτον τρέχουσι καὶ πρὸς μεσημβρίαν· μάλιστα γὰρ καὶ πλειστάκις οἱ ἐντεῦθεν πνέουσιν ἄνεμοι· νότος γὰρ καὶ βορέας ἀεὶ κατέχουσιν, οἱ δ' ἄλλοι πνέουσιν ὀλιγάκις.*

2.171. *Διὰ τί αἱ ἵπποι ἐὰν ὀχευθεῖσαι βρεχθῶσιν, ἀνοχεύονται; ἢ ὅτι συμβαίνει ψύχεσθαι ὑπ' αὐτοῦ καὶ ἡσθῆναι τὰ σώματα; φρίτττοντος οὖν τοῦ σώματος κωλύεται τοῦ συλλαμβάνεσθαι τὸ σπέρμα καὶ καταψυχόμενον ὥσπερ ἀποθνήσκει φθειρομένης αὐτοῦ τῆς δυνάμεως· διὰ τοῦτο πάλιν δεῖται ὀχείας.*

2.172. *Διὰ τί τοῖς μὲν ἵπποις καὶ τοῖς ὄνοις ἐκ τῶν οὐλῶν φύονται τρίχες, τοῖς δὲ ἀνθρώποις οὔ; ἢ ὅτι ἡ οὐλὴ κωλύει διὰ τὴν πύκνωσιν ἢ διὰ τὸ φθίνειν τὴν τροφήν; τοῖς μὲν οὖν ἀνθρώποις παντελῶς κωλύει διὰ τὴν*

τὴν[1]] *τὰ* G[ac] || *οἱ ἄρρενες* AFGKPT Buss.

2.169 (AFKMOPRSTbei*p* Buss. Us.; *ν* 3(5).33(35); 2.164 Buss.)

1 post *πονοῦσιν* add. *ἐὰν δὲ πολλὰς ἅμα, οὔ* K*ν* 2 *ἐκείνως* K*ν* || *τιλμὸς*] *τιλαὶ* M || *ἑλκομένης*] *ἑλκόμενον* P*ν* Us.: *ἑλκομένου* R[ac]: *ἐκκομένου* K 3 *τὸν ἀλγηδόνα* S || *ἀθρόαν* M 4 *ἑλκύσῃ*] *ἐκλύσῃ* T || *συνακολουθῇ* T

2.170 (AFKMOPRSTbei*p* Buss. Us.; *ν* 3(5).34(36); 2.165 Buss.)

1 *αἱ*] *οἱ* P || *ἐκτοπίζουσαι* Buss. Us.: *ἐκτομίζουσαι* codd. *p* || *θέωσι* Oi || *πρὸς τὴν ἕω* A Buss. || *πρὸς*[2]] *τὴν* A Buss. 2 *πρὸς*[2] om. AFPRTe Buss. 3 *τοῦτο* F || *γὰρ*] *δ'* K 4 *ὁ τὸ πάθος* sic A || *δέοντα* K 5-6 *καὶ πλειστάκις – ἄνεμοι*] *οἱ πρὸς ἐντεῦθεν πνέουσιν ἄνεμον· καὶ πλειστάκις* F 6 *πνέουσιν*] *πλέουσιν* A || *βοράς* A: *βορρᾶς* *p* Buss. || *οἱ δ' ἄλλοι* K*ν* Us.: om. cett. 7 *πνέουσιν* P Us.: *πνεῖν* AFP Buss.: *ποεῖν* MRe: *πρίειν* T: om. KOSbi*ν* || ante *ὀλιγάκις* add. *οἱ δ' ἄλλοι* Buss.

2.171 (AFKMOPRSTbei*p* Buss. Us.; 2.166 Buss.)

2 *ἡσθῆναι*] *ὕεσθαι* K 3 *κωλύεται* Us.: *κωλύει* codd. || *συλλαμβάνοντος* F || *σπέρμα*] *σῶμα* P 5 *τοῦτο*] *τὸ* T

2.172 (AFKMOPRSTbei*p* Buss. Us.; [Ar.] 10.29; *ν* 3(5).35(37); 2.167 Buss.)

1 *τοῖς*[2] om. P 2 *ὅτι*] *διότι* [Ar.] 3 *φθίνειν*] *φθείρειν* [Ar.]: sicut *φθάνειν* vertit *ν* || *κωλύει παντελῶς* P 4 *οὐ κεκώλυκε* AFMPRe[ac]*p* Buss. (*κεκώλυε* e[mg]): *οὐκ ἐκώλυκε*

latter by female sheep and goats.

2.169. Why are horses in distress if one pulls one or two or three hairs from them? Is it because then they are plucked, for when [a hair] is pulled singly, or two, the skin does not follow and produces pain by pulling back, but when one pulls [hairs] all together, no spasm occurs, because the skin follows?

2.170.[490] Why when mares on heat leave a place do they not run towards the east or the west, but towards the north or the south? Is it because in such a condition they are said to be impregnated by the wind? For they run towards the wind. The reason for this is that what happens to them is heat and they need cooling, so that it is reasonable that they run towards the north and the south, for the winds from there blow most and most often; for the south wind and the north wind always prevail, the others blow rarely.

2.171. Why do mares not conceive if they get wet after they have been mounted? Is it because their bodies are cooled and experience pleasure from [the water]; so when their body shivers the conception of the seed is prevented and, being chilled, it dies, as it were, when its power is destroyed. And for this reason another mounting is needed.

2.172.[491] Why do hairs grow out of the scars of horses and asses, but not of humans? Is it because the scar prevents it, because of its denseness or because the nourishment wastes away?[492] In humans it completely prevents it

---

[490] Similarly, as noted by Richter 1885, 31 (but without the explanation in terms of cooling) Ar., *HA* 6.18 572a13-17, cited by Aelian, *NA* 4.6 and thence by [Aristophanes of Byzantium], *HA Epit.* 2.613 150.4-5 Lambros.

[491] = [Ar.], *Probl.* 10.29. Cf. below, 2.189, and [Alex.], *Probl.* i.151, which gives an explanation different both from the present text and from 2.189, arguing that human skin is naturally denser than that of horses, as the nature of the hair shows, and so becomes denser still when scarred.

[492] See below, n.514 to 2.190.

ἀσθένειαν τῆς τριχός, τοῖς δ' ἵπποις οὐ κεκώλυκε μέν, ἔφθειρε δέ, διὸ καὶ λευκαὶ φύονται τούτοις· οὐκ ἐξήει δὲ διὰ τὸ πλῆθος. (ἐπισκέψασθαι εἰ καὶ τοῖς βουσίν.)

2.173. *Διὰ τί παραπλησίων ὄντων τῶν ἵππων καὶ τῶν ὄνων κατὰ τὸν χρόνον τῆς κυήσεως (ἄμφω γὰρ ἐνιαυτὸν φέρει καὶ μονότοκά ἐστι)· τὸν μὲν ἵππον ἀφιᾶσιν ἐπὶ τὴν ὀχείαν ἰσημερίας, τοὺς δὲ ὄνους πρὸ τροπῆς θερινῆς; ἢ διότι οἱ μὲν ἵπποι θερμοὶ τὴν φύσιν, οἱ δὲ ὄνοι δύσριγοι καὶ ψυχροί; σημεῖον δὲ τούτου· οὐ γὰρ γίνονται ἐν τῇ Σκυθίᾳ ὄνοι· ἐνιαυσιαίας οὖν τῆς κυήσεως οὔσης ἀνάγκη γεννᾶν εἰς τὴν αὐτὴν ὥραν· τοῖς μὲν οὖν ἵπποις τὸ ἔαρ σύμμετρον· οὔτε γὰρ σφόδρα ψυχρά, ἀλεεινὴ δέ, ὥστε διὰ ψῦχος μηδὲν πάσχειν· σκοποῦσι δὲ τὴν τῆς ὀχείας ὥραν τίς ἐπιτηδεία πρὸς τὴν τροφὴν τῶν τέκνων πάντες.*

2.174. *Διὰ τί τῶν μὲν ἵππων οἱ ὀρθὰ τὰ σκέλη καὶ μικρὸν ἔχοντες τὸ μεσοκύνιον ἢ μέγα λίαν χωλοὶ γίνονται ταχέως, οἱ δὲ μέσον ἀσφαλέστατοι; ἢ ὅτι μικρὸν μὲν ὂν οὐ παρέχει ἔνδοσιν, ἀλλ' οἷον πληγὴ γίνεται· ὅταν δὲ σύμμετρον ἐνδίδωσι, καὶ γίνεται βάσις, οὐ πληγή· ἐὰν δ' αὖ μεῖζον τοῦ καιροῦ, καταρρήγνυται πλείονος τοῦ βάρους ἐμπίπτοντος διὰ τὴν ἔνδοσιν; διὸ καὶ αἱ χελῶναι ἀνόμοιαι, τῶν μὲν μικρὸν ἐχόντων*

T: *οὐκ ἐκώλυε* KOSbi Us.: *οὐκ ἐκώλυσε* [Ar.] || *μέν* om. [Ar.] *v* || *διέφθειρε* [Ar.] || *δὲ* om. A 4-6 *διὸ – βουσίν* om. [Ar.] 5 post *τούτοις*: *διὰ τὸ δέρμα ἀσθενέστερον γεγονέναι καὶ ἡ λευκὴ θρίξ ἀσθενεστέρα p* || *ἐξήει*] *ἐξίοι* F: *ἔξεστι* K*v*: *ἐξίῃ* P: *ἐξίει* T || *ἐπισκήψασθαι* APSTbe 5-6 *ἐπισκέψασθαι – βουσίν* del. Us.

2.173 (AFKMOPRSTbei Buss. Us.; 4-5 tantum *p*; *v* 3(5).36(38); 2.168 Buss.)
2-3 *τὸν μὲν ἵππον*] *ὅτι τῶν μὲν ἵππων* AF: *ὅτι τὸν μὲν ἵππον* PT 3 *ἀμφίασιν* A || *ἐπὶ*] *εἰς* F || *ἰσημερίας* Buss.: *⟨ἀπ' ἐαρινῆς⟩ ἰσημερίας* Us.: *εἰς ἡμέρας* AKMOPRSTbei: *τῆς ἡμέρας* F: *sub tempus v* || *τοὺς* AFPTe Buss. Us.: *τὰς* KMORSbi 3-4 *πρὸ τροπῆς θερινῆς* Buss. Us.: *προτρέπεσθαι ῥίνας* A: *προτρέπεσθαι ῥίναις* MPRTe: *προτρέφεσθαι* K*v*: *προτρέπεσθαι* OSbi: *εἰς τὴν ὀχείαν τῆς νυκτὸς* F 5 *τούτου*] *τούτων* F || *ὄνοι*] *ὅμοιοι* F 6 *κυήσεως*] *κινήσεως* KS*v* 7 *καὶ τὸ ἔαρ* A || *ἀλεεινὴ*] *ἐλεινὴ* Oi: *ἀλεεινὸς* A 8 *ψῦχος*] *ψυχῆς* A || *πάσχει* Buss. || *σκοποῦσι*] *σκώπτουσι* F

2.174 (AFKMOPRSTbeim*p* Buss. Us.; 2.169 Buss.)
1-5 *Διὰ τί ... ἐμπίπτοντος* om. m qui 2.68 et 2.174 in unum coniungit (cf. Introd. §2.1.7) 1 *μικρὸν*] *μικρῶν* A: *μικρὰ* F || *τὸ* K Us.: om. cett. Buss. 2 *μεσοκύνιον* Us.: *ἐπισκύνιον* codd. Buss. *p* (*ἐπισκήνιον* F) || *μέσον*] *μέγα* A 3 *μικρὸν*] *μικρῶν* A: *μικρὸς* FMTe*p* || *ὂν*] *ὧν* A: *ὢν* FMPRe*p* 4 *σύμμετρον* K Buss. Us.: *σύμμετρος* F: *σύμμετροι* cett. || *βάσεις* A 4-5 *δ' αὖ*] *δὲ* KOi: *δ' οὖν* Sb 5 *μεῖζον* Buss. Us.: *μείζων* KOi: *μεῖζω* cett. || *πλέον ὡς* T || *καὶ διὰ* m 6 *ἀνόμοιοι* P*p* || *μικρῶν* AOST[ac]bi*p* 7 *παρεμπίπλαται* Us.: *παρεμπίπλανται p* Buss.:

because of the weakness of the hair, while in horses it has not prevented it, but corrupted it, and for this reason they grow white [hairs]. (Consider whether [this] also [happens] in cattle.)[493]

2.173.[494] Why, when horses and asses are similar in duration of pregnancy (for both carry the young for a year and produce one foal), do they set the stallion at stud after the equinox,[495] but asses before the summer solstice? Is it because horses are hot in nature, but asses are cold and do not endure cold well? An indication of this: there are no asses in Scythia.[496] So, since the pregnancy lasts a year, they necessarily give birth in the same season [as they conceived]. For horses the spring is suitable; for it is not very cold, but sunny, so that they do not suffer anything from the cold.[497] All [the herdsmen][498] consider what time of mating will be suitable for the rearing of the young.

2.174. Why do horses that have straight legs and small or excessively large pasterns[499] become lame swiftly, while those with a medium-sized [pastern] are safest [from this]? Is it because a small one does not provide any yielding, but [the step] is as it were a blow; when it is moderate it yields, and there is a support, not a blow; but if it is larger than is opportune, it breaks because too much weight falls upon it because of the yielding? For this reason also the frogs of the hoofs[500] are dissimilar, and of those which have a

---

493 This phrase was deleted by Usener. It is clearly in some sense a note, but it may go back very early in the tradition, possibly even to the original compiler; it is present in all the MSS which include this problem. We have therefore placed it in parentheses rather than between square brackets.

494 An expanded version of Ar., *GA* 2.8 748a28ff.: Richter 1885, 31. Similarly but much more briefly Pliny, *NH* 8.167.

495 Usener inserts a specific reference to the spring equinox in the actual text; but the author may have left this to be understood.

496 Ar., *HA* 8.25 605a21, 8.28 606b4, *GA* 2.8 748a25; Herodotus 4.28, 4.129; Strabo 7.3.18 (We are grateful to Ioana Ciuca and Simon Hornblower for the references to Herodotus, Pliny and Strabo).

497 The implication being: whereas it would still be too cold for asses.

498 The reference is presumably to those who have care of the animals, rather than to the animals themselves.

499 Usener's emendation for the MSS *episkunion*, the forehead above the eyebrows, which makes no sense in this context.

500 Reading *khelidones* for *khelônai*.

παρεμπίπλαται τὰ σκέλη καὶ ἀνοιδεῖ, τῶν δὲ μέγα καταρρήγνυται καὶ ἀφελκοῦται ὁ τόπος οὗτος.

2.175. *Διὰ τί ὁ ἵππος ἐὰν ἐμέσῃ ἀποθνήσκει; πότερον ὅτι παρὰ φύσιν; ἢ οὐ τοῦτο αἴτιον, ἀλλὰ σημεῖον ἑτέρου μείζονος; ὅτι γὰρ ἐπιθανάτως ἔχει ἐὰν ἐμέσῃ, οὐχ ὅτι ἐὰν ἐμῇ ἀποθνήσκει. ἢ ὅμοιον ὅσοι ἐν τοῖς εἰλεοῖς ἂν ἐμῶσι κόπρον; καὶ γὰρ οἱ ἵπποι ἐν τοῖς αὐτοῖς πάθεσι· τοῖς μὲν οὖν τὸ ἐκ τῆς κάτω κοιλίας σπάσαι παρὰ φύσιν (διὸ καὶ κρατοῦνται τοῖς πάθεσι), τοῖς δ᾽ ὅλως τὸ ἀνεμεῖσθαι.*

2.176. *Διὰ τί ὑπὸ διαρροίας ἐὰν ἁλῶσιν, ὡς ἐπὶ τὸ πολὺ ἀποθνήσκουσι; ἢ ὅτι μανὴ καὶ οὐ πυκνὴ ἡ σάρξ; σημεῖον δὲ καὶ τὸ εὐΐδρωτα εἶναι καὶ ὅλως τὸ εὐκίνητον· πλείων οὖν ἡ σύντηξις ὁρμῆς γενομένης παρὰ τοῖς ποσί.*

2.177. *Διὰ τί οἱ πρεσβῦται ἵπποι τὸ κάτω χεῖλος προτείνουσιν; ἢ ὅτι γηρασκόντων ἀνίεται ἡ φύσις, τὸ δὲ κάτω χεῖλος μεῖζον καὶ σαρκωδέστερον καὶ ἧττον πρὸς τῷ ὀστῷ;*

2.178. *Διὰ τί οἱ ἵπποι θᾶττον οὐροῦσιν ἐν τοῖς μαλακοῖς χωρίοις ἢ ἐν τοῖς στερεοῖς; πότερον ὅτι ῥαίνονται ἧττον, φεύγει δὲ τοῦτο σφόδρα; σημεῖον δέ τὸ διάγειν ἐπὶ πολὺ τὰ σκέλη, καὶ ὅταν ὑποστρωννύηται μέλλων καθεύδειν ἀεὶ οὐρεῖ· οὐ γὰρ τέγγεται ῥανίσιν. ἢ γίνεται μὲν καὶ τοῦτο, οὐ μὴν ἀλλ᾽*

περιπίμπλανται M: *παπίμπλανται* T: *παραπίμπλανται* cett.

2.175 (AFKMOPRSTbeim*p* Buss. Us.; *v* 3(5).37(39); 2.170 Buss; 4-5 Petrus ap. Olivieri 1988, 196-197)

2 *τοῦτο*] *τοῦτου* sic A: *τούτου* FORSTbim*p* 2-3 *ἐὰν ἐμέσῃ*] *ἂν ἐμέσῃ* Tm: *ἤμεσεν* K*v*: om. F 3 *ἐὰν ἐμῇ*] *ἐὰν ἐμεῖ* sic A: *ἂν ἐμῇ* PT: *ἀνεμεῖ* m: *ἐὰν* non vertit *v* (cf. ad 2-3) || *ἐν*] *ἂν ἐν* K: om. *v* || *ἰλεοῖς* FKPTm: *τοῖς εἰλεοῖς* om. *v* || *ἂν ἐμῶσι* MOPRi Us.: *ἀνεμοῦσι* em Buss.: *ἂν ἐμοῦσι* AFT: *ἐμῶσι* K*v*: *ἀνεμῶσι* Sb: *ἀνιμῶσι* *p* 4 *οἱ*] *ὡς* F || *τοῖς μὲν οὖν τὸ* om. F || *τοῖς*[2]] *τῆς* m 5 *κρατοῦνται*] *ὡμοίωνται* Buss. || *τοῖς δ᾽ ὅλως* Us.: *τοῖς δὲ ὅλως* K: om. F: *τοῖς ὅλως* cett. Buss. 6 *τὸ ἀνεμεῖσθαι*] *τὸ ἂν ἐμεῖσθαι* T: *τὸ ἄνω ἐμεῖν* K*v*: *μὴ ἀνεμεῖσθαι* A Buss.: om. F: *ἀνεμοῦσι* Buss. in var. lect.

2.176 (AFKMOPRSTbeim Buss. Us.; *v* 3(5).38(40); 2.171 Buss.)

1 *ἐὰν ἁλσῶσιν*] *cum triturent v* || *ἐπιπολὺ* MRe[ac] 2 *οὐ πυκνὴ* Me Buss. Us.: *rara v*: *πυκνὴ* tantum cett. 2 *εὐΐδρωτον* KOi 3 *εὐκίνητον* AKP Buss. Us.: *εὐκίνητα* FMORSTbeim*p* || *πλεῖον* T || *παρὰ*] *περὶ* FTm

2.177 (AFKMOPRSTbeim*p* Buss. Us.; *v* 3(5).39(41); 2.172 Buss.)

1 *προστείνουσιν* O: *προτείνωσιν* e

2.178 (AFKMOPRSTbeim*p* Buss. Us.; *v* 3(5).40(42); 2.173 Buss.)

2 *σημεῖον δέ*] *σημεῖον τόδε* APTm: *σημεῖον δὲ τούτου* F 3 *τὸ διάγειν*] *το δῆ ἄγειν* A: *ὅτι διάγει* Us. || *ὑποστρωννύηται*] *ὑποστρονίεται* sic A: *ὑποστρωννύεται* MORSTbeim: *ὑποστρωννύονται p* || *καθεύδει* F 4 *τέγγεται* Us.: *γί(γ)νεται* codd.

small one the legs fill up and swell, but of those which have a large one this place is broken and becomes ulcerated.

2.175. Why does a horse die if it vomits? Is it because this is unnatural? Or is this not the reason, but an indication of something more serious? It is because it is in a terminal condition if it vomits, not because it dies if it vomits. Or is it like those with intestinal obstruction if they vomit excrement? The horses have the same affliction. So for the former it is unnatural to draw [vomit] from the lower bowel, and for this reason also they are overwhelmed by their affliction, while for the latter vomiting generally [is unnatural].

2.176. Why if [horses] get diarrhoea do they for the most part die?[501] Is it because their flesh is rare and not dense? An indication is both that they sweat easily and in general that they move easily. So the colliquescence is greater because of the impulse in their feet.

2.177. Why do elderly horses thrust their lower lip forward? Is it because when they grow old their nature grows slack, and the lower lip is larger and more fleshy and less in contact with the bone?

2.178. Why do horses urinate more swiftly in soft places than in hard ones? Is it because they get splashed less, and they eagerly avoid this? An indication is their spreading their legs wide apart, and when they settle[502] intending to sleep they always urinate? For they are not wetted by the spray. Or is it that this does happen, but it seems that it is rather the relaxation of

[501] That prolonged diarrhoea in horses is dangerous is asserted at *Hippiatrica Berolinensia* 35.1 (193.1 Oder and Hoppe).

[502] Literally "spread out their bedding".

ἔοικε μᾶλλον ἡ ἄνεσις τοῦ σώματος αἰτία εἶναι; συντεταμένον γὰρ οὐ προΐεται τὸ σῶμα, καθάπερ οὐδὲ πονοῦντα, ἀλλ' ὥσπερ καὶ δυσουριᾷ, ἐνθένδ' ἀνίησιν ἐν τῷ μαλακῷ· τὸ δὲ στερεὸν ἀντίτυπον ὂν συντονίαν γεννᾷ· ἥ τε διαγωγὴ τῶν σκελῶν καὶ ἡ πρόσνευσις τῆς ὀσφύος εἰς τὴν πρόεσιν· ἅπαντά τε γὰρ σχεδὸν τοῦτο ποιεῖ, πλὴν ὅσα ῥᾳδίως οὐρεῖ· τὰ δὲ δύσουρα μᾶλλον εἰκότως τῶν ἄλλων.

2.179. Διὰ τί οἱ ἵπποι καὶ οἱ ὀρεῖς καὶ ὅλως τὰ λόφουρα ἥκιστα δύνανται πεινᾶν; διὸ καὶ ὅταν κατακλεισθῶσι, τὰς θύρας καὶ τὰς φλιὰς καὶ ὅλως πᾶν ξύλον κατεσθίουσιν. ὅτι ἔχουσι θηρίδια ἐν τῇ ἄνω κοιλίᾳ πολλὰ παρόμοια ταῖς πρασοκουρίσι· ταῦτ' οὖν ἐξαναλίσκει τε τὴν τροφήν, καὶ ὅταν μηκέτι ἔχῃ, τῇ κινήσει καὶ τῇ ξύσει τοῦ ἐντέρου πόνον παρέχει· παραπλήσιον δὲ τοῦ συμβαίνοντος τοῖς τὰς ἕλμινθας ἔχουσι τὰς πλατείας.

2.180. Διὰ τί αἱ ἵπποι ῥᾷον φέρουσιν ἐν τῇ κυήσει ἡμίονον, καίτοι ἰσχυρότερον ἡ ἡμίονος; ἢ ἴσως ἀληθὲς δι' αὐτὸ τοῦτο, καθάπερ καὶ αἱ γυναῖκες τὰ ἄρρενα τῶν θηλειῶν; ἰσχυρότερα γὰρ ὄντα μᾶλλον ἐξαναλίσκει τὰ περιττώματα, τὸ δὲ βαρύνεσθαι καὶ πονεῖν ἀπὸ τούτων.

Buss. *v* || ῥανίσιν] ῥάνισις Buss. 5 ἡ] ἢ m || συντεταμένον AP Buss. Us.: συντεταγμένον cett. *v* 6 προΐεται F*v* Buss. Us.: προσίεται cett. || πονοῦντα] πονοῦντι A Buss.: πονοῦν K*v* || δυσουριᾷ Buss. Us.: δυσουρίαν τινὰ F: δυσουρία γίνεται K*v*: δυσουρίαν cett. 7 ἐνθένδ' ἀνίησιν AP Buss.: del. Us.: ἔνθεν δ' ἀνίησι(ν) KM*v*: ἔνθεν δ' ἀνείησιν m: ἔθεν δ' ἀνίησι T: ἔνθεν δ' ἂν ἵησιν ORSbei: om. F || συντονίαν] συντονία AR 8 γεννᾶ Buss. Us.: τηνᾶ A: τινί M: τινὰ ποιεῖ K*v*: om. P: τινὰ cett. || πρόνευσις T || ὀσφρύος Re 9 πρόεσιν Buss. Us.: πρόθεσιν codd. *p*: *productionem v* || τε om. K

2.179 (AFKMOPRSTbeim*p* Buss. Us.: *v* 3(5).41(43); 2.174 Buss.

1 ἵππεις Oi || ὀρεῖς Buss. Us.: ὄρνεις FMPRSTem*p*: ὄρνις Obi: ὄρνοις sic A: ὄνοι K*v* || λόφουρα K Us.: λοφοῦρα Buss.: λοφῆρα sic A: λόφυρα P: λοφηρὰ T: *iumenta v*: λάφυρα cett. || post τὰ λόφουρα add. τὰ ἐκ τῶν λόφων A 2 φλοιὰς APT[ac] 3 ἢ ὅτι K (ἢ add. in marg.) || θηρίδα F 4 παρόμοια] παρομοιαται T: om. S || ταῖς om. T || πρασοκουρίσι] πρασικουρίσι F: *indagantibus v* || ταῦτ' οὖν] ταύταις A || ἐξαναλίσκει τε] ἐξαναλίσκεται A: ἐξαλίσκει τε Oi: ἐξαναλίσκει e[ac]: ἐξαναλίσκει τις e[pc] 6 τοῦ συμβαίνοντος] τὸ συμβαῖνον K Us. || τοῖς τὰς KM Us.: τοῖς A Buss.: τὰς cett. || τὰς[2] om. A Buss.

2.180 (AFKMOPRSTbeim*p* Buss. Us.; *v* 3(5).42(44); 2.175 Buss.)

1 ῥᾷον] *n o n facilius e q u o s q u a m mulas v* (cf. infra) || τῇ om. P || ἡμίονον Buss. Us.: ἵππους ἡμιόνων K*v*: ἡμιόνων cett. 2 ἰσχειότερον F: ἰσχυρότερος M || ἡ om. AFTm Buss. 2-3 *mulieribus, quae mares g r a v i u s ferunt quam feminas v* 4 ἀπὸ τούτου M

their body that is the cause? For when the body is taut it does not emit [it], just as it does not when labouring, but it experiences as it were difficulty in passing urine, [and] after this it releases [it] on the soft ground.[503] The hard ground, offering resistance, produces tautness. The spreading of the legs and inclination of the loins is for the emission; for almost all [creatures] do this, apart from those that pass urine easily; it is reasonable that those that do so with difficulty should do this more than the others.

2.179. Why are horses and mules and in general bushy-tailed animals least able to endure hunger? For this reason when they are shut up they eat the doors and the door posts and in general any wood. It is because they have little creatures in their upper bowel which are very like millipedes. These consume the food, and when they no longer have any, cause pain by moving and scraping the intestine. This is similar to what happens to those who have flatworms.

2.180. Why do mares carry a mule more easily[504] in pregnancy, although the mule is stronger? Perhaps it is true for just this reason, just as women carry males [more easily] than females? For being stronger [the males] use up the residues more, and the heaviness and discomfort come from these.

---

[503] Usener deletes the last clause.

[504] *sc.* than a foal.

2.181. Διὰ τί οὖν ῥᾷον ἀπαλλάττουσαι οὐ δύνανται συνεχῶς φέρειν ἡμίονον; διὸ καὶ διαλείπουσιν, ὥσπερ νεὸν λείποντες; ἢ ὅτι γηράσκει καὶ καταρρακοῦται τὰ σώματα θᾶττον ἰσχυροτέρων ὄντων, τοῦτο δὲ εὐλαβοῦνται; καὶ ἅμα τὸ μὴ δύνασθαι συνεχῶς τὸ ἰσχυρότερον ὥσπερ γῆν καρπὸν εὔλογον, ἀλλ᾽ ὅταν τέκῃ, κουφότερον ἀπαλλάττειν. τὸ δὲ καὶ κυΐσκεσθαι ῥᾷον ἡμίονον, ὥσπερ Ἀλκίμαχος ἔφη τοὺς κελτοὺς λέγειν, οὐκ ἀληθές· οὐ γὰρ προσδέχονται τοὺς ὄνους ὅτανπερ ἀκμάσῃ ἡ ὁρμή, δυσχεραίνουσαι τὴν ὀσμὴν καὶ τὸ ἀσύνηθες· διὸ καὶ πολλὰ δοκοῦσαι κύειν ἀναφαίνονται κεναί.

2.182. Διὰ τί δὲ ῥᾷστα καὶ τάχιστα ἐκτιτρώσκει ἡ ἵππος; ἢ ὅτι σφοδρότατον καὶ συντονώτατον, ὥσθ᾽ ὅ τι ἂν πονῇ, πᾶν μετὰ βίας; ἐκ δὲ τῶν τοιούτων ἐκτιτρωσμός.

2.183. Διὰ τί καὶ ἐνδίδωσι τάχιστα καὶ ἀναστρέφεται τὸ σῶμα; ἢ ὅτι μανὴ ἡ σάρξ; ταχὺ δὲ καὶ συμπίπτει καὶ ὀγκοῦται. ἥπερ τοῦτο αἴτιον, οἷον καὶ τὴν κύνα καὶ τὴν αἶγα καὶ εἴ τι ἄλλο εὐκίνητον ἐπ᾽ ἄμφω; ἢ εἰ μὴ ταὐτό ἐνδέχεται τοῖς ἑτέροις δι᾽ ἄλλας αἰτίας.

2.184. Διὰ τί οἱ μὲν ποικίλοι ἵπποι ὡς ἐπίπαν ἀγεννεῖς, τῶν δὲ φαλιῶν ἔνιοι

2.181 (AFKMOPRSTbeim Buss. Us.; *v* 3(5).43(45); 2.176 Buss.)
2 *ἡμίονον* F Buss. Us.: *ἡμιόνους* K*v*: *ἡμιόνων* cett. || *διὸ δὴ* KOSbi*v* || *διαλείπουσιν* K*v* Buss. Us.: *διαλυποῦσιν* cett. || *νεὸν* F Buss. Us.: *νέον* cett. || *λείποντες* M Us.: *λιπόντες* Buss.: *ληποῦντες* sic A: *λυποῦντα* P: *λυποῦντες* cett. 3 *καταρρακοῦνται* MRe^{ac} 4 *τῷ μὴ* OPRSTbeim || *τὸ ἰσχυρότερον συνεχῶς* MRe^{ac} 5 *ἀπαλλάττειν* P Us.: *ἀπαλάττει* F: *ἀπαλλάττει* cett. || *καὶ* om. F 6 *κυΐσκεσθαι*] *ἡνίσκεσθαι* F: *κυΐσκιται* sic R || *ἡμίονον* PTm Buss. Us.: *ἡμιόνων* cett. 7 *προσέσχονται* T || *ὅτανπερ ἀκμάσῃ* Us.: *ὅτι παρακμάσει* A: *ὅταν παρακμάσωσι* K*v* et Buss. in var. lect.: *ὅταν παρακμάσῃ* cett. || *ἡ ὁρμή* Us.: *πως* K*v* Buss. (qui tamen in var. lect. *πως* delendum censuit): *πως μὴ* cett. 8 *δυσχεραίνουσι* A || *καὶ*^{1}] *διὰ* K*v* || *πολλαὶ* FMPT: *πολλάκις* K*v* || *δοκοῦσι* FP || *κύειν*] *ἔχειν* A Buss.
2.182 (AFKMOPRSTbeim*p* Buss. Us.; *v* 3(5).44(46); 2.177 Buss.)
2 *σφοδρώτατον* S 3 *τῶν τοιούτων*] *τούτων* F
2.183 (AFKMOPRSTbeim Buss. Us.; *v* 3(5).45(47); 2.178 Buss.)
1 *Διὰ τί*] *διότι* e || *ἀνατρέφεται* K (sed *revertitur v*) 2 *μανὴ* MRe^{ac} Buss. Us.: *μόνη* cett. || *δὲ* om. KOSbi*v* || *συμπίπτει*] *συμπίνει* MRe^{ac} || *ἥπερ*] *ὅπερ* sic F: *εἴπερ* K: *ἥπερ* AM || *οἷον* KP*v* Buss. Us.: *οἵαν* AFMORSbeim: *οἳ ἂν* T 3 *εὐκίνητον* Us.: *συγκινητὸν* codd. Buss. 4 *ταὐτὸ* M Us.: *ταῦτα* T: *ταὐτ᾽* cett. || *ἐνδέχεται* FKOSbe^{mg}i Us.: *ἐνδέδεται* AMPRe^{ac}m Buss.: *ἐνδέδοται* T || *ἑταίροις* PR^{pc}Sbe || *δι᾽*] *δ᾽* F
2.184 (AFKMOPRSTbeim Buss. Us.; *v* 3(5).46(48); 2.179 Buss.)
1 *φαλιῶν* Buss. Us.: *φαλλιῶν* sic A: *φαλίων* cett. 2 *φαλιώτης* KOi || *ποικιλότης*]

2.181. Why, having been delivered more easily,[505] can they not continuously bear mules,[506] and for that reason stop, as if leaving land fallow? Is it because their bodies grow old and are reduced to shreds more swiftly because [the mules] are stronger, and they avoid this? And at the same time it is reasonable that they cannot continuously [bear] what is stronger, as the earth [cannot continuously bear] fruit, but when they have given birth, they are delivered more easily. That they also conceive a mule more easily, as Alcimachus[507] says the Celts say, is not true; for they do not receive the asses whenever they are on heat,[508] disliking their smell and their unfamiliarity; and for this reason many that seem to have conceived turn out to be empty.

2.182. Why does the mare miscarry very easily and very quickly? Is it because it is very eager and tense, so that whatever it does with effort, it does forcefully? And such [features] produce miscarriage.

2.183. Why does it both yield and withdraw its body very quickly? Is it because its flesh is rare? So it is quick both to contract and to swell. Or is the cause similar to that in the bitch and the she-goat and any other [creature] which is readily moved in both respects? Or if it is not the same, it may be for other reasons in the others.

2.184. Why are piebald horses generally ill-bred, but some with a white

---

505 As indicated in the previous problem.

506 Cf. Ar., *HA* 6.23 577b11-13.

507 Alkimachos (5b) in *RE* Supplbd. 1, 1903, 62. Not otherwise known.

508 See above, 2.163.

γενναῖοι, καίτοι ἡ φαλιότης ποικιλότης· ἢ ὅτι ἐπ' ἔλαττον ἡ ἐξαλλαγή; ἐν μικρῷ γὰρ τὸ φαλιόν, ὥστε ἡ ποικιλία διὰ τὸ ἀνώμαλον ἀγεννεῖς ποιεῖ· τὸ μικρὸν γὰρ οὐκ ἐνισχῦον ἧττον ἐξίστησιν.

2.185. *Διὰ τί οἱ πῶλοι κατακλίνονται μᾶλλον ἢ οἱ πρεσβύτεροι ἵπποι; ἢ ὅτι ῥᾷον καὶ ἀνίστανται καὶ κατακλίνονται οἱ νέοι τῶν πρεσβυτέρων· τοῦτο δ' ὅτι τοῖς μὲν πολὺ τὸ βάρος, τοῖς δ' ὀλίγον· τὰ μὲν γὰρ κῶλα οὐκ ἐπαυξάνεται τοῖς ἵπποις, τὸ δὲ σῶμα ἐπαυξάνεται· εὐφορώτερον οὖν τὸ σῶμα τοῖς νέοις.*

2.186. *Διὰ τί τὸ τῶν ἵππων περίττωμα ὅταν γυμνάζωνται δυσωδέστερον; πότερον ὅτι δυσέκκριτον κινεῖται ὑπὸ τῶν γυμνασίων, ὥσπερ ὁ ἱδρὼς ἔξω, οὕτως εἰς τὴν κοιλίαν φαῦλοι χυμοί· οἳ ποιοῦσι τὴν δυσωδίαν (ὥσπερ τοῖς καθαιρομένοις κάτω ὑπὸ φαρμάκου). ἢ ὅτι πλείω προσφέρονται τότε καὶ ἧττον πέττουσι, τὰ δὲ ἄπεπτα δυσωδέστατα καὶ σήπεται θᾶττον;*

2.187. *Διὰ τί ἵππος καὶ ὄνος οὐρεῖ μετὰ τὴν ὀχείαν; ἢ διὰ τὸν πόνον; πονούντων γὰρ συρρεῖ εἰς τὴν κύστιν, πονεῖ δὲ ὄνος μὲν καὶ ἵππος διὰ τὸ μέγεθος· βοῦς δὲ διὰ τὴν συντονίαν· τὸ γὰρ σύντονον ποιεῖ πόνον.*

2.188. *Διὰ τί ὄνος μὲν ἢ ἵππος ἐξουρεῖ τὸ σπέρμα, βοῦς δὲ οὐκ ἐξουρεῖ; ἢ ὅτι τοῖς μὲν πόρρω ἐξακοντίζεται διὰ τὴν συντονίαν, τοῖς δὲ ἐγγύς (ἅμα γὰρ προΐεται καὶ ἐξάγει);*

ποικιλία Us. 3 *φαλιόν* Buss. Us.: *φαλίον* codd. || *ἀγενεῖς* AST Us. 4 *ἐξανίστησιν* KORSbei

2.185 (AFKMOPRSTbeim*p* Buss. Us.; *v* 3(5).47(49); 2.180 Buss.)

1 *ἢ ὅτι* K*v* Us.: om. cett. 2 *ῥᾷον* T Us.: post *κατακλίνονται* exhibent K*v*: om. cett. || *τοῦτο δ'*] *τοῦδ'* S 4 *ἐπαύξεται* bis K || *εὐφορώτερον*] *εὐφορώτατον* KOSbi*v*

2.186 (AFKMOPRSTbeim*p* Buss. Us.; *v* 3(5).48(50); 2.181 Buss.)

1 *τὸ* om. Oi 2 *κινεῖται* Us.: *γίνεται* AF Buss.: *τε γί(γ)νεται* MPRSTbem: *ἐστιν* KOi || *γυνμασιῶν* Us. || *ὁ ἱδρὼς ἔξω*] *ἐν ἄλλῳ τόπῳ* A Buss. 3 *οὕτως εἰς τὴν κοιλίαν*] *εἰς παγκαλίαν* F || *φαῦλοι ⟨γὰρ⟩* Buss. || *οἱ χυμοὶ* M || *οἳ*] *ἢ* T || *οἳ ποιοῦσι τὴν δυσωδίαν* in hunc locum transpos. Us.: post *φαρμάκου* (4) exhibent codd. *v* Buss. 4 *φαρμάκου*] *φαρμάκων* F*p*: *φαρμάκω* T || *φέρονται* F || *τότε*] *ὥστε* F

2.187 (AFKMOPRSTbeim*p* Buss. Us.; *v* 3(5).49(51); 2.182 Buss)

1 *ὁ ἵππος* P || *ἡ ὄνος* P 2 *πονούντων* om. m || *συρρεῖ γὰρ* m || *πονεῖ δὲ*] *πονεῖ μὲν* F: *πονεῖ* MRe^{ac}: ante *πονεῖ* add. *πόνος δὲ* e^{mg} || *μὲν* om. F

2.188 (AFKMOPRSTbeim Buss. Us.; *v* 3(5).50(52); 2.183 Buss)

1 *ἢ*^{1} FKMOPRSbeim*v*: *καὶ* AT*p* Buss. Us. 1-3 *ἢ – ἐξάγει* om. M 2 *τοις*^{1}] *τῷ* Us. 3 *προΐεται* Us.: *emittunt* *v*: *ποιεῖται* AFPTm Buss.: *ποιεῖ* KORSbei

mark well-bred, although having a white mark is being piebald? Is it because the departure [from type] is to a lesser extent? For the white mark is in a small [area]. So being piebald makes them ill-bred because of the unevenness; for what is small does not have strength and makes them depart [from type] less.

2.185. Why do foals lie down more than older horses? Is it because the young both stand up and lie down more easily than the old? This is because the weight of the latter is great, that of the former small. For horses' legs do not grow, but their bodies do.[509] So the young carry their bodies more easily.

2.186. Why does the excrement of horses smell worse when they exercise? Is it because what is excreted with difficulty is moved by the exercise, as sweat to the exterior, so bad humours into the bowel, and these produce the bad odour (as with those who are purged downwards by a drug)? Or is it because they consume more then and concoct it less, and what is unconcocted has the most evil odour and rots more quickly?

2.187. Why do the horse and ass urinate after mounting? Is it because of the effort? For when they make an effort it runs together into the bladder, and the horse and the ass make an effort because of their size. The bull because of its vehemence; for vehemence produces effort.

2.188. Why do the [she-]ass or mare[510] expel the seed together with the urine, but the cow does not? Is it because in the case of the latter it is projected a long way because of the vehemence [of the bull], but in that of the former [only to the place that is] nearby (for [the males] emit it and withdraw at the same time)?

---

509 See above, 2.136.

510 The Greek does not indicate the sex of any of the animals in this sentence. But see below, 2.192.

2.189. Διὰ τί τοῖς μὲν ἵπποις καὶ τοῖς ὄνοις, ἐκ τῶν οὐλῶν φύονται τρίχες, τοῖς δ' ἀνθρώποις οὔ; ἢ ὅτι τοῖς μὲν ἄλλοις ζῴοις ἐστὶ τὸ δέρμα μέρος τῆς σαρκός, ἀνθρώπων δὲ καθάπερ πάθος σαρκός; δοκεῖ γὰρ αὐτῆς τὸ ἐξ ἐπιπολῆς καταψυχόμενον στερεώτερον γίνεσθαι, καθάπερ τῶν ἑφθῶν ἀλεύρων αἱ γραῖαι καλούμεναι· ἐπεὶ οὖν κἀκεῖναί εἰσιν ἄλευρον ἑφθόν, καὶ τὸ τοῦ ἀνθρώπου καλούμενον δέρμα σάρξ ἐστι· τραυματισθέντος δὲ ἢ τριβέντος ἀνθρώπου συμβαίνει τὴν σάρκα φθείρεσθαι καὶ πυκνοῦσθαι· διὸ ἠλλοιωμένης τῆς ἐπιπολῆς σαρκὸς οὐ τὴν αὐτὴν ἐπίφυσιν λαμβάνει τὰ τραύματα ἥνπερ καὶ ἐκ γενετῆς· ἀλλοιωθείσης δὲ αὐτῆς μὴ γίνεσθαι ἔτι τὸ ἐξ αὐτῆς οὐδὲν ἄτοπον, καθάπερ καὶ ἐκ τῶν ἀλωπεκιῶν καλουμένων· καὶ γὰρ τῆς ἐπιπολῆς σαρκὸς φθοραὶ καὶ ἀλλοιώσεις εἰσί· τῶν δ' ὑποζυγίων τριβέντων καὶ ὑγιαζομένων πάλιν ἀναπληροῦται τὰ νοσώδη μέρη τοῦ σώματος τῶν αὐτῶν μέν, ἀσθενεστέρων δὲ ἢ ἐξ ἀρχῆς· ἐπειδὴ δ' ἐστὶ καὶ τὸ δέρμα μέρος αὐτῶν, γίνοιτο ἂν καὶ φύοιντο τρίχες, λευκαὶ δὲ φύονται διὰ τὸ δέρμα ἀσθενέστερον γεγονέναι τοῦ ἐξ ἀρχῆς καὶ τὴν λευκὴν ἀσθενεστάτην εἶναι τρίχα.

2.190. Διὰ τί εἰ ἐπὶ τὰ τῶν ἵππων ἕλκη κριθὰς κατακαύσαντες καὶ

2.189 (AFKMOPRSTbeim Buss. Us.; [Ar.] 10.27; *ν* 3(5).51(53); 2.183bis Buss) 1 *οὐλῶν*] *αὐλῶν* F: *οὔλων* T 1-2 *Διὰ τί – οὔ* om. M (cf. ad 2.188.1-3) 2 *ὅτι*] *διότι* [Ar.] 2-3 *τῆς σαρκὸς μέρος* m 3 *ἀνθρώπων*] *ἀνθρώπῳ* FM [Ar.] Us.: *ἀνθρώποις* Kmv ‖ *πάθος σαρκὸς* [Ar.] Us.: *πάγος σαρκὸς* Buss.: *μέρος σαρκὸς ἢ πάχος* A: *πάχος σαρκὸς* cett. ‖ *αὐτῆς*] *αὐτοῖς* AFKm et [Ar.]Ap.C[a]X[a] 3-4 *ἐξ ἐπιπολῆς*] *ἐπιπολῆς* tantum [Ar.] 5 *ἀλεύρων*] *ἀλλ' εὗρον* T ‖ *γραῖαι*] *ἀγρίαι* [Ar.] codd. ‖ *οὖν* om. T 6 *ἐστι* KOSbe[mg]i*ν* Us.: *ἐστιν* Buss.: *εἶναι* AFMPRTe[ac]m: *ἂν εἴη* [Ar.] ‖ *ἢ*] *καὶ* A Buss. 7 *τοῦ ἀνθρώπου* K ‖ *ἀνθρώπου μὲν τὴν σάρκα συμβαίνει* [Ar.] ‖ *φθείρεσθαι καὶ* om. [Ar.] 7-8 *διὸ ἠλλοιωμένης* [Ar.] Us.: *διηλλοιωμένης* codd. Buss. 8 *τῆς*] *δὲ τῆς* K*ν* Buss.: om. F ‖ *οὐ* AFPTm [Ar.] Buss. Us.: om. KMORSbei*ν* ‖ *αὐτὴν* om. P ‖ *ἐπίφυσιν*] *ἐπὶ φύσιν* R: *ἐπίφυσις* T: *φύσιν* tantum [Ar.] 9 *ἥνπερ*] *ὅνπερ* F: *ἧπερ* [Ar.] 10 *τὸ ἐξ αὐτῆς* M [Ar.] Us.: *ἐξ αὐτῆς* codd. Buss. ‖ post *ἐξ αὐτῆς* add. *τρίχας* K*ν* ‖ *ἐκ τῶν*] *αἰτιῶν* M 11 *καὶ γὰρ αὗται* [Ar.] 12 *ἀναπληροῦνται* F ‖ *νοσώδη* KOSbe[mg]i Us.: *νενοσηκότα* [Ar.] Buss.: *νοσήματα* APT: *νοσήσαντα* m: *μνησάσαντα* F: *δυσώδη* MRe[ac] 13 *ἐξ ἀρχῆς ἦν* [Ar.] 13-14 *ἐπειδὴ δ'*] *ἐπεὶ δ'* K [Ar.]: *ἐπειδὴ δὲ* T 14 *αὐτῶν*] *αὐτῆς* P ‖ *γίνοιτο ἂν*] *γίνοιτό τ' ἂν* [Ar.] ‖ *φύοιντο* [Ar.] Us.: *φύοι* KP*ν*: *φῦον* Buss.: *φύον* cett. ‖ *τρίχες* P [Ar.] Us.: *τρίχας* cett. ‖ post *τρίχες*: *ἐκ δέρματος γὰρ φύονται τρίχες* [Ar.] 15 *φύονται*] *γίνονται* m ‖ *τὸ τὸ* AKe[pc] et [Ar.] codd. plerique 16 *ἀσθενεστέραν* K*ν*

2.190 (AFKMOPRSTbeim*p* Buss. Us.; *ν* 3(5).52(54); 2.184 Buss)
1 *εἰ* M Buss. Us.: *ἐὰν* K*ν*: om. cett. 2 *λευκαὶ*] *λευκὰς* M Us. ‖ *ἀναφύονται*]

2.189.[511] Why do hairs grow from the scars of horses and asses, but not of humans? Is it because in the other animals the skin is a part of the flesh, but in humans it is like something that happens to the flesh? For it seems that the part of it that is on the surface is chilled and becomes more solid, as with what is called the crust of boiled wheatmeal. For this [crust] too *is* boiled wheatmeal, and what is called the skin of a human is flesh. So when a human is wounded or chafed the result is that the flesh is destroyed and compressed; and for this reason, when the flesh on the surface has been changed, wounds do not have the same growth over them that [they did] from birth. It is not at all strange that when it has been changed what came from it no longer does so, just as from alopecia; for this too is a corruption and alteration of the surface skin. But when yoke-animals are chafed and heal the affected parts of the body are filled again with the same things, but weaker than in the beginning. Since their skin is a part of them, hairs will occur and grow, but they are white because the skin is weaker than it was originally and white hair is the weakest.

2.190.[512] Why, if they burn and grind up barley and sprinkle it on the

---

[511] = [Ar.], *Probl.* 10.27. Cf. above, 2.172, with a different explanation.

[512] Cf. [Alex.], *Probl.* i.150 (similar explanation in different words).

λεπτύναντες ἐπιπάττουσιν, οὐ λευκαὶ ἀναφύονται αἱ οὐλαί, ἀλλ' ὁμοιόταται ταῖς πρότερον, ἄν τε μέλας ὢν τύχῃ ἄν τε ξανθὸς ἢ ὁποιοσοῦν ᾖ; ἢ ὅτι λευκὴ διαφθειρομένης τῆς τροφῆς καὶ τῆς ἀρχῆς ὅθεν ἡ ἔκφυσις, αὕτη δὲ φθείρεται διυγραινομένη καὶ λεπτύνουσα τὸ δέρμα, τὸ δὲ τῆς κριθῆς κατάκαυμα ξηρὸν ὄν, ἐξαίρει καὶ ἀναλαμβάνει ταύτην, ὅθεν ἐξομοιοῖ τὸ δέρμα τῷ ἄλλῳ, συνεξομοιωθέντος δὲ καὶ ἡ ἔκφυσις ὁμοία; καὶ γὰρ τοῦ δέρματος ἡ ἔκφυσις.

2.191. *Διὰ τί οἱ ἵπποι κατὰ τὴν ἕδραν ἐν τοῖς ποταμοῖς ἀποπνίγονται; ἢ ὅτι εὐρὺ τὸ ἔντερον ἔχουσι τὸ ἀπὸ τοῦ ἀρχοῦ; δηλοῖ δὲ ὅτι εὐμαρῶς ἀφοδεύουσι βαδίζοντες· πληθύοντα ⟨δὲ⟩ τὴν ἕδραν διαδίδωσιν εἰς τὴν κοιλίαν, ἧς ἐπαιρομένης θλιβομένη ἡ ἀρτηρία ἀποπνίγει.*

2.192. *Διὰ τί τὰς ἵππους καὶ τὰς ὄνους ὅταν ὀχευθῶσι, διώκουσιν; ἢ ὅτι οὐροῦσιν εὐθύς; συμβαίνει οὖν ἐξουρεῖν τὴν γονήν.*

ἀναφύουσιν F Us.: *ἀναφαίνονται* APTm: *φαίνονται p*: *γίνονται ἀναφυόμεναι* Se[mg] || *αἱ* om. A Buss. 3 *ὁμοιοτάτας* Us. || *ταῖς*] *τῶν* A || *πρότερον*] *προτέρον* A: *προτέραις* OP || *ὁποιοσοῦν*] *ὁποῖος οὖν* AMRe: *ὁποῖος ἂν* F 4 *ἢ* KT*v* Us.: om. cett. Buss. || *λευκὴ*] *ἡ λευκὴ* K: *λευκαὶ* Us. || *τῆς*[2] om. F || *ἔκφυσις γίνεται* K*v* 5 *αὕτη* AFKMP Buss. Us.: *αὐτὴ* ORSTbeim*v* 7 *ἐξομοιοῖ* AKe*p* Buss. Us.: *ἐξομοιοῖ* cett. || *τῷ ἄλλῳ*] *τὸ ἄλλον* m || *καὶ*[1] om. MRe 8 *ἡ* om. A

2.191 (AFKMOPRSTbeim*p* Buss. Us.; *v* 3(5).53(55); 2.185 Buss.)

1 *ἀναπνίγονται* M 2 *τὸ ἔντερον εὐρὺ* Buss. || *τοῦ ἀρχοῦ*] *τῆς ἀρχῆς* A Buss. || *εὐμαρῶς*] *ἀβαρῶς* Buss. 3 *βαδίζοντες*] *βαδίζοντα* FORSTbei: om. Buss. || *πληθύοντα ⟨δὲ⟩* Us.: *πολὺ οὖν τὰ κατὰ* Buss.: *πολὺ ὄντα* AMOPRSTbei: *πολὺ ἔχοντες* K*v*: *ἀπολύονται* m: om. F || *διαδίδωσιν* Buss.: *διαδώσει* Us.: *διαδῶσιν* A: *διαδώσειν* FMOPRSTbeim: *πλατεῖαν διαβαίνουσα* K*v* 4 *θλιβομένη* om. F

2.192 (AFKMOPRSTbeim*p* Buss. Us.; *v* 3(5).54(56); 2.186 Buss)

1 *τοὺς ἵππους* AFOTi || *τοὺς ὄνους* F || *ὅταν ὀχευθῶσι*] *ὀχευθείσας* MRe 2 *συμβαίνει*] *σημαίνει* m

wounds of horses, do the scars not grow white,[513] but very like those before, whether black or brown or whatever? Is it because [a hair] is white when the nourishment and the principle from which it grows is destroyed, and it is destroyed when the skin becomes moist and thin,[514] and the burned barley, being dry, draws out and receives [the moisture], and as a result of this it makes the skin like the rest, and when it is made like the rest the outgrowth is similar too? For the outgrowth [of hair] is from the skin.

2.191. Why do horses suffocate through their rump in rivers? Is it because their intestine is wide from the anus? This is shown by the fact that they excrete solid matter easily while walking.[515] The rump fills up [with water] and sends it into the intestine, which swells up and crushes the windpipe so that they suffocate.

2.192.[516] Why do they chase mares and she-asses when they have been mounted? Is it because [otherwise] they urinate at once? So the result is that they expel the seed with the urine.

---

513 i.e., grow white hairs. With Usener's text this is explicitly stated; but his reading involves selection of words from a minority of MSS, and the reading of the majority of MSS, that the scars (themselves) do not grow (intransitive) white, though not entirely logical, is clear enough in the context.

514 The reference here, by contrast with 2.172, seems to be to the state of the wound *before* scar tissue forms. See above, n.331 on 2.71. There and in 2.172 it is the *density* of the scar that hinders the growth of hair; here the point is that the skin is (initially) thinner where there has been a wound, and thinness of skin goes with lack of hair, because there is less matter to make up the hair.

515 See above, 2.131.

516 Similarly Ar., *HA* 6.23 577a23, *GA* 2.8 748a21 (only of she-asses). Richter 1885, 31. See above, 2.188.

<titulus deest>

3.1. ⟨Διὰ τί ἐὰν⟩ τις καρύῳ ἐγγράψας γράμματα φυτεύσῃ, τὸ ἐξ αὐτοῦ φυὲν δένδρον ταὐτὰ ἔχοντα γράμματα οἴσει τὰ κάρυα; ὅτι καὶ ἐξ ἀνθρώπου οὐλὴν ἔχοντος γέγονεν ἔγγονος ἐν τῷ αὐτῷ τόπῳ ἔχων οὐλήν.

3.2. Διὰ τί ἐπὶ τῶν μνημάτων μάλιστα τὰς καρύας φυτεύουσιν; ἢ ὅτι τὸν καρπὸν φέρουσι προσεοικότα τοῖς σοροῖς; ἢ ὅτι δάκρυον μάλιστα ἀφίησι τῶν δένδρων ὃ τοῖς τελευτῶσιν ἀποδιδόαμεν γέρας;

3.3. Διὰ τί τὰ ἀμύγδαλα μόνα τῶν καρπῶν καὶ σκληρὰ καὶ μαλακὰ καὶ γλυκέα καὶ πικρά; ἢ ὅτι μάλιστα ἥμερά ἐστι τῶν καρπῶν, μὴ τυχόντα δ' ἐπιμελείας εἰς τὸ πικρὸν μεταβάλλει;

3.4. Διὰ τί ἡ τῆς ὑὸς κόπρος ἐξ ὀξείας γλυκερὰν τὴν ῥόαν ποιεῖ; ἢ ὅτι θερμοτάτη ἐστίν; δῆλον δ' ὅτι εὖ ἐξίκμασται. ἔχει οὖν ἐν αὐτῇ ὥσπερ ἡ κονία πῦρ· τοῖς δ' ἄλλοις φυτοῖς ἀσύμμετρος διὰ τὸ καυστικὴν ⟨εἶναι⟩.

3.5. Διὰ τί εἴ τις ἐξέλοι τοῦ κλήματος τὴν ἐντεριώνην, γίνονται οἱ βότρυες γίγαρτα μὴ ἔχοντες; ἢ ὅτι ὥσπερ ἡ ἰλύς ἐστιν ἐν τῷ οἴνῳ, ἀπὸ τῶν γιγάρτων γίνεται, οὕτω καὶ ἐν τῇ ῥαγὶ αὐτῇ τὸ μετὰ τῆς τροφῆς τῆς ὑγρᾶς συνεισιὸν γεῶδες εἰς μέσον ὑφίσταται; μεῖζον δ' οἱ πυρῆνες πάντων τῶν ἀγρίων εἰσίν.

3.6. Διὰ τί τὸ πήγανον βασκανίας φασὶ φάρμακον εἶναι; ἢ βασκαίνεσθαι δοκοῦσι π⟨νευμάτων ὁ⟩ρμώντων οἱ λάβρως καὶ πολλὰ ἐσθίοντες; τὸ πήγανον οὖν προεδεσθὲν θερμαντικὸν ὂν φύσει ἠραίωσε τὸ δ⟨εχόμεν⟩ον

3.1 (A Buss.)
1 Διὰ τί ἐὰν add. Buss.: A legi non potest || αὐτοῦ A 2 ἔχον τὰ sic A 3 ἔγγονος] ἔγγονον sic A
3.2 (A Buss.)
3.3 (A Buss.)
3 μεταβάλει A^{ac}
3.4 (A Buss.)
1 ἐξοξίας sic A 2 θερμοτάτει A || ὥσπερ] ὄσπερ, i.e. ὥσπερ, A: ὡς Buss. 3 εἶναι suppl. Buss.
3.5 (A Buss.)
3 αὔτη A 4 μεῖζον δ' οἱ πυρῆνες Buss.: μείζων δὲ καὶ οἱ πυρίναις A
3.6 (A Buss.; [Ar.] 20.34 926b20-21 et 28-31)
1 ἢ διότι βασκαίνεσθαι [Ar.] 2 πνευμάτων ὁρμώντων Buss.: π ..... ρῶντων A: om. [Ar.] || οἱ om. [Ar.] || καὶ πολλὰ om. [Ar.] 3 τῇ φύσει [Ar.] || δεχόμενον [Ar.] Buss.: δ .... ων A 4 τὰ σίτια ἀγγεῖον Buss.: τὰ σίτια ἀγγεί A: ἀγγεῖον τὰ

<NATURAL PROBLEMS>

<BOOK III>

3.1. <Why if> one carves letters on a nut and plants it, does the tree that grows from it bear nuts with the same letters? Because a man who has a scar, too, begets a child with a scar in the same place.[518]

3.2. Why do they plant nut-trees[519] especially on tombs? Is it because they bear fruit that is like caskets? Or is it because it most of all trees sheds tears, which we give as a offering to the dead?

3.3. Why are almonds alone of fruit both hard and soft and sweet and bitter? Is it because they are the most domesticated of fruits, and if they are not looked after, they change to bitterness?[520]

3.4. Why does the manure of pigs make the pomegranate sweet instead of sharp?[521] Is it because it is very hot? It is clear that it dries well. For it has fire in it, like quicklime; but for other plants it is disproportionate because of its burning effect.

3.5. Why, if one takes out the pith of the branch, do the grapes grow without pips?[522] Is it because as the dregs are in the wine, [for] they come from the pips, so in the grape itself the earthy matter that enters in along with the moist nourishment settles into the middle? For the kernels are the greater [part] of all wild fruits.

3.6.[523] Why do they say that rue[524] is a charm against witchcraft? Is it that people seem to be bewitched when they suffer wind because they eat much greedily? So rue eaten beforehand, having a natural heating effect, makes

---

[518] Both the claim in the problem and that in the solution are of course false.

[519] *karua*: probably hazel, *Corylus avellana* L. The word is also used for the sweet chestnut, *Castanea sativa* L. (but the cases of its nuts do not much resemble funerary urns) and for the walnut (*Juglans regia* L.).

[520] Cf. Theophrastus, *CP* 2.15.1.

[521] Cf. Theophrastus, *HP* 2.2.11, *CP* 2.14.2.

[522] Cf. Theophrastus, *CP* 5.5.1.

[523] A shortened version of [Ar.], *Probl.* 20.34.

[524] *Ruta graveolens* L.

τὰ σιτία ἀγγεῖ⟨ον⟩· διὸ ἐξίεσθαι τὸ ἐγκ⟨αταλαμ⟩βανόμενον πνεῦμα συμβαίνει.

3.7. Διὰ τί ⟨οἱ ἄναλοι ἄ⟩ρτοι πλείονα σταθμὸν ἕλκουσι τῶν ἡλισμένων, καίτοι οἱ ἅλες βαρύτεροι ὕδατος [ὄντες]; ἢ ὅτι ξηραίνουσιν οἱ ἅλες.

3.8. Διὰ τί οἱ ψυχροὶ ἄρτοι ἐὰν βρεχθέντες ἅψωνται ἀλλήλων, οὐ συνέρχονται, οἱ δὲ θερμοί; ἢ ὅτι οἱ ψυχροὶ μετὰ τῆς ἀτμίδος ἀφῆκαν τὸ ἐν αὐτοῖς γλίσχρον ὑγρόν· οὗ ἐξελθόντος οὐ κολλῶνται· οἱ δὲ θερμοὶ ἔχουσι τινὰ γλισχρότητα;

3.9. Διὰ τί οἱ ἡμιπλήρεις πίθοι ὀξίζουσιν; ἢ διότι τὸ ὄξος ὥσπερ ⟨τις πέψις⟩ οἴνου ἐστί, σήπεται δ' ὑπὸ τοῦ ἐκτὸς ἀέρος;

3.10. Διὰ τί ὁ οἶνος χείρων γίνεται μεταντλούμενος ἐν τοῖς νοτίοις ἢ ἐν τοῖς βορείοις; ὅτι ἐν ταῖς ἀλέαις μᾶλλον [.........]ανται καὶ χωρίζεται τῆς τρυγὸς καὶ ἀσθενέστερος γίνεται, ἡ δ' ἄντλησις χωρίζει τὴν τρύγα;

3.11. Διὰ τί δήποτε ὅταν ὁ οἶνος εἰς ἀσκὸν ἀντληθῇ, ἐξ ἀνάγκης μὲν τὸν

σίτια καὶ τὸ ἄλλο σῶμα [Ar.] 4-5 *διὸ – συμβαίνει* Buss.: *διὸ εξίεσθαι τὸ ἐγκ .... βανόμενων, πνεύμα συμβένει* sic A: *διὸ ἔξωθεν* (*ἐξίεσθαι* Gaza in [Ar.]) *τὸ ἐγκαταλαμβανόμενον πνεῦμα συμβαίνει* [Ar.]

3.7 (A Buss.; [Ar.] 21.5 927a35-38)

1 *⟨οἱ ἄναλοι ἄ⟩ρτοι* suppl. Buss. ex [Ar.]: evanuerunt in A || *ἕλκουσι*] *ἔχουσι*[Ar.] 2 *τοῦ ὕδατος* [Ar.] || *ὄντες* del. Buss.: *εἰσιν* [Ar.] || *ἢ ὅτι* A [Ar.]: *ὅτι* Buss.

3.8 (A Buss.; [Ar.] 21.6 927b6-11)

1 *οἱ μὲν ψυχροὶ* [Ar.] 2 *συνέχονται* [Ar.] || *οἱ μὲν ψυχροὶ* [Ar.] || *ἀφῆκαν*] *ἀφιᾶσι* [Ar.] 4 *τινὰ* A [Ar.]: om. Buss.

3.9 (A Buss.; cf. Macrobius, *Sat.* 7.12.11-12)

1 *τις πέψις* add. Buss. || *πέψις*] *σῆψις* coni. Buss. in var. lect. 1-2 *ἢ διότι – τοῦ ἐκτὸς ἀέρος*] *in illud enim vacuum quod superne liquido caret aer advena incidit, qui tenuissimum quemque humorem elicit et exorbet: eo siccato vinum quasi spoliatum viribus, prout ingenio imbecillum aut validum fuit, vel acore exasperatur vel austeritate restringitur* Macrobius 2 *ἐκτὸς* A: *ἔξω* Buss.

3.10 (A Buss.)

3.11 (A Buss.)

1 *ἀντληθῇ*] *ἀντλησθεῖ* sic A 3 *συ⟨νηθέστερον⟩* suppl. Buss. || *οὕτως*] *οὗτος* sic A

the vessel that receives the food rarer in texture, and the result is that the trapped wind is released.

3.7.[525] Why do unsalted loaves of bread weigh more than salted, although salt is heavier than water? Because the salt dries them.

3.8.[526] Why if cold loaves which have been wetted touch each other do they not coalesce, but hot ones do? Is it because the cold ones have emitted the sticky moisture in them along with the steam, and when this has departed they do not stick, but the hot ones have a certain stickiness?

[On Wine and Olive-Oil][527]

3.9.[528] Why do half-full [wine]-jars turn to vinegar? Is it because vinegar is like a sort of fermenting of wine, but it decays because of the external air?

3.10. Why does wine become worse if decanted during a south wind than a north wind?[529] Because in hot weather ... more ... and is separated from the lees and becomes weaker, and the decanting separates the lees?

3.11.[530] Why indeed when wine is decanted into a wineskin, [although] [the

---

525 A very much compressed version of [Ar.], *Probl.* 21.5.

526 A shortened version of [Ar.], *Probl.* 21.6.

527 The title of a work by Theophrastus: **384** no.2f FHS&G, probably regarded in antiquity as book 7 of the *De causis plantarum*; cf. Sharples 1995, 132-133, and references there.

528 Cf., with Linke 1880, 54 n.2, Macrobius, *Sat.* 7.12.11-12 = Ar., fr.224 Rose^3, contrasting wine with oil which is said to improve in such circumstances; similarly, but answering the problem why honey is best at the bottom of the jar, wine in the middle but oil at the top, Plutarch, *Quaest. conviv.* 7.3. 701E-702B (also cited by Rose): the dense part of the honey sinks to the bottom, wine is best in the middle because it is not exposed to the air, oil is best at the top because it is in contact with the air. Plutarch cites Aristotle in connection with the last point. Our problem appears to be an abridged remnant of a fuller original discussion.

529 Cf. the problematic passage on the effect of the south wind on storage jars in Hippocrates, *Morb. sacr.* 16.

530 Cf., with Bussemaker, [Ar.], *Probl.* 25.8, which is however concerned with the question why wine in wineskins takes up less space than when it was in a jar, and offers the explanation that air in the wine escapes in the process of transfer to the skins.

οἶνον ἔσω δέχονται, καὶ ἔτι κενὸν περίεστιν; ἤγουν ἀφαίρεσις γίνεται τοῦ οἴνου, ἢ συστολή. συ⟨νηθέστερον⟩ τοῦτό πως συμβαίνει. οὕτως Ἀριστοτέλης.

3.12. *Διὰ τί ὁ γλυκὺς οἶνος οὐ γίνεται ἀνθοσμίας, ὁ δ' αὐστηρός; ὅτι ἀποκρινομένου τοῦ γεώδους ἡ ὀσμὴ γίνεται, γλυκὺς δὲ γίνεται τοὐναντίον [διὰ τί] ἀποκρινομένου τοῦ ὑδατώδους;*

3.13. Πότερον *ὁ οἶνος, ἢ τὸ ὕδωρ ἀπεπτότερον; πῇ μὲν γὰρ δόξει τοῦτο, πῇ δὲ ἐκεῖνο· καὶ γὰρ ἔξω τὰ ἐμβαλλόμενα εἰς αὐτὸ διαχεῖται ἑψόμενα καὶ σήπεται ἑώμενα· ὁ δ' οἶνος στύφει καὶ ταριχεύει, καὶ ὅλως τὸ μὲν λεπτὸν ὥστε διαδῦνον διαχεῖν, ὁ δ' οἶνος παχύτερος, ὥστε μὴ ὁμοίως εἰσδύνεσθαι εἰς τοὺς πόρους, τὴν δὲ πέψιν εἶναι διάχυσιν καὶ σῆψιν. εἰ δ' ὁ οἶνος θερμός, τὰ δὲ θερμὰ πεπτικά, ἔχει δ' ἀπορίαν, εἰ θερμός, πῶς οὐχὶ μαραίνει, ἢ μαραίνεται, ὥσπερ πῦρ ἐπὶ πυρί· τὸ μὲν οὖν λεπτὸν εἶναι τὸ ὕδωρ ἀληθές, ἀλλὰ φύσει ψυχρὸν ἔτι. τὸ μὲν ὕδωρ ἄπεπτον εἰσέρχεται καὶ πολλῆς ἐργασίας δεόμενον, ὁ δ' οἶνος πεττόμενος ὑπὸ τοῦ ἡλίου. συμβάλλεται δὲ τὸ ὕδωρ εἰς τὴν πέψιν, οὐ τῷ εὔπεπτον εἶναι, ἀλλὰ τῷ ἄτροφον· διὸ καὶ*

3.12 (A Buss.)
1 *οὐ*] *οὐ* sic A 2 *τοὐναντίων* sic A 3 *διὰ τί* deleviimus

3.13 (A Buss.)
1 *δόξη* sic A ‖ *πῇ*^{2} Buss. in var. lect.: *εἶ* A 2 *αὔτο* sic A 4 *ὁ*] *οἱ* A 5 *εἰκός* post *σῆψιν* addendum coni. Buss. in var. lect. 6 *εἰ*] *ἡ* A 9 *πεττόμενος*] *πεμπόμενος* A 10 *οὐ τῷ*] *οὗτο* sic A ‖ *τῷ*^{2}] *τὸ* A 11 *οἶον*] *ὡς* Buss.

wineskins] receive the wine under pressure, yet there is still a void? Either there is a loss from the wine [by evaporation], or it contracts. This happens in some way more usually. Thus Aristotle.

3.12.[531] Why does sweet wine not develop fine fragrance, but harsh wine does? Because the fragrance occurs when the earthy part is separated out, but [wine] becomes sweet in the opposite way when the watery part is separated out.[532]

3.13. Is wine or water harder to digest? In a way it will seem that the latter is, in a way the former: for things placed into [the water] dissolve when boiled and decay if left; but wine has an astringent and pickling effect. And in general what is rare enters in and separates, but wine is thicker, so that it does not enter into the pores in a similar way, and concoction is a separation and decay. But if wine is hot, and hot things effect concoction, there is a difficulty as to why it does not either weaken or get weakened, as fire applied to fire. Well, it is true that water is rare, but it is by nature cold. So water enters in in an unconcocted state needing much work on it, while wine has been concocted by the sun.[533] Water contributes to concoction not

---

531 Mismatch between odour and flavour generally is discussed by Theophrastus, *CP* 6.9.4, 6.14.5 (linking sweet flavour in fruit to earthiness), 6.16.2 (sweetness in ripened fruit comes from removal of the watery element), 6.16.6-7 (mentioning wine), *Od.* 3, 5, and **431** FHS&G. See also below, 3.13. What *CP* 6.16.6-7 (cf. 6.7.4) makes clear, while our passage does not, is that we are dealing with two successive but overlapping stages of a process, the relation between which will depend on the initial constitution of the wine; the fragrance is produced when the earthy part (the lees) separate out and as the watery part evaporates, the sweetness when what is left is fluid rather than earthy, but also no longer watery.

532 The MS has "...way. Why when the watery part is separated?", so that the problem seems to end with the statement of a further problem, and no solution. In the light of Theophrastus, *CP* 6.16.2 (see the previous note) it seems that "Why" should be deleted.

533 Galen attributes to Aristotle (fr. 226 Rose^3) and to Theophrastus (**425** FHS&G) the view that heat from the sun retained in the grapes continues to concoct and overpower the watery element, using this himself to explain why some new wine appears sweet because of the extra acquired heat in it. He also attributes to them (fr. 221 Rose^3/**427** FHS&G) the view that additional heat and movement hasten the ripening of naturally warm wines but put inferior ones to the test and accelerate the processes – by implication, of spoiling – which were going to occur anyway.

πλεῖον ἐσθίουσιν οἱ ὑδροπόται, οἷον ἀφῃρημένου τινὸς ἀπὸ τῆς τροφῆς.

3.14. Διὰ τί ὁ οἶνος θερμὸς ὢν παύει μᾶλλον τὴν δίψαν ἢ τὸ ὕδωρ ψυχρὸν ὄν; ἢ διότι διὰ τὴν θερμότητα καὶ ὑγρότητα αἱ ἐκθερμανθεῖσαι * * δίψαν ποιοῦσιν;

3.15. Διὰ τί ὁ οἶνος ὁ παριστάμενος καὶ ὁ ἐξιστάμενος λαμβάνει ἄνθος; ἢ τὸ μὲν πάθος τὸ αὐτό, ἐφ' οἷς δὲ γίνεται, ἐναντίως ἔχουσιν; ὁμονύμως δὲ λέγεται σαπρός, καὶ ὅταν ὁ καρπὸς ᾖ τετελεσμένος· οὐ γὰρ τὸ αὐτό ἐστι καὶ ἐπὶ τοῦ οἴνου· τὸ μὲν γὰρ διεφθαρμένον ἄνθος ἐστί, τὸ δὲ πεπεμμένον.

3.16. Διὰ τί εἰς φυλακὴν συμφέρει τοῖς οἴνοις ἐν μὲν τῷ χειμῶνι ἀλεεινὸς τόπος, ἐν δὲ τῷ θέρει ψυχρός; ἢ ὅτι αἱ ὑπερβολαὶ ἐξιστῶσιν; τὸ λίαν οὖν ψῦχος τὴν ἐν οἴνῳ θερμότητα σβέννυσι, καὶ ἡλίου ἀλέα, ὥσπερ ἔνια σήπει, οὕτω καὶ τὸν οἶνον· ἔστι δ' οἴνου σαπρότης ἡ ὀξύτης.

3.17. Διὰ τί ὑπὸ τοῦ χαλκοῦ τὸ μὲν ὕδωρ οὐκ ἰοῦται, ὁ δ' οἶνος ἰοῦται; πότερον τὸ μὲν ἄχυμόν ἐστι, τὸ δ' ἔχει χυμόν; ἢ ὅτι τὸ μὲν ὕδωρ λεπτὸν καὶ κοῦφον, ὁ δ' οἶνος παχὺς καὶ βάρος ἔχει, ὥστε καὶ ἀπατμίζειν τὸ ἰῶδες κωλύει διὰ τὸ πάχος ἀπὸ τοῦ χαλκοῦ καὶ τῷ βάρει τρῖψιν ποιεῖ, πάντα δὲ τὰ τρίβοντα ἰοῦται; ἔτι ψαθυρὸν τὸ ὕδωρ, καὶ οὐχ ὥσπερ τὸ ἔλαιον γλίσχρον, ὥστ' οὐχ ὑπομένει· ὁ δ' οἶνος καὶ τὸ ἔλαιον διὰ τὸ βάρος καὶ διὰ τὴν γλισχρότητα ἐπιμένοντα καὶ ποιεῖ καὶ πά⟨σ⟩χει ἐπὶ τοῦ χαλκοῦ.

3.18. Διὰ τί τὸ ὄξος ἀσαπές; ἢ ὅτι κωλύει διίστασθαι τοὺς χυμοὺς διεσθίον;

3.14 (A Buss.)
2 verbum post *ἐκθερμανθεῖσαι* omissum coni. Buss. in var. lect.
3.15 (A Buss.)
1 *λαμβάνει*] *ποιεῖ* Buss. 2 *αὐτό* A 3 *οὐ*] *οὐ* sic A || *αὐτό*] *αὐτό* A
3.16 (A Buss.)
2 *ὑπερβολαὶ*] *ὑποβολαὶ* A 3 *ψῦχος*] *ψυχρὸν* Buss.
3.17 (A Buss.)
1 *ὕδωρ* deletum in cod. restituit Buss. 2 *τὸ*[2]] *ὁ* Buss. in var. lect. 7 *γλισχρότιταν* A || *πάσχει* Buss.: *πάχει* A
3.18 (A Buss.)

because it is easily concocted but because it provides no nourishment; and for this reason those who drink water eat more, because something has been taken away from their nourishment.

3.14.[534] Why does wine, which is hot, quench thirst more than water, which is cold? Is it because the ... that have been made hot by the heat and moisture produce thirst?

3.15. Why does wine that is ready and that which has deteriorated acquire a bloom? Is it that the effect is the same, but that what it happens to is an opposite condition? Fruit is called by the same name, "squashy",[535] also when it is fully ripe. It is different in the case of wine too; for one bloom is from corruption, the other from concoction.

3.16. Why does it benefit wines to be kept in a warm place in winter and a cool one in summer? Is it because excesses change its nature? So excessive cold quenches the heat in the wine, and, as the heat of the sun corrupts some things, so it does wine. The corruption of wine is acidity.[536]

3.17.[537] Why is water not corrupted by bronze, but wine is? Is it that one has no flavour and the other does? Or because water is rare and light, wine is dense and has heaviness, so that by its density it stops the corruption from the bronze from departing as vapour and by its weight causes friction, and all things that undergo friction corrupt? Moreover water is brittle, and not viscous like olive oil, and so does not remain [in contact]; but the wine and the oil because of their heaviness and stickiness remain and both act on and are acted on by the bronze.

3.18. Why does vinegar not suffer corruption? Is it because it prevents the

---

[534] See above, Introduction, n.54.

[535] Literally "rotten", but in English that has only a bad sense. Cf. LSJ, *sapros* II.3.

[536] Galen attributes to Aristotle (fr. 222 Rose^3) and to Theophrastus (**426** FHS&G) the explanation of the change of wine to vinegar by the replacement of the innate heat of the wine by the alien heat due to corruption. He also makes the point that both excess heat and excess cold are harmful in the context of reporting their views on the ripening of good wine and, by implication, the spoiling of inferior wine (Aristotle fr. 221 Rose^3/Theophrastus **427** FHS&G).

[537] Rose 1863, 219 compares Plutarch, *Pyth. Orac.* 3 395F, where Aristotle is cited as explaining the fact that oil creates rust (patina) on bronze by arguing that other liquids disperse the rust but the oil because of its density does not do so. See above, Introduction, n.73.

3.19. *Διὰ τί τὸ ὄξος τὸν χαλκὸν καὶ τὸν σίδηρον ἰοῖ μάλιστα; ἢ ὅτι διαδοτικὸν καὶ διαδῦνον δὲ ἀναζυμοῖ καὶ ἐξάγει τὸ γεῶδες;*

3.20. *⟨Διὰ τί⟩ τὸ ὄξος τῷ μέλιτι, οἱ δ' ἅλες τῷ ἐλαίῳ ἁρμόττουσιν; ἢ ὅτι τὸ μὲν ἔλαιον διὰ τὸ κοῦφον εἶναι ἐπιπολαστικόν ἐστι καὶ πλησμονὴν παρέχει, μετὰ δὲ τοῦ ἁλὸς κάτω φέρεται· τὸ δὲ μέλι κοῦφον μὲν οὐκ ἔστι [καὶ] ἀλλὰ βαρύσταθμον, ἅμα δὲ τρόφιμον καὶ ἄσηπτον, τὸ δ' ὄξος ἄτροφον καὶ διαλυτικὸν πάντων, ὥστε καὶ τοῦ μέλιτος ἀφαιρεῖται τὴν ὑπεροχήν, εἰ λυπηρόν ἐστιν;*

3.21. *Διὰ τί ὁ λευκὸς οἶνος οὐ γίνεται ὄξος; ὅτι ἀσθενέστερος.*

3.21a. *Διὰ τί τῶν μὲν ἄλλων ὑγρῶν ἡ ὕλη καὶ ἡ ὑπόστασις καὶ ἡ ἰλὺς [τοῦ] κάτω ὑφίσταται, τοῦ δὲ μέλιτος ἄνω; ἢ ὅτι τὰ μὲν ἄλλα κουφότερα τῆς γῆς ἐστι, καὶ ἔλαιον καὶ ὕδωρ καὶ οἶνος, τὸ δὲ μέλι βαρύτατον καὶ ὑγρότατον τῶν ὑγρῶν, καὶ τὸ βέλτιον βαρύτερον τοῦ χείρονος; ἀνάγκη δὲ τοῦ μέλιτος κάτω φερομένου ἄνω τὴν ἀχλὺν εἶναι διὰ τὸ κουφοτέραν εἶναι.*

3.22. *Διὰ τί ὁ μὲν οἶνος παλαιούμενος γίνεται ἡδίων, τὸ δὲ μέλι οὔ; ἢ ὅτι ξηρὸν τῇ φύσει ἐστίν (σημεῖον δέ, ὅτι ἑψόμενον παχύτερον γίνεται)· ὁ δ' οἶνος ὑγρόν; τὸ δὲ μέλι ὑγρὸν οὐ γίνεται παλαιούμενον, ἀλλὰ ξηρόν, ὥστ' ἄχυμον.*

3.23. *Διὰ τί τὸ ἔλαιον αὐτὸ μὲν μόνον ἥκιστα ἐκκάεται, μιχθὲν δ' ἄλλοις μάλιστα; διὰ θερμότητα συνεκκάει τὸ ὑγρὸν μετὰ τῆς ἀπὸ πυρὸς θερμότητος, αὐτὸ διὰ γλισχρότητα ἀκολουθεῖ τὰ ἐξατμίζοντα, οἷον ἐν ῥεύματι.*

3.19 (A Buss.)
3.20 (A Buss.)
1 *Διὰ τί* add. Buss. || *τῷ*[1]] *τὸ* A || *τῷ*[2]] *τὸ* A 2 *ἐπιπολαστικόν*] *ἐπιπολετικὸν* sic A 3 *καὶ* del. Buss.
3.21 (A Buss.)
1 *οὐ*] *οὑ* A
3.21a (A Buss.)
1 *διὰ τί* Buss. in var. lect.: *καὶ* A || *ἰλὺς*] *υλις* sic A 2 *τοῦ* del. Buss. 3 *καὶ ἔλαιον καὶ ὕδωρ καὶ οἶνος* tacite om. Buss. || post *μέλι* add. *πάντων* Buss. 5 *ἀχλὺν*] *ἰλύν* coni. Buss. in var. lect. || *κουφότερον* perperam Buss.
3.22 (A Buss.; cf. Macrobius, *Sat.* 7.12.9-10)
1 *ἡδίων, τὸ δὲ* Buss.: *ἰδίων, το δε* sic A: *optima* Macrobius
3.23 (A Buss.)
1 *αὐτὸ* A 2 *ἀπὸ*] *ὑπο* A 3 *αὐτὸ* A || *διὰ γλισχρότητα ἀκολουθεῖ τὰ* A: *δ' ἀκολουθεῖ καὶ διὰ γλισχρότητα* (*τὰ*?) sic Buss. || *οἷον*] *οἰόιον* sic A

flavours from separating by corroding them?

3.19. Why does vinegar corrupt bronze and iron especially? Is it because it is penetrative and by penetrating it ferments and brings out the earthiness?

3.20. Why does vinegar go with[538] honey, and salt with oil? Is it because the oil on account of being light floats on the surface and produces satiety, but is carried downwards with the salt; the honey is not light but heavy in weight, and at the same time nourishing and free from corruption, while the vinegar does not provide nourishment but dissolves everything, so that it also takes away the excess of the honey, if this is burdensome.

3.21. Why does white wine not turn to vinegar? Because it is weaker [than red wine].

3.21a. Why does the matter and sediment and dregs of other liquids settle downwards, that of honey upwards? Or because the others, oil, water and wine, are lighter than earth, but honey is the heaviest and moistest of liquids, and the better is heavier than the worst? So it is necessary that as the honey moves downwards the "haze" is on top because it is lighter.

3.22.[539] Why does wine become sweeter as it ages but honey not? Is it because [honey] is dry by nature (an indication is that it becomes thicker when boiled) but wine is moist? Honey does not become moist as it ages, but dry, so that it lacks taste.[540]

3.23. Why does olive oil on its own burn least, but mixed with other things most? Because of its heat it burns away moisture, together with the heat from the fire, and itself, because of its stickiness, follows what vaporizes, as in a stream.

---

538 I.e. "suit", "fit with" (*harmottein*), in diet as the sequel shows. – That there is an oily element in salt water is asserted in [Ar.], *Probl.* 23.9 and 15, and similarly for salt itself by Pliny, *NH* 31.91 in a section that may derive from Theophrastus; cf. Steinmetz 1964, 316-317, and Sharples 1998b, 221 and 224.

539 Cf. [Alex.], *Probl.* ii.70, which however argues that *both* honey and wine lose sweetness with moisture as they age. Macrobius, *Sat.* 7.12.9-10, compared with the present text by Linke 1880 54 n.2, draws a similar contrast to that here between wine and honey, but argues for the dryness of honey from its medicinal effects rather than from the effect of boiling.

540 The implication being that the wine becomes sweeter as the watery element in it is removed; cf. above, 3.12 and n.531.

3.24. Διὰ τί τὸ ἔλαιον τῶν ἐναντίων ἐστίν; ἧττον γὰρ ῥιγῶσιν οἱ ἀλειφόμενοι, καὶ ἧττον καίονται ὑπὸ τοῦ ἡλίου. ἢ τὸ μὲν διὰ τὸ φύσει θερμὸν εἶναι, τὸ δὲ διὰ τὸ θερμὸν ἅπαν τοῦ ψυχροῦ μᾶλλον αἰσθητικώτερον εἶναι;

3.25. Διὰ τί τινα τῶν ὑγρῶν ξηραίνεται βραδύτερον; ἢ ὅτι τὸ ξηραίνεθαί ἐστι τὸ ψύχεσθαι, ψύχεται δὲ θᾶττον τὰ ψυχρότερα; τὸ δ' ἔλαιον θερμότατον.

3.26. Διὰ τί τὸ ἔλαιον τοῖς μὲν ἔξω συμφέρει τοῦ σώματος, τοῖς δ' εἴσω οὔ; ἢ ὅτι μαλάσσει; διὸ τὴν κηρωτὴν ἔνιοί φασιν ἄριστον εἶναι τοῖς ἰατροῖς εὑρημάτων. τοῖς δ' εἴσω οὐ συμφέρει ὅτι ἄπεπτον.

3.27. Διὰ τί τὰ ἐλαιωθέντα βιβλία εὐαναγνωστότερα μᾶλλον ὄπισθεν ἢ ἔμπροσθεν; ἢ ὅτι διὰ τὸ ἀραιόπορα γίνεσθαι, ὥστε διέρχεσθαι τὴν ὄψιν ῥᾷον;

3.28. Διὰ τί τὸ ἔλαιον ἐπιχεόμενον σχῆμα στρογγύλον ἄνωθεν ἔχει πλέον τῶν ἄλλων ὑγρῶν χεομένων; ἢ ὅτι μόνον ἐλαφρότερον ὕδατος· στρογγύλον δέ, ὅτι πάντῃ ἐπιχεόμενον ὅμοιον; συνωθεῖ οὖν πάντῃ ὁμοίως ὡς ἀδιάσπαστον· τὸ δὲ πάντῃ ὅμοιον σχῆμα στρογγύλον.

3.29. Διὰ τί τὸ ἔλαιον τῇ θαλάσσῃ ἐπιχεόμενον γαληνιᾷ; ἢ ὅτι λεῖόν ἐστιν; ἀπολισθαίνει τοίνυν τὸ πνεῦμα, καὶ οὐ ποιεῖ φρίκην.

3.24 (A Buss.)
3.25 (A Buss.)
3.26 (A Buss.)
3.27 (A Buss.)
3.28 (A Buss.)
1 *πλέων* A 2 *ἄλλων* add. sup. lin. A 3 *πάντῃ*] *πάντι* A 4 *πάντῃ*] *πάντι* A || *ὁμοίος* sic A || *ὁμοιόσχημα* Buss. in var. lect.
3.29 (A Buss.)
1 *θαλάσσῃ* add. sup. lin. A 2 *οὐ ποιεῖ φρίκην* sic A: *φρίκην οὐ ποιεῖ* Buss.

3.24. Why does olive oil [produce] opposites? Those who are anointed with it feel the cold less, and are less burned by the sun. Is the former because it is naturally hot, the latter because everything that is hot is more aware of the cold?[541]

3.25. Why do some moist things[542] dry more slowly? Is it because to be dried is to be chilled, and colder things are chilled faster? But olive oil is very hot.

3.26. Why is olive oil beneficial for the external parts of the body, but not for the internal? Is it because it softens? And for this reason some say that the wax-ointment is the best of doctors' discoveries. But it does not benefit the internal parts because it is not easily concocted.

3.27. Why are books treated with olive oil read more easily from the back than the front?[543] Is it because their pores become looser-textured, so that the sight passes through more easily?

3.28. Why when olive oil is poured does it have a round shape above[544] more than other liquids that are poured? Is it that it alone is lighter than water, and round because it is similar everywhere when it is poured? So it pushes in every direction similarly, as inseparable [from itself]; and the shape that is similar in every direction is round.

3.29.[545] Why is oil that is poured on the sea calm? Is it because it is smooth? So the wind slips off it, and does not create roughness.

- - - - -

---

[541] And therefore less affected by the heat? But this seems directly to contradict the first half of the answer.

[542] The solution suggests that olive-oil specifically is meant.

[543] Papyrus rolls, as opposed to codices, generally had writing on one side only.

[544] The solution suggests that the reference is not to the convexity of the surface of the oil, but to its spreading out in a circle when poured onto the top of water. In either case surface tension due to viscosity is the explanation, as our author indicates by describing the oil as "inseparable", *sc.* from itself.

[545] See below, 3.47. Plutarch, *Prim. frig.* 13 950b attributes this view to Aristotle, as noted by Rose 1863, 219 and Flashar 1962b, 360 n.1; Plutarch himself rejects it, arguing rather that *any* added moisture dissipates the waves.

3.30. Διὰ τί ἐν τῷ γάλακτι οὐ φαίνονται αἱ ὄψεις; ὅτι οὐχ ὁμαλόν ἐστιν· ἀφρῶδες γάρ ἐστι, καὶ μικραί εἰσι πομφόλυγες.

3.31. Διὰ τί τὸ γάλα γλυκύτερον τῶν μελάνων προβάτων; πότερον ὅτι γλυκύτερον, διότι τὸ αἷμα τῶν λευκῶν προβάτων θερμότερον ὂν ἀναλίσκεται τὸ γλυκύ; ἢ ὅτι παχύτερόν ἐστι τὸ τῶν μελάνων; καὶ ἐπὶ τῶν γυναικῶν παχύτερόν ἐστι τὸ τῶν μελαινῶν, ὅτι καὶ τὸ αἷμα. λέγομεν οὖν ὅτι τὰ μὲν λευκὰ πέττει διὰ θερμότητα τὴν τροφήν, τὰ δὲ περιττωματικά, οἷον τὰ μέλανα, οὔ· ψυχρότερα γάρ.

3.32. Διὰ τί τὰ ἐρίφεια τῶν ἀρνείων βελτίω, καὶ τὰ αἴγεια τῶν προβάτων; ἢ ὅτι τῶν μὲν προβάτων τὸ γάλα παχὺ καὶ λιπαρόν, τὸ δ' αἴγειον τοὐναντίον;

3.33. Διὰ τί ὕστερα τὰ δυσπεπτότερα προσφέρεται, τὰ δ' εὐπεπτότερα πρῶτον, οἷον κρέα ἰχθύων, καίτοι ἔδει ἀποδιδόναι πλείονα χρόνον τοῖς δυσπέπτοις; ἢ ὅτι τὸ προπεπεμμένον ⟨δεῖ⟩ ὑπάγεσθαι, ἵνα μὴ ἐμποδίζῃ ἄλληλα;

3.34. Διὰ τί αἱ γαστέρες κνισωθεῖσαι ἥδισταί εἰσι τῶν αἰγῶν, καὶ Ὅμηρος,

γαστέρες αἵδ' αἰγῶν κέατ' ἐν πυρί;

ἢ ὅτι κνισωδέστατόν ἐστι τὸ αἴγειον στέαρ; λιπαρώτατον γάρ ἐστιν· διὸ καὶ οἱ ἰατροὶ πρὸς τὰ μαλάγματα αὐτῷ χρῶνται μάλιστα· κνισῶδες δ' ὂν καὶ λιπαρὸν ἄρτυμα γίνεται τῆς κοιλίας· τὰ δ' ἠρτυμένα ἡδίω.

3.35. Διὰ τί τὸ ἔλαιον οὐ πήγνυται, ἀλλὰ συνίσταται, τὸ δὲ στέαρ πήγνυται; ὅτι λιπαρώτερόν ἐστι τὸ ἔλαιον τοῦ στέατος. ἥκιστα οὖν πήγνυται· τὸ δ' ὕειον στέαρ ὑδαρέστερον· ἔχει γὰρ χυμοὺς ἐν αὐτῷ

3.30 (A Buss.)

3.31 (A Buss.)
1 *τὸ* om. Buss. 2 *ὂν*] *ὧν* A

3.32 (A Buss.)
1 *προβατείων* Buss. in var. lect. 3 *τοὐναντίον*] *τοὐναντίων* A

3.33 (A Buss.)
3 *προπεπεμένον* A^{ac} || *δεῖ* add. Buss.

3.34 (A Buss.)
3 *κέατ' ἐν* Buss. ex Homero: *καὶ αὐτ' εν* A 4 *κνυσσώδεστατον* A 5 *μαλάγματα* Buss.: *μαλακά* A || *αὐτῷ* A 6 *τοῖς κοιλίας* A

3.35 (A Buss.)
2 *λιπαρῶτατον* sic A 4 *οὐ* delevimus: exhibet A, se tamen ipsum per

3.30. Why do images not appear in milk? Because it is not even; for it is frothy, and there are little bubbles.

3.31. Why is milk from black sheep sweeter? Is it because it is sweeter because the blood of white sheep is hotter and consumes the sweetness? Or because [the blood] of black [sheep] is denser? In the case of women too the [milk] of dark ones is thicker, because the blood is too. So we say that [animals] that are white concoct the nourishment because of their heat, but those with residues, like black ones, do not, because they are colder.

3.32. Why is [the meat?] of kids better than that of lambs, and of goats than of sheep? Is it because sheep's milk is rich and thick, goats' [milk] the opposite?

3.33. Why are things that are harder to digest served afterwards, ones that are easier to digest first, like the flesh of fish, though those that are hard to digest need more time? Is it because what is digested first needs to be got out of the way, so that they do not impede one another?

3.34. Why are goats' paunches steamed in suet most pleasant, as Homer says "these stomachs of goats are resting in the fire"?[546] Is it because goats' suet steams best? For it is most rich; and for this reason doctors use it most as an emollient. Being steamy and rich it is a seasoning for the stomach; for seasoned things are more pleasant.

3.35. Why does olive oil not become solid, but thickens, while suet becomes solid? Because oil is more fatty than suet. So it becomes solid least. The suet of pigs is more watery; for it has juices from residues in it, and such

---

[546] *Odyssey* 18.44.

περιττωματικούς· οἱ δὲ τοιοῦτοι εἰκότως [οὐ] πήγνυνται.

3.36. Διὰ τί τοῦ σαπέρδου φαύλου μὲν ὄντος ἡ κεφαλὴ σπουδαία ἐστί, σπουδαίου δ' ὄντος μοχθηρά; οὗτος δ' ὁ ἰχθύς ἐστιν ὁ λεγόμενος κορακῖνος.

3.37. Διὰ τί τὰ ὑπηνέμια ᾠὰ οὐκ ἀποσήπεται, ἄψυχά γε ὄντα, ὥστε ἔδει, ὥσπερ καὶ τὰ ἄλλα μέρη στερισκόμενα ψυχῆς σήπεται, σήπεσθαι καὶ αὐτά; ἢ ὅτι πηρώματά ἐστι, καὶ οὐχ ὥσπερ νεκρός· σήπεται δ' ἐν οἷς ἐκλείπει ψυχή, ἀλλ' οὐκ ἐν οἷς μηδ' ἐγεννήθη ὅλως;

3.38. Διὰ τί τῶν ᾠῶν τὸ ἕτερον ἄκρον σκληρότερον; ἢ ὅτι μία ἀρχὴ ἐν παντί, ὥστε καὶ τοῦ ᾠοῦ; τὸ οὖν ὄστρακον φυλακῆς χάριν, ἡ δὲ φυλακὴ περὶ τὴν ἀρχὴν ἀεὶ τέτακται, [κατὰ δὲ] κατὰ τὸ ὀξὺ δὲ γίνεται τοῦ ᾠοῦ ὁ νεοσσός, ἐὰν μὴ παραπέσῃ ἐν τῇ γενέσει· ἅπαξ γὰρ ὤφθη.

3.39. Διὰ τί τὰ ᾠὰ κατὰ μῆκος μὲν θλιβόμενα ταῖς χερσὶν οὐ κατάγνυται, ἐγκλινόμενα δέ; ὅτι κατὰ μῆκος μὲν θλιβόμενα πρὸς ὀρθὴν τῇ χειρὶ κείμενα συμβαίνει θλίβεσθαι, ἡ δ' ὀρθὴ ἄκαμπτος διὰ τὸ ἴσον καὶ τὸ εὐθύ· ἐὰν δὲ ἐγκλιθῇ, ἡ ἑτέρα ὀξυτέρα δὲ γίνεται γωνία, ὥστε ταύτῃ ἐνδίδωσιν. ἢ ὅτι ὁμοίως πάντα θλίβεται τότε; πάντα γὰρ τὰ ἴσον ἀπέχοντα μόρια τοῦ ἀνωτάτω ὅμοια ἀλλήλοις· ἐὰν δ' ἐγκλιθῇ, οὐκέτι πρὸς τοῦτο τὸ σημεῖον ἅπαντα ἔχει ὁμοίως.

3.40. Διὰ τί, ἐάν τις σπόγγον ὑπερίσχῃ τῶν ἑψημάτων, ἀφαιρεῖται τὸ ὄξος

---

coniecturam addidisse credidit Buss. praef. xviii.

3.36 (A Buss.)

3.37 (A Buss.)

1 *οὐκ ἀποσήπεται*] *οὐκαθὸ σήπεται* A || *ἔδει*] *δῆ* A 3 *αὐτά* A 4 *ἐκλείπει ψυχή*] *ἐκλύπει ψυχῇ* sic A

3.38 (A Buss.)

1 *διὰ τί* coni. Buss. in var. lect.: *ἢ* A || *ἀρχῇ* sic A || *ἐν* A 3 *κατα δὲ* del. Buss

3.39 (A Buss.)

1 *ταῖς*] *τε* A

3.40 (A Buss.)

1 *ὑπερίσχει* A 2 *ἐξ αὐτῶν* om. Buss. (per errorem? exhibet transl.) || *σικῦαι* A

[juices], reasonably, become solid.[547]

3.36. Why is the head of an inferior *saperdês* good, but of a good one bad? This is the fish that is called *korakinos*.[548]

3.37. Why do wind-eggs not decay, when they are lifeless, so that they should decay themselves in the same way as other parts do when deprived of soul? Is it because they are a deformity, and not like a corpse, and those things decay in which soul has ceased, not those in which it never was at all?

3.38. Why is one end of an egg harder? Is it because there is one first principle in everything, so that there is in an egg too? The shell is for the sake of protection, and the protection is always arranged around the principle,[549] and the chick is at the pointed end of the egg—if it does not shift as it is generated; that was seen once.

3.39.[550] Why do eggs that are squeezed along their length by the hands not break, but [do if they are squeezed] at an angle? Because if they are squeezed along their length the result is that they are squeezed at a right angle against the hand, and the right angle does not bend because it is equal and straight; but if [the egg] is at an angle, one angle becomes more acute, so that it yields on that side. Or is it because [in the first case] everything is squeezed equally? For all the parts that are equidistant from the topmost one are similar to one another; but if it is at an angle, no longer are all [the parts] in a similar relation to this point.

3.40. Why, if someone holds a sponge over boiled wine-must, is the vinegar

---

547 MS A has "do not become solid"; Bussemaker claimed, perhaps as the result of a slip, that he had added "not" himself. He interprets the text to mean "such juices *on the other hand* reasonably do not become solid", referring not to the juices just mentioned but to the oil; but this seems awkward.

548 The term *korakinos* is used for a whole range of fish: *Corvina nigra* (*Sciaena umbra* L.); *Umbrina cirrosa* Cuv.; *Chromis castanea*; *Tilapia Nilotica*; or *Labeo niloticus*. *Saperdes* and *korakinos* are identified by Dorion ap. Athenaeus 3.87 118b and Euthydemus ibid. 7.81 308e. Cf. Thompson 1947, 123, 226. As Bussemaker 1857, xix notes, the solution to the problem is missing.

549 So Ar., *GA* 3.2 752a14-15. Peck 1942, 282 n.(c) cites Platt as observing that 'Aristotle must have been a "little-endian", for the germ always floats up to the top whichever way the egg is placed.' If so, it seems possible that what we have here is a correction of Aristotle – though, perhaps characteristically, Aristotle's view of the matter is regarded as the norm.

550 Cf. [Alex.], *Probl.* ii.48. The second explanation given here is the correct one.

ἐξ αὐτῶν; ὥσπερ σικύα ἕλκει ὁ σπόγγος· ὁ γὰρ ἐν τρήμασιν ἀὴρ θερμαινόμενος ἀντὶ πυρὸς γίνεται.

3.41. *Διὰ τί, τῶν ᾠῶν ῥηγνυμένων ἐν τῷ ὀπτᾶσθαι, τοῦ κελύφους πρότερον τὸ ὠχρὸν τοῦ λευκοῦ ἔξω; ἢ ὅτι τὸ μὲν [λευκὸν] ⟨ὠχρὸν⟩ οὐσίᾳ ῥεῖ μᾶλλον ἢ τὸ λευκὸν [ὠχρόν], καίτοι γε μᾶλλον ἐντός ἐστι τὸ ὠχρὸν τοῦ λευκοῦ; διὸ καὶ ὁ νεοσσὸς ἐξ αὐτοῦ γίνεται, τὸ δ' ὠχρὸν καθάπερ τρύξ. ὥσπερ οὖν ἐν τοῖς οἴνοις ἡ μὲν τρὺξ ἀπαθής ἐστιν, ὁ δ' οἶνος μεταβάλλει ἑκάστοτε καὶ χρώματα καὶ ὀσμήν, οὕτω καὶ τοῦ ᾠοῦ τὸ λευκὸν μεταβλητικώτερον τοῦ ὠχροῦ· διὸ καὶ τὸ πυρούμενον στερεὸν γίνεται, καὶ ἐπῳαζόμενον. ἀλλὰ καὶ ἐάν τις ᾠὸν τρυπῇ, [τὸ ᾠῶν] τὸ λευκὸν ἐκχεῖται, τὸ δὲ πυρρὸν ὑπομένει διὰ τὸ μὲν λευκὸν ἔνυγρον σῶμα εἶναι, τὸ δὲ καθάπερ ἐπιπεπασμένον ἔοικεν.*

3.42. *Διὰ τί τῶν ἀγγείων οἱ πυθμένες θερμαινομένου τοῦ ὕδατος, ἕως ἂν ᾖ ψυχρὸν τὸ ὕδωρ, θερμότεροί εἰσιν· ἐπὰν δὲ ζέῃ, ψυχρότεροι γίνονται; ἢ διότι ψυχροῦ μὲν ἔτι ὄντος τοῦ ὕδατος, ἐγκατακλείεται τὸ θερμὸν καὶ*

3.41 (A Buss.)
1 *τοῦ κελύφους*] *τὸ κέλυφον* A 2 *λευκὸν* delevimus et *ὠχρὸν* addidimus (cf. infra) 3 *ἢ*] *εἶ* sic A || *ὠχρόν* seclusimus: *ὠχρόν* potius quam *λευκὸν* maluit Buss. in var. lect. 8 *ἐπωαζόμενων* sic A || *ᾠὸν*] *ωὼν* A || *τρυπῇ*] *τριπήσας* A || *τὸ ᾠῶν* del. Buss.: corr. ex *τὸ ὦν* (*ῶ* superscr.) A
3.42 (A Buss.; = [Ar.] 24.8)
2 *ἐπὰν – γίνονται* om. [Ar.] || *διότι* A [Ar.]: *ὅτι* Buss. 5 *γινομένου* [Ar.] 6

drawn off from it? The sponge has as it were cupping-instruments; for the air in its pores, being heated, serves instead of fire.

3.41. Why, when eggs break when being roasted, does the yolk [come] out of the shell before the white? Is it because the yolk is a runnier substance than the white,[551] although the yolk is further inside than the white? For this reason the chick comes from it; and the yolk[552] is like the dregs. As in wine the dregs are not affected, but the wine changes constantly both in colour and in odour, so in the egg the white changes more than the yolk; and for this reason what is heated by the fire becomes solid, and also when incubated. But also if someone pierces an egg, the white pours out, the yolk remains, because the white is a moist body, the [yolk] is like something sprinkled over it.[553]

- - - - -

3.42.[554] Why, when water is being heated, are the bases of the vessels warmer as long as the water is cold, but become colder when it boils? Is it because when the water is still cold the heat is trapped and compressed

---

551 The MS has "the white is a runnier substance [literally; "runnier in its substance"] than the white yolk". Bussemaker conjectured "the white is a runnier substance than the yolk", but the argument requires the reverse. It appears that a corruption resulting in the presence of "white" in both places was corrected, perhaps by the insertion of "yolk" in the margin, but "yolk" was then restored in the wrong place.

552 Bussemaker prints "yolk" in his Greek text with the MS, but translates by "the white". "Yolk" is however correct; the sequel shows that the yolk is like the dregs in not changing, the white like the wine in being susceptible to change. The analogy, if pressed, is however imperfect in so far as the yolk is not the dregs *of the white*, but of the process that produces the chick; moreover, the white is affected more than the yolk by boiling whether the egg is fertile or not.

553 As with grain sprinkled on wine: cf. Plato, *Republic* 405e referring to *Iliad* 11.638ff. As our text fails to make completely clear, the real explanation for the difference between the two cases of roasting the egg and piercing is that in the former the white has been solidified by heat, in the former not; moreover a pin-hole in an egg will only allow what is nearest to run out first.

554 = [Ar.], *Probl.* 24.8; cf. id. 24.5, which argues that the base of a vessel containing boiling water is cooled by the water.

ἀντιπεριίσταται εἴσω κωλυόμενον ἐξιέναι· ὅταν δὲ διαθερμανθῇ τὸ ἐνὸν ὕδωρ, οὐκέτι ἀποστέγοντος, ἀλλὰ διαπνέοντος καὶ ἐλάττονος γενομένου τοῦ πυρός, ψυχρότερος γίνεται ὁ πυθμὴν ὁμοίως ὥσπερ καὶ τὰ βαλανεῖα χειμῶνος θερμότερα ἢ τοῦ θέρους;

3.43. Διὰ τί τὸ ἑφθὸν ὀπτᾶν οὐ νόμος, ὀπτὸν δὲ ἕψειν νόμος; πότερον ⟨δὲ⟩ διὰ τὰ λεγόμενα ἐν τῇ τελετῇ; ἢ ὅτι ὕστερον ἔμαθον ἑφθὰ ἢ ὀπτά; τὸ γὰρ ἀρχαῖον ὤπτων πάντα. ἐπανιέναι γοῦν ἐγίνετο τὸ ἑφθὸν καὶ ὀπτᾶν· διὸ ἀπεδοκίμασαν· τὸ δ' ὀπτὸν ἐφέψειν οὐκ ἐκώλυσαν. ἢ ὅτι βελτίω ἑψηθέντα; ξηρότερα γὰρ τὰ ἑφθὰ τῶν ὀπτῶν· ἐὰν δ' ὀπτὰ ὄντα, καὶ ἔτι μᾶλλον.

3.44. Διὰ τί οἱ Θρᾷκες σολοικίζουσιν ἀπογηράσκοντες; ἢ διὰ τὸ τὰ μὲν κατὰ φύσιν ἅπαντα ῥᾴδιον εἶναι ποιεῖν, τὰ δὲ παρὰ φύσιν χαλεπόν· κἂν δυναμένῳ δή, εἴ τι μέλλει τις ποιῆσαι τῶν παρὰ φύσιν; ἡ μὲν οὖν οἰκεία διάλεκτος ὥσπερ φύσει αὐτοῖς γίνεται· ἡ δ' ὑστέρα παρὰ φύσιν· διὸ καὶ καταφέρονται εἰς τὴν πρώτην διάλεκτον, γηράσκοντές τε καὶ οἰνούμενοι.

3.45. Διὰ τί ἡ Ὑρκανία λίμνη πότιμος οὖσα πλύνει καὶ ῥύπτει τὰ ἱμάτια; πλύνει μὲν τῷ γλυκεῖ, ῥύπτει δὲ τῷ πικρῷ· ἅμα δ' οὐχ οἱ ....

βαλανεῖα· καὶ γὰρ ταῦτα [Ar.] 7 τοῦ χειμῶνος [Ar.] || post θέρους haec habet [Ar.]: διὰ τὸ ἐγκατακλείεσθαι τὸ θερμὸν ἐν τῷ χειμῶνι μᾶλλον ἢ ἐν τῷ θέρει ὑπὸ τοῦ περιέχοντος ἀέρος ὄντος ψυχροῦ.

3.43 (A Buss.)

1 δὲ add. Buss. 2 ἑφθὰ] ἕψα A 3 ὤπτων] ὠπτὸν A || ὀπτᾶν] ὠπτῶν A 4 βελτίων A

3.44 (A Buss.)

2 παρὰ Buss.: περι A || χαλεπῶν A 2-3 κἂν δυναμένῳ] καὶ δυνάμεως A 3 δή Buss.: δῆ A || μέλει A 4 αὐτοῖς A 5 οἰνούμενοι] ὀνοῦμενοι A

3.45 (A Buss.; = [Ar.] 23.40 935b34-36)

1 ἡ Ὑρκανία Buss.: η ἀρκανία A: πᾶσα [Ar.] codd.: Παῖσα Gaza [Ar.] vertens: παλαστίνη, coni. Louis, Ar. *Meteor.* 2.3 359a17 conferens || πλύνει] πλήνει A 2 πλύνει] πλήνει A || μὲν γὰρ [Ar.] || γλυκεῖ Buss.: γλυκῦ A: τὸ γλυκύ [Ar.] || τὸ πικρόν [Ar.] || post οἱ deficit A: οἷόν τε ἔχειν ταῦτα [Ar.]

within [the base of the vessel],[555] being prevented from emerging; but when the water which is inside gets thoroughly hot, since the fire no longer retains [the heat][556] but evaporates and becomes less, the base becomes colder, in the same way that baths are hotter in winter than in summer.

3.43. Why is it not the custom to roast what is boiled, but it is the custom to boil what is roasted? Is it because of what is said in the mystery rite? Or because people learned of boiled things after roast things? For in ancient times they roasted everything. So to roast what is boiled would be to regress, so they rejected it; but they did not prevent boiling in addition what has been roasted. Or is it because they are better when boiled? For boiled things are drier than roast ones,[557] and if [they are boiled] when they are [already] roasted, even more.

3.44. Why do the Thracians make mistakes in speech[558] when they grow old? Is it because it is easier to do everything that is natural, difficult [to do] what is unnatural? And so even for the person who can do it, if someone is going to do what is unnatural? So their own dialect is like what is natural for them, the one [acquired] later unnatural; and for this reason they return to their original dialect when they grow old and when they are drunk.

3.45.[559] Why does the Hyrcanian[560] lake, which is drinkable, wash and cleanse clothes? [Water] washes with its sweetness, and cleanses with its bitterness; but <both can>not <be present> at the same time ...

\- - - - - -

555 For the principle that cold compresses and thus increases heat cf. above on 2.119; Ar., *Meteor.* 1.12 348b2-5; [Ar.], *Problems* 1.29, 14.3, 33.6; Theophrastus, *Ign.* 12-18; Flashar 1962b, 329; Sharples 1998, 122-123; as an explanation for the bath which is hotter in winter at Theophrastus, *Ign.* 13.

556 Following Forster's interpretation of [Aristotle].

557 See above, Introduction, n.72.

558 I.e., revert to their local dialect rather than speaking "proper Greek".

559 = [Ar.], *Probl.* 23.40, init., going on to suggest that the detergent effect is in fact due not to bitterness but to stickiness which usually, but not in this case, is associated with the bitterness of soda; for the connection between oiliness and soda cf. Theophrastus, fr. **223**.25-6 FHS&G, and above, n.538 to 3.20.

560 [Aristotle] as emended by Sylburg has "the lake at Paesus", near Lampsacus. The Hyrcanian *sea* is the Caspian.

3.46. Διὰ τί τὰ τῶν ἀετῶν πτερὰ ἄλλοις πτεροῖς ἐπιτιθέμενα φθορὰν αὐτοῖς ἐμποιεῖ; ὅτι σηπεδόνος τινὸς μετέχουσι φυσικῆς· δυσωδέστατον γὰρ πτηνῶν ἁπάντων ὁ ἀετός· ὅθεν οὐδὲ τῆς ἑαυτοῦ χθιζῆς θήρας τὰ λείψανα αὖθις ἐπῆλθε ποτέ.

3.47. Διὰ τί τὸ ἔλαιον τῇ θαλάσσῃ ἐπιρραινόμενον γαλήνην ποιεῖ; ὅτι γλίσχρον ὂν καὶ ὑγρὸν διολισθαίνειν ποιεῖ τὰ κύματα πρὸς ἑαυτὰ καὶ ἐνδιδόναι τῆς οἰκείας ὁρμῆς καὶ φορᾶς.

3.48. Διὰ τί ὁ ἄνεμος ἀρχόμενος ἀμυδρὸς πνεῖ, προκόπτων δὲ καὶ ἐπὶ ἄλλην γῆν φερόμενος, γίνεται σφοδρότερος; διὰ τὸ συγκαταμίγνυσθαι αὐτῷ ἄλλας ἀναθυμιάσεις ἐκ τῶν κατὰ μέρος.

3.49. Πόθεν δῆλον ὅτι ὅσον παχύτατόν ἐστι τὸ ὕδωρ, τοσοῦτον γίνεται ἁλμυρώτατον; ἰδού φημι ἡ ἐν Παλαιστίνῃ νεκρὰ μυθευομένη θάλασσα παχυτάτη οὖσα, ἁλμυρωτάτη πάντων ἐστὶ τῶν ὑδάτων. καὶ πόθεν δῆλον ὅτι παχυτάτη ἐστίν; εἰ γάρ τις ἄνθρωπον, ἢ ἕτερόν τι ζῷον, καταδήσας ἐμβάλοι εἰς τὸ τῆς νεκρᾶς θαλάσσης ὕδωρ, ἀνωθεῖται ὑπὸ τοῦ ὕδατος, καὶ οὐ γίνεται ὑποβρύχιον διὰ τὸ εἶναι παχὺ πάνυ καὶ βαστάζειν τὸ ἐμπεσόν.

3.46 (BDHJLW Buss.)
1 *τὰ τῶν ἀετῶν πτερὰ*] *τῶν ἀετῶν τὰ πτερὰ* H: *τὰ τῶν ἀετῶν τὰ πτερὰ* W || *πτεροῖς* om. BHL Buss. 2 *αὐτοῖς* om. BL Buss. || *γὰρ* om. H 3 *πτηνῶν ἁπάντων*] *πάντων πτηνῶν* W: *πτηνὸν ἁπάντων* H[ac] || *ὅθεν*] *ὥστε* W || *θήρας* om. W 4 *αὖθις* om. Buss. || *ἐπῆλθε*] *ἐπέλθῃ* J: *ἐπέλθοι* DHW[pc]: *ἐπέλθῃ* W[ac]
3.47 (BDHJLW Buss.)
2 *ὂν* om. BL || *διολισθαίνει* BL || *ἑαυτὰ* D Buss.: *αὐτὸ* HJW: *αὐτῇ* BL 3 *ἐνδιδῶναι* J || *φορᾶς*] *φοράς* W: *φθορᾶς* B
3.48 (DGYo Buss.; = Olympiodorus, *In Meteor.* 178.31-34)
1 *διὰ τί*] *τί φησιν; ὅτι* Ol. || *ὁ ἄνεμος ἀρχόμενος* G: *ὁ ἄνεμος ἐν ἀρχῇ* D Buss.: *ἀρχόμενος μὲν ὁ ἄνεμος πνεῖν* Ol.: *ὁ ἄνεμος* tantum Yo || *ἀμυδρὸς* GY Ol.: *ἀμυδρὸν* D Buss.: *ἀμυδρῶς* o || *δὲ καὶ*] *δ' ὡς* Ol. 2 *γίνεται σφοδρότερος*] *σφοδρότερος γίνεται* D Buss. 3 *ἐκ τῶν κατὰ μέρος*] *ἐξ ὧν διέρχεται μερῶν* D Buss.: *ἐκ τῶν κατὰ μέρος, ἐφ' ἃς ἐπέρχεται* Ol.
3.49 (DG Buss.; = Olympiodorus, *In Meteor.* 163.17-32)
1 *πόθεν δῆλον ὅτι*] *ὅτι γὰρ* Ol. || *τοσοῦτον γίνεται*] *τοσοῦτον ἐστὶ καὶ* D: *τοσοῦτόν ἐστι καὶ* Buss. 2 post *ἁλμυρώτατον*: *δῆλον* Ol. || *ἰδού φημι* G: *ἰδοὺ γάρ*: *φασὶν ὡς* D Buss.: *ἰδοὺ γὰρ* Ol. || *μυθευομένη* om. D Buss. 3 *τῶν* om. D Buss. || *πόθεν δῆλον* om. Ol. 4 *ἐστίν*] *ἐστί, δῆλον* Ol. || *ὅτι εἰ γὰρ* G || *ἢ ἕτερόν τι ζῷον*] *ἢ ζῷον ἕτερον* D Buss. 5 *ἐμβάλοι* Ol.: *ἐμβάλει* DG: *ἐμβάλῃ* Buss. || *ὑπὸ*] *ἀπὸ* G 7 *καὶ*

3.46. Why do eagle's feathers, when placed on other feathers, make them corrupt?[561] Because they share in some natural rottenness. For the eagle is the most evil-smelling of all birds. As a result it never even approaches again the remains of its previous day's hunting.[562]

3.47.[563] Why does oil poured on the sea produce calm? Because being viscous and moist it makes the waves slip over each other and give up their proper impulse and motion.[564]

3.48.[565] Why does the wind blow weakly at the beginning, but as it advances and travels to another land it becomes more vehement? Because there are mixed in with it other exhalations[566] from the parts [it traverses].[567]

3.49.[568] How is it clear that the more water is very dense, the more it comes to be very salty? Look, I say that the Dead Sea that is told of in Palestine, being very dense, is the saltiest of all waters. And how is it clear that it is very dense? Because, if one were to tie up a man, or some other animal, and throw him into the water of the Dead Sea, he is pushed up by the water, and is not submerged, because the water is very dense and supports what falls

---

561 Aelian, *NA* 9.2 has the alleged fact, but explains it rather by the eagle's kingly nature which the feathers of the other birds cannot endure.

562 One might rather expect that *no other* creature should approach the remains; but presumably the eagle's smell is so awful that it cannot even endure it itself.

563 See above, 3.29, where the explanation is rather that the oil makes the *wind* slip over it. Rose 1863, 219 regards 3.47 as deriving from 3.29; that seems questionable.

564 See above, Introduction, § 1.2 .

565 This text is closely parallel to Olympiodorus, *In Meteor.* 178.31-34.

566 If we press "other", this follows Aristotle's view that wind is formed from the dry exhalation, as opposed to Theophrastus' view that wind is moving air. Cf. Sharples 1998, 146-152.

567 "It traverses" is explicit in the text of Olympiodorus but omitted in MSS GYo of our text; it is present, but in different words from those of Olympiodorus, in MS D.

568 This text is closely parallel to Olympiodorus, *In Meteor.* 163.16-32, commenting on Aristotle, *Meteor.* 2.3 359a16-22, with minor variations in wording. After "and supports what falls into it" Olympiodorus has "Let no-one say that the creature swims by the capacity of swimming; for it is established that it is tied up", omitted in our text. The occurrences of "look" (*idou*), unique in these *Problems*, are also in the text of Olympiodorus, and are characteristic of it (80 occurrences in the entire commentary).

καὶ πόθεν δῆλον ὅτι ἁλμυρώτατόν ἐστιν; ἐπειδὴ ῥυπτικόν ἐστι παντὸς ῥύπου· σμήχει γὰρ τὸν ῥύπον, ἤπερ ἄλλο τι σμῆγμα, τὸ δὲ ῥύπτειν ἄκρας ἐστὶν ἁλμυρότητος· ἰδοὺ γάρ· ἅλες μὲν καὶ αὐτοὶ ῥύπτουσι, νίτρον δὲ πλέον, ὡς ἐπιτεταμένην ἔχον τὴν ἁλμυρότητα ἤπερ οἱ ἅλες· ὅθεν καὶ διὰ τὴν ἐπίτασιν ὑπόπικρόν ἐστι τὸ νίτρον· ἔπειτα καὶ τὸ εἶναι ἄγονον τὸ ὕδωρ τοῦτο δῆλον ἐστὶ ὑπερβαλλούσης ἁλμυρότητος· οὐδὲ γὰρ ἰχθύς· ἢ ἄλλό τι ζῷον ἐγγίνεται τῇ νεκρᾷ θαλάσσῃ διὰ τὴν ἁλμυρότητα καὶ πικρότητα.

(Usener, appendix 45) Διὰ τί οὐ φαίνεται πληθυνομένη ἡ θάλασσα καίτοι πάντων τῶν ποταμῶν εἰς αὐτὴν ἐμβαλλόντων; ὅτι διὰ τὸ ἐκτετάσθαι ἐπὶ τὸ πλάτος τὴν θάλατταν πολὺ ἀπ' αὐτῆς διαφορεῖται καθ' ἡμέραν· ἄνωθεν μὲν ὑπὸ τοῦ ἡλίου ἐξατμιζομένης· κάτωθεν δὲ ὑπὸ τῆς γῆς ἀναπινομένης· διὰ γοῦν τὴν πολλὴν αὐτῆς διαφόρησιν οὐχ ὁρᾶται πλήθουσα· εἰ καὶ πάντες οἱ ποταμοὶ εἰς αὐτὴν ἐμβάλλουσιν.

(Usener, appendix 46) Πῶς ὁ Ἀριστοτέλης ποτὲ μὲν ὑγρὰν καὶ θερμὴν τὴν ἀτμίδαν λέγει, ποτὲ δὲ ὑγρὰν καὶ ψυχράν; καὶ λέγω ὅτι ὑγρὰ καὶ ψυχρά ἐστι κατὰ φύσιν (τοιοῦτον γὰρ καὶ τὸ ὕδωρ)· ὑγρὰ δὲ καὶ θερμὴ κατὰ συμβεβηκός· ἐπειδὴ ἐπίκτητον ἔχει τὴν θερμασίαν ὥσπερ τὸ ὕδωρ.

πόθεν – ἐπειδὴ] aliter Ol. 8 *ῥύπον τῶν ἱματίων* Ol. ‖ *ἤπερ* G et Ol. cod. G: *εἴπερ* D Ol. cod.V Buss.: *ἤπερ* Ol. cod. a: *ἤπερ* Ideler ad Ol.: ⟨*μᾶλλον*⟩ *ἤπερ* coni. Stüve ad Ol. 9 *ἀρμυρότητος* G 10 *ὡς*] *ὥσπερ* Buss. 10-13 *ὅθεν – πικρότητα* om. D Buss. 12 *ὑπερβαλλούσης* Ol.: *ὑπερβαλλούμενον* G

Usener, appendix 45 (Go Us.)

1 *καίτοι*] *καίπερ* Us. 2 *τῶν* om. o Us. ‖ *ἐπὶ* Us.: *περὶ* codd. 3 *τὴν θάλατταν* Us.: *τῆς θαλάσσης* codd. ‖ *πολὺ ἀπ' αὐτῆς* G: *πολὺ δι' αὐτῆς* o: *πολύ τι αὐτῆς* Us. 4 *ἐξατμιζομένης ... ἀναπινομένης* Us.: *ἐξατμιζομένη ... ἀναπινομένη* codd. 5 *γοῦν*] *γὰρ* o Us.

Usener, appendix 46 (Go Us.; = Olympiodorus, *In Meteor*. 172.8-11)

1 Πῶς *ὁ Ἀριστοτέλης ποτὲ μὲν*] *πῶς γὰρ ἀνωτέρω ἔλεγεν αὐτὴν* Ol. 1-2 *τὴν ἀτμίδαν λέγει, ποτὲ δὲ*] *νῦν δὲ* Ol. 2 *καὶ λέγ(ω) ὅτι* Go: *καὶ λέγομεν ὅτι* Us.: *καὶ λέγομεν* Ol. ‖ *ὑγρὰ μὲν* Ol. 3 *δὲ* om. o ‖ *θερμή ἐστι* Ol. 4 *θερμασίαν ὥσπερ τὸ ὕδωρ*] *θερμασίαν* tantum, deinde nullis verbis interpositis proximum lemma *Ὥσπερ ὕδωρ θερμανθέν* Ol.

into it. And how is it clear that it is very salty? Because it cleanses all dirt; for it washes off dirt more than[569] any other unguent [does], and to cleanse is characteristic of extreme saltiness; for look, salt itself cleanses, and soda more, as having a more intense saltiness than salt does. And for this reason soda is rather bitter because of the intensity. Next, the fact that this water is sterile is characteristic of extreme saltiness; for neither fish nor any other creature occurs in the Dead Sea, because of its saltiness and bitterness.[570]

Usener appendix 45.[571] Why does the sea not apparently increase, although all the rivers flow into it? Because, on account of the sea being spread out widely, much is dispersed from it every day, upwards as it is evaporated by the sun, and downwards as it is absorbed by the earth. At any rate, on account of the great amount that is dispersed from it it is not seen to increase, even though all the rivers run into it.

Usener appendix 46.[572] How is it that Aristotle sometimes says mist is moist and hot, and sometimes that it is moist and cold? I say that it is moist and cold by nature, for that is what water too is like, but moist and hot *per accidens*, since it has acquired heat, like water.

---

569 Stüve conjectured that "more" should be added in the text of Olympiodorus; but "than", the reading of MS G here and of Stüve's MS G in Olympiodorus, can be used idiomatically to mean "more than" (cf. LSJ s.v. *ê* B.1 ad fin., though here the form is *êper*). MS D here, and MS V in Olympiodorus, have "if (*eiper*) any other unguent [can]"; Ideler conjectured in Olympiodorus "in the way in which (*hêiper*) any other unguent can".

570 For 3.50-54 Bussemaker, not included here, see above, Introduction, at nn. 46-49.

571 Cf. Ar., *Meteor.* 2.2 355b20-32, but without the point about absorption by the earth, absent also from Alexander, *In Meteor.* 75.1-13; it is however included in Olympiodorus, *In Meteor.* 141.11-20, which in other respects too is closer to our passage.

572 = Olympiodorus, *In Meteor.* 172.8-11, on 2.4 360a23, which describes mist as moist and cold; that it is moist and hot is the reading of some MSS at *Meteor.* 1.3 340b28, followed by Olympiodorus (*In Meteor.* 29.22). See Lee 1952, 20 n. (a), who cites *GC* 2.3 330b4 for mist being moist and hot, while doubting it actually implies this.

Index of passages cited

General Index

For frequently occurring terms this index includes only the more significant occurences. Where a term appears both in the main text and in a footnote on the same page, the footnote is not listed separately. For entries marked * the Index of passages cited should also be consulted.

Bei Fragen zur Produktsicherheit wenden Sie sich bitte an:
If you have any questions regarding product safety,
please contact:

Walter de Gruyter GmbH
Genthiner Straße 13
10785 Berlin
productsafety@degruyterbrill.com